T0134851

Structural Stability and Morphogenesis

Structural Stability and Morphogenesis:

An Outline of a General Theory of Models

RENÉ THOM
Institut des Hautes Etudes Scientifiques

Translated from the French edition, as updated by the author, by

D. H. FOWLER
University of Warwick

Advanced Book Program

CRC Press is an imprint of the
Taylor & Francis Group, an **informa** business
A CHAPMAN & HALL BOOK

Structural Stability and Morphogenesis:
An Outline of a General Theory of Models

Originally published in 1972 as
Stabilité structurelle et morphogénèse
Essai d'une théorie générale des modèles
by W.A. Benjamin, Inc.

Published 1989 by Westview Press

Published 2018 by CRC Press
Taylor & Francis Group
6000 Broken Sound Parkway NW, Suite 300
Boca Raton, FL 33487-2742

© 1989 by Taylor & Francis Group, LLC
CRC Press is an imprint of Taylor & Francis Group, an Informa business

No claim to original U.S. Government works

ISBN 13: 978-0-201-40685-6 (pbk)

This book contains information obtained from authentic and highly regarded sources. Rea-son-able efforts have been made to publish reliable data and information, but the author and publisher cannot assume responsibility for the validity of all materials or the consequences of their use. The authors and publishers have attempted to trace the copyright holders of all material reproduced in this publication and apologize to copyright holders if permission to publish in this form has not been obtained. If any copyright material has not been acknowledged please write and let us know so we may rectify in any future reprint.

Except as permitted under U.S. Copyright Law, no part of this book may be reprinted, reproduced, transmitted, or utilized in any form by any electronic, mechanical, or other means, now known or hereafter invented, including photocopying, microfilming, and recording, or in any information storage or retrieval system, without written permission from the publishers.

For permission to photocopy or use material electronically from this work, please access www.copyright.com (http://www.copyright.com/) or contact the Copyright Clearance Center, Inc. (CCC), 222 Rosewood Drive, Danvers, MA 01923, 978-750-8400. CCC is a not-for-profit orga-niza-tion that provides licenses and registration for a variety of users. For organizations that have been granted a photocopy license by the CCC, a separate system of payment has been arranged.

Trademark Notice: Product or corporate names may be trademarks or registered trademarks, and are used only for identification and explanation without intent to infringe.

Visit the Taylor & Francis Web site at
http://www.taylorandfrancis.com

and the CRC Press Web site at
http://www.crcpress.com

Library of Congress Cataloging-in-Publication Data

Thom, René, 1923-
Structural stability and morphogenesis.

(Advanced book classics series)
Translation of: Stabilité structurelle et morphogénèsis.
Bibliography: p.
1. Biology--Mathematical models. 2. Morphogenesis --Mathematical models. 3. Topology. I. Title.
QH323.5.T4813 1988 574'.0724 88-19428
ISBN 0-201-09419-3 (H) ISBN 0-201-40685-3 (P)

This unique image was created with special-effects photography. Photographs of a broken road, an office building, and a rusted object were superimposed to achieve the effect of a faceted pyramid on a futuristic plain. It originally appeared in a slide show called "Fossils of the Cyborg: From the Ancient to the Future," produced by Synapse Productions, San Francisco. Because this image evokes a fusion of classicism and dynamism, the future and the past, it was chosen as the logo for the Advanced Book Classics series.

Publisher's Foreword

"Advanced Book Classics" is a reprint series which has come into being as a direct result of public demand for the individual volumes in this program. That was our initial criterion for launching the series. Additional criteria for selection of a book's inclusion in the series include:

- Its intrinsic value for the current scholarly buyer. It is not enough for the book to have some historic significance, but rather it must have a timeless quality attached to its content, as well. In a word, "uniqueness."
- The book's global appeal. A survey of our international markets revealed that readers of these volumes comprise a boundaryless, worldwide audience.
- The copyright date and imprint status of the book. Titles in the program are frequently fifteen to twenty years old. Many have gone out of print, some are about to go out of print. Our aim is to sustain the lifespan of these very special volumes.

We have devised an attractive design and trim-size for the "ABC" titles, giving the series a striking appearance, while lending the individual titles unifying identity as part of the "Advanced Book Classics" program. Since "classic" books demand a long-lasting binding, we have made them available in hardcover at an affordable price. We envision them being purchased by individuals for reference and research use, and for personal and public libraries. We also foresee their use as primary and recommended course materials for university level courses in the appropriate subject area.

The "Advanced Book Classics" program is not static. Titles will continue to be added to the series in ensuing years as works meet the criteria for inclusion which we've imposed. As the series grows, we naturally anticipate our book buying audience to grow with it. We welcome your support and your suggestions concerning future volumes in the program and invite you to communicate directly with us.

Advanced Book Classics

V.I. Arnold and A. Avez, *Ergodic Problems of Classical Mechanics*

E. Artin and J. Tate, *Class Field Theory*

Michael F. Atiyah, *K-Theory*

David Bohm, *The Special Theory of Relativity*

P.C. Clemmow and J. P. Dougherty, *Electrodynamics of Particles and Plasmas*

Ronald C. Davidson, *Theory of Nonneutral Plasmas*

P.G. deGennes, *Superconductivity of Metals and Alloys*

Bernard d'Espagnat, *Conceptual Foundations of Quantum Mechanics, 2nd Edition*

Richard Feynman, *Photon-Hadron Interactions*

Dieter Forster, *Hydrodynamic Fluctuations, Broken Symmetry, and Correlation Functions*

William Fulton, *Algebraic Curves: An Introduction to Algebraic Geometry*

Kurt Gottfried, *Quantum Mechanics*

Leo Kadanoff and Gordon Baym, *Quantum Statistical Mechanics*

I.M. Khalatnikov, *An Introduction to the Theory of Superfluidity*

George W. Mackey, *Unitary Group Representations in Physics, Probability and Number Theory*

A. B. Migdal, *Qualitative Methods in Quantum Theory*

Philippe Nozières and David Pines, *The Theory of Quantum Liquids, Volume II* - new material, 1990 copyright

David Pines and Philippe Nozières, *The Theory of Quantum Liquids, Volume I: Normal Fermi Liquids*

F. Rohrlich, *Classical Charged Particles - Foundations of Their Theory*

David Ruelle, *Statistical Mechanics: Rigorous Results*

Julian Schwinger, *Particles, Source and Fields, Volume I*

Julian Schwinger, *Particles, Sources and Fields, Volume II*

Julian Schwinger, *Particles, Sources and Fields, Volume III* - new material, 1989 copyright

Jean-Pierre Serre, *Abelian ℓ-Adic Representations and Elliptic Curves*

R.F. Streater and A.S. Wightman, *PCT Spin and Statistics and All That*

René Thom, *Structural Stability and Morphogenesis*

Vita

René Thom

Professor at I.H.E.S., Bures-sur-Yvette since 1963, he was born in Montbéliard, France on September 2, 1923. Professor Thom studied at the Ecole Normale Supérieure of Paris from 1943-46, and obtained his Ph.D. in Mathematical Sciences in 1951. After a year spent at a graduate college of Princeton University, he was Professor at the Faculty of Sciences of Strasbourg University from 1954-1963. In 1954 Professor Thom invented and developed the theory of cobordism in algebraic topology. This classification of manifolds used homotopy theory in a fundamental way and became a prime example of a general cohomology theory. For this work, Professor Thom received the Field Medal in 1958. Later on at I.H.E.S., he originated, with E.C. Zeeman, the celebrated Catastrophe Theory. Professor Thom is a member of the American Academy of Arts and Sciences, Académie des Sciences de Paris, Deutsche Akademie der Naturforscher Leopoldina (DRG), and the Academy of Sciences of Brazil.

Special Preface

It is now some fifteen years since *Structural Stability and Morphogenesis* was first published in the French edition. The purely mathematical theory of singularities of smooth maps has developed tremendously in that time, especially after the work of V. I. Arnold and his school. However, the main interest of this book lies not in its mathematics, but in the methodological perspective opened up by Catastrophe Theory models.

Of course the book enjoyed no small success around 1973-74 when the work of C. Zeeman and his Warwick school brought Catastrophe Theory to the wider public. Undoubtedly it had some influence in re-directing the attention of physicists from the Quantum World to Classical Dynamics and our usual "macroscopic" Physical World. This trend is remarkable owing to the present emphasis on Chaos and Turbulence Theories. But it seems fair to say that many of the teachings of *Structural Stability and Morphogenesis* are still awaiting full recognition and possible applications. The hopes raised by C. Zeeman's models in the "soft" sciences dwindled somewhat after the great controversy of the years 1975-76. It became clear that Catastrophe Theory formalism is not able, in general, to predict phenomena, at least in the usual sense of quantitative modeling. But the possibility remains of a qualitative understanding, a fast interpretation of many dynamical situations involving opposite trends.

The views developed in this book about possible applications in many domains, especially in Biology, still await workers to develop them and to confront them with the "hard" facts of experimentation. In that respect, I very much hope that reprinting in the Addison-Wesley Advanced Book Classics series does not mean the book is now no more than a museum curiosity. Like D'Arcy Thompson's book *On Growth and Form*, it still contains many ideas that may burgeon in the mind of a curious and intellectually active reader. I sincerely hope that will happen.

René Thom
April, 1988

Contents

Foreword

to the English Edition

When I was asked, several years ago, to write a Foreword to the French version of René Thom's *Stabilité Structurelle et Morphogénèse*, his work was very little known among biologists, and since Thom often insisted on its importance for biology it seemed not inappropriate that a biologist should introduce it. By now, many more biologists have at least heard of Thom's theories and are ready to consider their import. This translation, so admirably and clearly made by David Fowler, will make them readily available to a wide audience. Readily—but perhaps not easily; it is important to emphasize that Thom's work is a part of mathematics, not in a direct way a part of biology. And it is difficult mathematics. At the same time, it is a branch of mathematics developed with a subject matter in view which is a definite aspect of reality. For the sake of dealing with this subject matter, Thom is ready to forego an insistence on all the possible niceties of rigour. For instance, on p. 43 he writes "...when the catastrophes are frequent and close together, each of them, taken individually, will not have a serious effect, and frequently each is so small that even their totality may be unobservable. When this situation persists in time, the observer is justified in neglecting these very small catastrophes and averaging out only the factors accessible to observation." A scientist is likely to feel, on reading such a passage, that it was written by one of his own colleagues rather than by a mathematician of the traditional kind he is accustomed to meeting.

What, then, is this subject matter, to which Thom will devote such arduous and subtle thought, even at the sacrifice of some mathematical purity? It is nothing less than one of the major problems of epistemology. Thom defines it in his first few sentences: "...it is indisputable that our universe is not chaos. We perceive beings, objects, things to which we give names. These beings or things are forms or structures endowed with a degree of stability; they take up some part of space and last for some period of time." And, Thom emphasizes, they have boundaries; a boundary implies a discontinuity; and the mathematics used in almost all science so far is based on the differential calculus, which presupposes continuity. This is the lacuna which Thom is attempting to fill.

I stress again that his attempt is that of a mathematician, not of the experimental scientist whose imagination is largely confined to the real four-dimensional world of observables. Quite early in his exposition, Thom points out that "we therefore endeavor in the program outlined here to free our intuition from three-dimensional experience and to use more general, richer, dynamical concepts, which will in fact be independent of the configuration spaces." The biologist or other scientific reader must continually strive to ensure that he is not interpreting Thom's words in too simple and conventional a manner. For instance, Thom speaks much of "morphogenesis," a word biologists commonly employ to refer to that shaping of developing tissues into recognizable forms such as those of particular bones, muscles, and the like. Thom, however, is using the word in a considerably wider sense: "Whenever the point m meets K, there will be a discontinuity in the nature of the system which we will interpret as a change in the previous form, a *morphogenesis* "(p. 7). Thus the "morphogenesis" of Thom's title is a more inclusive concept than that familiar to the biologist. The reader must not be surprised to find himself confronted with discussions not only of physics—a subject to which we have all had to try to acclimatize ourselves—but also of human communication and linguistics, in the last chapter, one of the most stimulating in the book.

Undoubtedly Thom sees in biology, and particularly in developmental biology, the main area for the application of his ideas. The manner of his writing is not, perhaps, always persuasive of the need to view his theories as abstract mathematical statements rather than as straightforward descriptions of the geometry of solid bodies. It is easy to accept that the chemical composition of differentiating cells needs for its description a multidimensional function space, and that the sudden transitions and sharp boundaries between one tissue and another are examples of Thomian catastrophes. His fundamental theorem, that in a four-dimensional world there are only seven basic type of elementary catastrophe, which he proceeds to describe, provides many very provocative ideas in connection with the known processes of embryonic differentiation. I personally feel that I have not adequately thought through the argument whereby Thom moves swiftly from a four-dimensional space of chemical composition to the four-dimensional world of extended material structures in time. The key point is perhaps most clearly stated on p. 169: "Suppose that, in the model, the space U into which the growth wave is mapped, and which parametrizes the average biochemical state of each cell, is a four-dimensional state $\mathbf{R}^4$ identified with space-time. This is not so restricted an assumption as it may seem at first sight, for if U actually has many more dimensions and we suppose that the growth wave $F(B^3, t)$ describing the evolution of the embryo is an embedding, the only effective part of U (in normal epigenesis) will be a four-dimensional domain." I think that much of the debate about the particularities of Thom's biological suggestions will turn on the justification of that remark.

The mentally straitlaced and orthodox should perhaps also be warned that Thom clearly takes a certain mischievous pleasure in provoking them. "This attitude," he remarks on p. 205, "is defiantly Lamarckian; it supposes that, on the whole, *the function creates the organ....*" In an evolutionary context, there is nothing in this notion that need upset any believer in natural selection unless he has been brainwashed into accepting that Lamarckism is a dirty word; the organ is in fact evolved by selection pressures acting in directions dictated by the function. Here I suspect that Thom himself is not quite clear in his own mind whether he is talking about the evolution or the embryonic development of the organ; the fault is perhaps to some extent with the French language, which is very uncertain in making this important distinction. There are many other examples of the highly individual nature of Thom's thought processes, characterized often by extremely long intuitive leaps.

I have mentioned some of the oddities of Thom's writings to warn readers that they will not find only what they already expect, and to advise them as earnestly as I can not to allow themselves to be sidetracked from the main argument. I am convinced that Thom's book is one of the most original contributions to the methodology of thought in the last several decades, perhaps since the first stirrings of quantum and relativity theories. In the particular field of embryological morphogenesis, which is so central to it, it is in my opinion more important than D'Arcy Thompson's great work, *On Growth and Form.* Thompson's contribution was to apply well-known types of mathematical thinking where they had not been applied before, whereas Thom invents not only the applications, but the mathematics as well. Just as much of the detail in D'Arcy Thompson has turned out to be invalid, or at best incomplete, so it is likely that not all Thom's suggestions will prove acceptable; but in neither case does that constitute any reason to overlook the importance of the new insights which these authors have given us.

C.H. Waddington

Foreword

to the French Edition

I am honoured to have been invited to write a *préface* to Dr. René Thom's *Stabilité structurelle et morphogénèse*. I cannot claim to understand all of it; I think that only a relatively few expert topologists will be able to follow all his mathematical details, and they may find themselves less at home in some of the biology. I can, however, grasp sufficient of the topological concepts and logic to realize that this is a very important contribution to the philosophy of science and to theoretical general biology in particular.

The state of biology at the present time seems to call for concerted attempts, from several sides, to develop with greater rigour and profundity the concepts and logical systems in terms of which we can usefully consider the major characteristics of life processes, at all levels. It is a remarkable fact that, although theoretical physics is a well-recognized discipline, served by its own journals and with separate academic departments in many universities, there is no similar acceptance of any discipline called theoretical biology. Is this, perhaps, because no such subject exists or is called for? It might be argued that, in their essence, living beings are no more than very complex physicochemical systems, and do not call for the development of any general theory other than that which can be borrowed from the physical sciences. Proponents of such a view would admit that there are many special aspects of biological processes which require the elaboration of bodies of appropriate special theory. It is obvious that the hydrodynamics of body fluids, the membrane permeabilities involved in nerve impulses and kidney function, the network systems of nervous connections, and many other biological phenomena require theoretical developments in directions which have not previously been forced on students of nonliving things, but which physicochemical theorists can accommodate without much difficulty in the types of thinking to which they are accustomed. These are, however, clearly only theories of parts of biology, not of general biology as a whole.

Could one go further and argue that biology does not demand the development of any general theoretical biology, but can be dealt with adequately by a number of separate theories of biological processes, each no more than an extrapolation and development of some appropriate physical theory? In the recent past, the consensus in practice has amounted to an acceptance of this point of view, since little attempt has been made to construct any more general theory. Until the last decade or so, the orthodox view was that the most fundamental characteristic of living things was the existence of a metabolism in apparent (but only apparent) conflict with thermodynamics; but it was held that this was merely a rather elaborate chemistry, demanding little addition to existing theory except the recognition that many proteins can act as enzyme catalysts. More recently geneticists have urged that heredity and evolution are still more basic living processes, but the brilliant discoveries of molecular biology have made it possible to argue that these phenomena essentially fall within the realm of a chemistry expanded only by a few additional theorems about the template activities of nucleic acids.

There is a good deal of weight behind these arguments. Indeed, so long as we regard living things as scientific objects, that is to say, as objectively observable, neglecting any considerations of subjectivity and consciousness, in a certain sense they must be "mere physics and chemistry." There is nothing else but the physical and chemical raw materials out of which they could be constructed; and if at any time our knowledge of the physical sciences is not enough to account for some biological phenomenon, such as the catalytic activities of polyaminoacids or the template properties of polynucleotides, then what we have to do is to develop further our physical and chemical theories. But, all this being granted, there is still a case to be made for the desirability of constructing a general theoretical biology in addition.

This case is most obvious in certain areas of biology which deal with phenomena a long distance removed from the physicochemical processes that must in the last resort be their ultimate base. The evolution of higher organisms is an outstanding example. It would be outrageously clumsy to try to express the problems directly in physicochemical terms, and we have in fact developed quite an elaborate theory of population and evolutionary genetics. There are similar, though not quite so obtrusive, needs for specifically biological theories in other major areas, such as development and metabolism, though as yet these are not so definitely formulated as evolutionary theory. Such biological theories might perhaps be compared with the general theories of particular types of physical phenomena, such as aerodynamics, electronic circuitry, and optics. They would still not demand the formulation of any general theoretical biology, were it not for the fact that they share certain characteristics; and it is the exploration of the common characteristics of the various types of biological theory that constitutes the main topic of theoretical biology.

When a category of biological processes, such as evolution or development, leads to the formation of an appropriate and specifically biological body of theory, it does so because it exhibits two characteristics; it involves entities which have a certain global simplicity and definiteness of character (e.g., a given species of animal or plant, an organ such as the heart or liver, or a cell type such as a muscle or nerve cell), but if one attempts to analyze these entities into basic constituents, such as genes or molecules, they turn out to be of unmanageable complexity. The logical structure of important biological concepts is almost always an actual simplicity (exhibited in their relations to other concepts in the theoretical scheme), included within which is an extreme complexity (revealed on reductional analysis). If there was no simplicity, there would be nothing to make a theory about; if the complexity remained manageable, physicochemical theories would suffice.

Because of their inherent analytical complexity, biological concepts in general imply a multidimensionality. To specify a liver cell, the kidney, or an evolving population would require—if it could be done—the enumeration of a large number of parameters—10^2? 10^3? 10^4? We do not know. Such concepts can therefore be properly related to one another only within a logicomathematical framework which can handle a multiplicity of dimensions. Statistical mechanics is one such system, but the most general is topology. I hope I may be excused for remarking that as long ago as 1940, in my book *Organisers and Genes*, I urged the need to develop a topology of biology. In the intervening years, as not even an amateur mathematician, I have been quite unable to follow my own prescription. I am all the more grateful, therefore, to René Thom, who has now entered the field with such a strong and extended effort. Thom has tried to show, in detail and with precision, just how the global regularities with which biology deals can be envisaged as structures within a many-dimensioned space. He not only has shown how such ideas as chreods, the epigenetic landscape, and switching points, which previously were expressed only in the unsophisticated language of biology, can be formulated more adequately in terms such as vector fields, attractors, catastrophes, and the like; going much further than this, he develops many highly original ideas, both strictly mathematical ones within the field of topology, and applications of these to very many aspects of biology and of other sciences.

It would be quite wrong to give the impression that Thom's book is exclusively devoted to biology. The subjects mentioned in his title, *Structural Stability and Morphogenesis*, have a much wider reference; and he relates his topological system of thought to physical and indeed to general philosophical problems. I have little competence to make any comments on these aspects of the book. In biology, Thom not only uses topological modes of thought to provide formal definitions of concepts and a logical framework by which they can be related; he also makes a bold attempt at a direct comparison between topological structures within four-dimensional space-time, such as catastrophe hypersurfaces, and the physical structures found in developing embryos. I have not yet made up my mind whether I think that, in these efforts, he is allowing his desire to find some "practical results" of his theories to push him further than he ought to go in literal interpretation of abstract formulations. Whether he is justified or not in this makes little difference to the basic importance of this book, which is the introduction, in a massive and thorough way, of topological thinking as a framework for theoretical

Preface

translated from the French Edition

This book, written by a mathematician, is hopefully addressed to biologists and specialists of other disciplines whose subjects have so far resisted mathematical treatment. Although the radically new methods that are here developed will not require more than a very crude mathematical formalism, they do demand close acquaintance with concepts and objects basic both to differential topology and classical mechanics: differential manifolds, vector fields, dynamical systems, and so forth. I will not deny that communication will be difficult, all the more so since there is as yet no accessible modern introduction to these ideas. The nonmathematical reader can approach them through the mathematical summary at the end of this book and omit at first reading the more technical portions, Chapters 3 and 7. This may do little to ease communication but my excuse is an infinite confidence in the resources of the human brain!

Although no one can say that this book lacks originality, many will surely point to its precursors, both known and unknown to me. I would like to mention the following that I know: the most classical, *On Growth and Form* by D'Arcy Thompson, of which this book attempts a degree of mathematical corroboration; the books of C. H. Waddington, whose concepts of the "chreod" and the "epigenetic landscape" have played a germinal part in the formation of my theory; and among physiologists there is J. von Uexküll (*Bedeutungslehre*) and K. Goldstein (*Der Aufbau des Organismus*).

In a work with such comprehensive aims as this, the text should not be weighed down by all the possible references. Accordingly, I have retained only those involving necessary details of specific technical facts and have refrained from citing any reference where general or philosophical statements are concerned; for this I apologize to any writers who feel slighted.

I cannot express sufficient gratitude to the many colleagues who helped me with my task, especially Ph. L'Héritier, Etienne Wolff, and C. H. Waddington, who gave so much of their time and their valued conversation. I should also like to thank my colleagues at the University of Strasbourg, P. Pluvinage and his assistant, M. Goeltzene, who helped me in their laboratory to produce the photographs of caustics for this book.

René Thom

REFERENCES

D'A. W. Thompson, *On Growth and Form*, Cambridge University Press, 1917; second edition, 1942; abridged edition, 1961.

J. von Uexküll, *Bedeutungslehre*, J. A. Barth, 1940. French translation: *Mondes animaux et monde humain*, Gonthier, 1965.

K. Goldstein, *Der Aufbau des Organismus: Einführung in die Biologie unter besonderen Berücksichtigung der Erfahrungen am kranken Menschen*, Nijhoff, 1934. English translation: *The Organism: a Holistic Approach to Biology Derived from Pathological Data in Man*, Beacon Press, 1963.

Structural Stability and Morphogenesis

Translator's Note

The translation has been vastly improved by the time and effort that M. Thom devoted to it. He removed blunders, obscurities, and *contresens horribles* with which it was infested, brought some of the material up to date, and added some new sections and notes. I should like here to express my gratitude. I am also indebted to my wife, who is still patiently trying to teach me French, to Christopher Zeeman, who encouraged and helped me with this translation, to the many other people who provided help, in particular Klaus Jänich and Stephen Stewart, and finally to those who typed the various versions, most particularly Mademoiselle A. Zabardi, who reduced a mountain of scribble to a beautiful typescript.

David Fowler

CHAPTER 1

INTRODUCTION

The waves of the sea, the little ripples on the shore, the sweeping curve of the sandy bay between the headlands, the outline of the hills, the shape of the clouds, all these are so many riddles of form, so many problems of morphology...

D'ARCY THOMPSON
On Growth and Form

1.1. THE PROGRAM

A. The succession of form

One of the central problems studied by mankind is the problem of the succession of form. Whatever is the ultimate nature of reality (assuming that this expression has meaning), it is indisputable that our universe is not chaos. We perceive beings, objects, things to which we give names. These beings or things are forms or structures endowed with a degree of stability; they take up some part of space and last for some period of time. Moreover, although a given object can exist in many different guises, we never fail to recognize it; this recognition of the same object in the infinite multiplicity of its manifestations is, in itself, a problem (the classical philosophical problem of concept) which, it seems to me, the Gestalt psychologists alone have posed in a geometric framework accessible to scientific investigation. Suppose this problem to be solved according to naive intuition, giving to outside things an existence independent of our own observation.[1]* Next we must concede that the universe we see is a ceaseless creation, evolution, and destruction of forms and that the purpose of science is to foresee this change of form and, if possible, explain it.

B. Science, and the indeterminism of phenomena

If the change of forms were to take place at all times and places according to a single well-defined pattern, the problem would be much

*Superior numbers refer to notes at the end of the chapter.

easier; we could then set out, once and for all, the necessary order of the change of form (or systems of forms) in the neighborhood of any point (e.g., as a table or a graph), and this would then be at least an algorithm giving a prediction of phenomena, if not an explanation. Very probably the mind would become used to considering this necessary order of change of forms as imposed by a causality or even a logical implication

The fact that we have to consider more refined explanations—namely, those of science—to predict the change of phenomena shows that the determinism of the change of forms is not rigorous, and that the same local situation can give birth to apparently different outcomes under the influence of unknown or unobservable factors. It is ironical to observe here that this science which, in principle, denies indeterminism is actually its ungrateful offspring, whose only purpose is to destroy its parent! Thus classical mechanics, a strictly quantitative and deterministic theory, was created in order to remove the indeterminism found in all instances of moving bodies (e.g.: Is this bullet going to hit its target or not? Will this weight stay in equilibrium or not?); conversely, if some disciplines, like social sciences and biology, resisted mathematical treatment for so long, even if they have succumbed, this is not so much because of the complexity of their raw material, as is often thought (all nature is complicated), but because qualitative and empirical deduction already gives them sufficient framework for experiment and prediction.

1.2. THE THEORY OF MODELS

A. Formal models

In those ambiguous or catastrophic situations where the evolution of phenomena seems ill determined, the observer will try to remove the indeterminacy and thus to predict the future by the use of *local models*. The single idea of a spatiotemporal object already implies the idea of a model (this is discussed in Chapter 2). From this point of view, we say that a system of forms in evolution constitutes a *formalizable* process if there is a formal system P (in the sense of formal logic) satisfying the following conditions:

1. Each state A of the phenomenological process under consideration can be parameterized by a set of propositions a of the formal system P.

2. If, in the course of time, state A is transformed into state B, then B can be parameterized by a set b of P such that b can be deduced from a in P.

In other words, there is a bijective map h from some or all of the propositions of P onto the set of forms appearing globally in the process, and the inverse of this map transforms temporal into logical succession.

Such a model is not necessarily deterministic, for a set a of premises of P can, in general, imply a large number of formally different conclusions, and so the model is not entirely satisfactory for, being indeterministic, it does not always allow prediction. Atomic models, in which all forms of the process under consideration arise by aggregation or superposition of elementary indestructible forms, called atoms, and all change is a change in the arrangements of these atoms, are to some extent models of this type. But the theory will not be satisfactory unless it allows for prediction, and for this it is necessary, in general, to construct a new theory, usually quantitative, which will be the thermodynamical theory governing the arrangement of these particles.

All models divide naturally in this way into two a priori distinct parts: one *kinematic*, whose aim is to parameterize the forms or the states of the process under consideration, and the other *dynamic*, describing the evolution in time of these forms. In the case envisaged above of a formalizable process, the kinematic part is given by the formal system P together with the bijective map h of P onto the forms of the process. The dynamic part, if it is known, will be given by the transition probabilities between a state A parameterized by a set of propositions a and a state B parameterized by b, a consequence of a. In this way the kinematical theory, if formalizable, implies a restriction on the dynamical theory, because the transition probability between states A and C must be zero if $h^{-1}(C)$ is not a consequence of $a = h^{-1}(A)$ in P. It is altogether exceptional for a natural process to have a global formalization; as we shall see later, it is a well-known experience that initial symmetries may break in some natural processes.[2] As a result we cannot hope for a global formalization, but local formalizations are possible and permit us to talk of cause and effect. We can say that the phenomena of the process are effectively explained only when P is effectively a system of formal logic; in most known cases P has a less rigid structure, with only a preordering in the place of logical implication. Dropping the natural restriction that P contains a countable number of elements (parameterized by symbols, letters, etc.), we obtain quantitative or continuous models.

B. Continuous models

It is quite natural to make P into a topological space with the convention that, if the point representing a system lies outside a certain closed set K of P (the set of catastrophe points), the qualitative nature of the state does not vary for a sufficiently small deformation of this state. Each type, each form of the process then corresponds to a connected component of

P-K. If P also has a differentiable structure (e.g., Euclidean space $\mathbf{R}^m$, or a differential manifold) the dynamical structure will be given by a vector field X on P. Existence and uniqueness theorems for solutions of differential equations with differentiable coefficients then give what is without doubt the typical paradigm for scientific determinism. The possibility of using a differential model is, to my mind, the final justification for the use of quantitative methods in science; of course, this needs some justification: the essence of the method to be described here consists in supposing a priori the existence of a differential model underlying the process to be studied and, without knowing explicitly what this model is, deducing from the single assumption of its existence conclusions relating to the nature of the singularities of the process. Thus postulating the existence of a model gives consequences of a local and qualitative nature; from quantitative assumptions but (almost always) without calculation, we obtain qualitative results. It is perhaps worthwhile tc discuss this question in greater detail.

1.3. A HISTORICOPHILOSOPHICAL DIGRESSION

A. Qualitative or quantitative

The use of the term "qualitative" in science and, above all, in physics has a pejorative ring. It was a physicist who reminded me, not without vehemence, of Rutherford's dictum, "Qualitative is nothing by poor quantitative." But consider the following example. Let us suppose that the experimental study of phenomenon Φ gives an empirical graph g with equation $y = g(x)$. To explain Φ the theorist has available two theories θ_1 and θ_2; these theories give graphs $y = g_1(x)$ and $y = g_2(x)$, respectively. Neither of these graphs fits the graph $y = g(x)$ well (see Figure 1.1); the graph $y = g_1(x)$ fits better quantitatively in the sense that, over the interval considered, $\int|g - g_1|$ is smaller than $\int|g - g_2|$, but the graph g_2 has the same shape and appearance as g. In this situation one would lay odds that the theorist would retain θ_2 rather than θ_1 even at the expense of a greater quantitative error, feeling that θ_2, which gives rise to a graph of the same appearance as the experimental result, must be a better clue to the underlying mechanisms of Φ than the quantitatively more e: :ct θ_1. Of course this example is not a proof, but it illustrates the natural tendency of the mind to give to the shape of a graph some intrinsic value; it is this tendency that we shall develop here to its ultimate consequences.

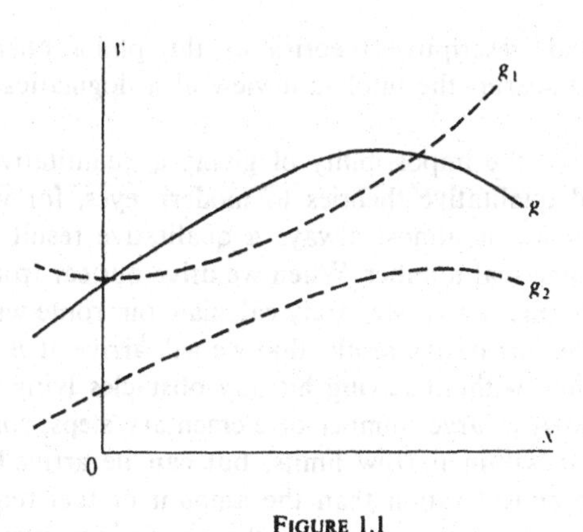

FIGURE 1.1

B. The shadow of history

History gives another reason for the physicist's attitude toward the qualitative. The controversy between the followers of the physics of Descartes and of Newton was at its height at the end of the seventeenth century. Descartes, with his vortices, his hooked atoms, and the like, explained everything and calculated nothing; Newton, with the inverse square law of gravitation, calculated everything and explained nothing. History has endorsed Newton and relegated the Cartesian constructions to the domain of curious speculation. The Newtonian point of view has certainly fully justified itself from the point of view of its efficiency and its ability to predict, and therefore to act upon phenomena. In the same spirit, it is interesting to reread the introduction to Dirac's *Principles of Quantum Mechanics*, wherein the author rejects as unimportant the impossibility of giving an intuitive context for the basic concepts of quantum methods.[3] But I am certain that the human mind would not be fully satisfied with a universe in which all phenomena were governed by a mathematical process that was coherent but totally abstract. Are we not then in wonderland? In the situation where man is deprived of all possibility of intellectualization, that is, of interpreting geometrically a given process, either he will seek to create, despite everything, through suitable interpretations, an intuitive justification of the process, or he will sink into resigned incomprehension which habit will change to indifference. In the case of gravitation there is no doubt that the second attitude has prevailed, for we have not, in 1975, less reason to be astonished at the fall of the apple than had Newton. The dilemma posed all scientific explanation is this: magic or geometry.[4] From this point of view, men striving for understanding will never show toward the qualitative and descriptive theories of the philosophers from the pre-Socratics to Descartes the intolerant view of a dogmatically quantitative science.

However, it is not the impossibility of giving a quantitative result that condemns the old qualitative theories to modern eyes, for what matters most for everyday use is almost always a qualitative result and not the precise value of some real number. When we drive our car from town A to town B a hundred miles away, we rarely calculate our route with precision. What matters is the qualitative result: that we will arrive at B after a finite and reasonable time without having hit any obstacles lying in our path. Such a result follows a large number of elementary steps, some of which must be computed within narrow limits, but can be arrived at after no more quantitative consideration than the amount of fuel required at the start. What condemns these speculative theories in our eyes is not their qualitative character but the relentlessly naive form of, and the lack of precision in, the ideas that they use. With the exception of the grandiose, profound, but rather vague ideas of Anaximander and Heraclitus, the first pre-Socratic philosophers, all these theories rely on the experience of solid bodies in three-dimensional Euclidean space. This intuition, however natural and innate in our development and use of our original tools, is insufficient for a satisfactory account of most phenomena, even on a macroscopic scale.

C. An extension of our basic intuition

We may then ask whether we cannot, by a refinement of our geometric intuition, furnish our scientific investigation with a stock of ideas and procedures subtle enough to give satisfactory qualitative representations to partial phenomena. It is necessary to emphasize one point: we can now present qualitative results in a rigorous way, thanks to recent progress in topology and differential analysis, for we know how to define a *form* and can determine whether two functions have or have not the same form or topological type. We therefore endeavor in the program outlined here to free our intuition from three-dimensional experience and to use much more general, richer, dynamical concepts, which will in fact be independent of the configuration spaces. In particular, the dimension of the space and the number of degrees of freedom of the local system are quite arbitrary—in fact, the universal model of the process is embedded in an infinite-dimensional space. This is obviously necessary; there is no doubt that the closer the study approaches the infinitesimal, the more degrees of freedom are needed, so that all qualitative representation of microscopic phenomena will require the use of an infinite-dimensional function space.

One essential feature of our use of local models is that it implies nothing about the "ultimate nature of reality"; even if this is ever revealed by analysis complicated beyond description, only a part of its manifestation, the so-called observables, are finally relevant to the macroscopic description of the system. The phase space of our dynamical model is defined using only these observables and without reference to any more-or-less chaotic underlying structures.

To each partial system, relatively independent of the environment, we assign a local model that accounts qualitatively and, in the best cases, quantitatively for its behavior. But we cannot hope, a priori, to integrate all these local models into a global system. If it were possible to make such a synthesis, man could justifiably say that he knew the ultimate nature of reality, for there could exist no better global model. For myself, I think that this would be extravagant pretension; the era of grand cosmic synthesis ended, very probably, with general relativity, and it is most doubtful that anybody will restart it, nor would it seem to be useful to attempt to do so.

1.4. THE CONSTRUCTION OF A MODEL

A. The catastrophe set

We propose the following general model to parameterize the local states of a system: the space of observables M contains a closed subset K, called the *catastrophe set*, and as long as the representative point m of the system does not meet K, the local nature of the system does not change. The essential idea introduced here is that the local structure of K, the topological type of its singularities and so forth, is in fact determined by an underlying dynamic defined on a manifold M which is in general impossible to exhibit. The evolution of the system will be defined by a vector field X on M, which will define the macroscopic dynamic. Whenever the point m meets K, there will be a discontinuity in the nature of the system which we will interpret as a change in the previous form, a *morphogenesis*. Because of the restrictions outlined above on the local structure of K we can, to a certain extent, classify and predict the singularities of the morphogenesis of the system without knowing either the underlying dynamic or the macroscopic evolution defined by X. In fact, in most cases we proceed in the opposite direction: *from a macroscopic examination of the morphogenesis of a process and a local and global study of its singularities, we can try to reconstruct the dynamic that generates it.* Although the goal is to construct the quantitative global model (M, K, X), this may be difficult or even impossible. However, the local dynamical interpretation of the singularities of the morphogenesis is possible and useful and is an indispensable preliminary to defining the kinematic of the model; and even if a global dynamical evolution is not accessible, our local knowledge will be much improved in the process.

The method dealt with here puts emphasis above all on the morphogenesis of the process, that is, on the discontinuities of the phenomenon. A very general classification of these changes of form, called *catastrophes*, will be given in Chapter 4.

B. The independence of the substrate

That we can construct an abstract, purely geometrical theory of morphogenesis, *independent of the substrate of forms and the nature of the forces that create them*, might seem difficult to believe, especially to the seasoned experimentalist used to working with living matter and always struggling with an elusive reality. This idea is not new and can be found almost explicitly in the classical book of D'Arcy Thompson, *On Growth and Form*, but the theories of this innovator were too far in advance of their time to be recognized; moreover, they were expressed in a geometrically naive way and lacked the mathematical justification that has only been found in the recent advances in topology and differential analysis.

This general point of view raises the following obvious question: if, according to our basic hypothesis, the only stable singularities of all morphogenesis are determined solely by the dimension of the ambient space, why do not all phenomena of our three-dimensional world have the same morphology? Why do clouds and mountains not have the same shape, and why is the form of crystals different from that of living beings? To this I reply that the model attempts only to classify local accidents of morphogenesis, which we will call *elementary catastrophes*, whereas the global macroscopic appearance, the form in the usual sense of the word, is the result of the accumulation of many of these local accidents. The statistic of these local accidents and the correlations governing their appearance in the course of a given process are determined by the topological structure of their internal dynamic, but the integration of all these accidents into a global structure would require, if we wanted to pursue the application of the model, a consideration of catastrophes on spaces of many more dimensions than the normal three. It is the topological richness of the internal dynamics that finally explains the boundless diversity of the external world and perhaps even the distinction between life and inert matter.

C. Biological and inert forms

It is here important to note a generally neglected situation: for centuries the form of living beings has been an object of study by biologists, while the morphology of inert matter seems only accidentally to have excited the interest of physicochemists. There are many disciplines of obvious practical use (e.g., meterology and the forms of clouds, structural geology and the geomorphology of the forms of terrestial relief) for which there is a valuable body of experimental observation and occasionally some satisfactory local dynamical explanations, but scarcely any attempt at global integration beyond a purely verbal description. A similar situation exists in astronomy in the study of the morphology of stellar objects (and, in particular, spiral nebulae, which we shall consider in Chapter 6), as well as in the study of dislocations in crystalline lattices, a research inspired by the needs of practical metallurgy. However, the central problem of the study of the geometrical partition of a substance into two phases has never been systematically attacked; for example, there is no model to explain the dendritic growth of crystals.

The reason for this neglect by physicochemists is clear:[5] these phenomena are highly unstable, difficult to repeat, and hard to fit into a mathematical theory, because the *characteristic of all form, all morphogenesis, is to display itself through discontinuities of the environment*, and nothing disturbs a mathematician more than discontinuity, since all applicable quantitative models depend on the use of analytic and therefore continuous functions. Hence the phenomenon of breakers in hydrodynamics is little understood, although it plays an important role in morphogenesis in three-dimensional space. D'Arcy Thompson, in some pages of rare insight, compared the form of a jellyfish to that of the diffusion of a drop of ink in water;[6] it may happen that biological morphogenesis, which is better known, which takes place slowly, and which is controlled strictly, may help us to understand the more rapid and fleeting phenomena of inert morphogenesis.

D. Conclusion

Finally, the choice of what is considered scientifically interesting is certainly to a large extent arbitrary. Physics today uses enormous machines to investigate situations that exist for less than 10^{-23} second, and we surely are entitled to employ all possible techniques to classify all experimentally observable phenomena. But we can at least ask one question: many phenomena of common experience, in themselves trivial (often to the point that they escape attention altogether!)—for example, the cracks in an old wall, the shape of a cloud, the path of a falling leaf, or the froth on a pint of beer—are very difficult to formalize, but is it not possible that a mathematical theory launched for such homely phenomena might, in the end, be more profitable for science?

The pre-Socratic flavor of the qualitative dynamics considered here will be quite obvious. If I have quoted the aphorisms of Heraclitus at the beginnings of some chapters, the reason is that nothing else could be better adapted to this type of study. In fact, all the basic intuitive ideas of morphogenesis can be found in Heraclitus: all that I have done is to place these in a geometric and dynamic framework that will make them some day accessible to quantitative analysis. The "solemn, unadorned words," like those of the sibyl that have sounded without faltering throughout the centuries, deserve this distant echo.

APPENDIX

The notion of an object. Let U denote the space of all possible positions of my body in the external world; in each position $u \in U$, the set of all my sense data decomposes, in principle, into distinct entities, *forms*, each of which persists during a small displacement of u. In this way to each $u \in U$ is associated a discrete space F_u, and the union $\cup F_u$ can be given a topology making it a space $\hat{U}$ stacked (*étalé*) over U. When V is an open subset of U, let $G(V)$ be the set of sections of the map $\hat{V} \to V$ induced by $\hat{U} \to U$; there is a canonical restriction map $G(V') \to G(V)$ for $V' \supset V$. We define an *object* c to be a maximal section of $\hat{U} \to U$ for the restriction operation $G(V') \to G(V)$, and with each object associate a *domain of existence* $J(c) \subset U$. Such a domain is necessarily connected; for example, if form A is a sheet of paper, and I crumple this sheet, giving it form B, then A can be continually deformed into B and, conversely, corresponding to smoothing the paper, B into A. But if I now tear the sheet, giving it form C, this produces a new object because the transition $A \to C$ is not reversible.

Another postulate underlying the usual idea of an object is the following: *every object c is characterized by its domain of existence $J(c)$*. Equivalently, if two objects have the same domain, they are identical. Consequently, if one describes a loop k in U and extends the section of $\hat{U} \to U$ along k, these sections will never interchange; the principle of indiscernibility of particles in quantum mechanics shows the shortcomings of this postulate. In this case, when U is the space corresponding to a field of k isolated particles, there will be a closed subspace K corresponding to the state when two or more particles are in collision; U-K is connected, and if u and v are two points of U-K corresponding to states $(a_1, \ldots, a_k)$ and $(b_1, \ldots, b_k)$ of the particles, all distinct, each path q from u to v will, by lifting, give an identification of $(a_1, \ldots, a_k)$ and $(b_1, \ldots, b_k)$, depending, in general, on the homotopy class of the path q. It may happen that two objects A and B are bound together in the sense that every transformation $A \to A'$ in $\hat{U}$ can be extended to transformation $B \to B'$, but there can exist transformations $B \to B''$ not affecting A; we then call B a *reflection* of A. This pathology is avoided for solid bodies in the outside world only because of the solidarity of matter, which allows each body to be manipulated in isolation. It seems probable that the properties of connection, reversibility, and "indecomposibility," which define the space of the same object, have as origin not the physical properties of the outside world but the constraints of the dynamic of our brain, which prevent us from thinking of more than one thing at a time.

NOTES

1. The most convinced solipsist, when living and going about his business, must adjust to the world outside and admit its structural invariance in his use of it; does this not amount to admitting the existence of a certain reality?

2. We are referring here to the phenomenon of *generalized catastrophe*, described in Chapter 6, which is connected in principle with the breaking of a symmetry.

3. ". . . The main object of physical systems is not the provision of pictures, but the formulation of laws governing phenomena, and the application of these laws to the discovery of new phenomena. If a picture exists, so much the better; but whether a picture exists or not is a matter of only secondary importance." P. M. Dirac, *Principles of Quantum Mechanics*, third edition, Oxford, 1961, p. 10.

4. We shall see in Chapter 13 that classical Euclidean geometry can be considered as magic; at the price of a minimal distortion of appearances (a point without size, a line without width) the purely formal language of geometry describes adequately the reality of space. We might say, in this sense, that *geometry is successful magic*. I should like to state a converse: is not all magic, to the extent that it is successful, geometry?

5. This lack of interest in inert morphogenesis has probably a biological origin, for all our perceptual organs are genetically developed so as to detect the living beings that play a large role, as prey or predators, in our survival and the maintenance of our psychological equilibrium; by contrast, an inanimate object excites only a passing glance, and that only insofar as its form suggests that of a living being.

6. D'A. W. Thompson, *On Growth and Form*, abridged edition, Cambridge University Press, 1961, pp. 72–73.

CHAPTER 2

FORM AND STRUCTURAL STABILITY

Pour nous autres Grecs, toutes choses sont formes....

PAUL VALÉRY, Eupalinos

2.1. THE STUDY OF FORMS

A. The usual sense of form or shape

What is form? It is far from an easy task to give a definition to cover all common uses. A mathematical definition is immediate: two topological spaces X and Y have the same form if they are *homeomorphic* (this means that there is a map from X onto Y which is bijective and bicontinuous). If we restrict our attention to the shape of objects in three-dimensional Euclidean space $\mathbf{R}^3$, it is clear that this idea of homeomorphism is far too general to cover the usual definition of the form of an object. Let us observe first that what is usually called an object in space is, topologically, a closed subset (and usually compact, for objects rarely extend to infinity). Then we might propose a new definition: two objects A and A' in Euclidean space have the same shape if there exists a displacement D such that $A' = D(A)$. Now the definition is too restrictive, however, for two objects cannot have the same shape without being metrically equal, whereas it will not be obvious (to the eyes of a Gestalt psychologist, for example) that a square with horizontal and vertical sides has the same shape as the same square with sides inclined at 45° to the horizontal (see Figure 2.1). Thus the group of displacements is not suitable and must be replaced by a pseudogroup G of local homeomorphisms, leaving the form invariant. An a priori formal definition of this pseudogroup is not easy; it must leave unchanged the horizontal and vertical, and contain both translations and a neighborhood of the identity in the affine group

$x' = kx$, $y' = k'y$ of the Oxy-plane.[1] A unique pseudogroup G valid for all forms probably does not exist, and it is quite likely that each subjectively defined form F admits its own particular pseudogroup G_F of equivalences.

It is clear that some forms have special value for us or are biologically important, for example, the shapes of food, of animals, of tools. These forms are genetically imprinted into our understanding of space, and their invariant pseudogroups G_F are narrowly and strictly adapted to them. A precise definition of this pseudogroup G_F is in general difficult or impossible: to define the spatial form of an animal, for example, is equivalent to defining what is called in biology the phenotype of this animal, and nothing is more difficult to define exactly, as we shall see in Chapter 8. We must therefore tackle the problem of the classification of spatial forms from a different, more Gestalt point of view.

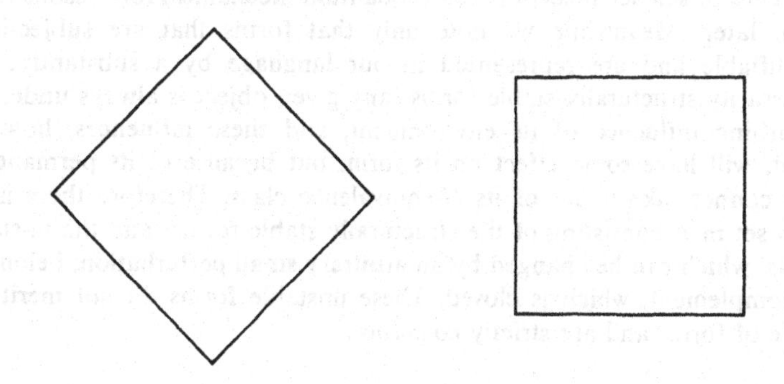

FIGURE 2.1

B. The space of forms

We shall consider the space Y of compact subsets of $\mathbf{R}^3$ with its Hausdorff metric: this space is, to a large extent, divided into basins of attraction, each attractor corresponding to a biologically important type of form. This division into basins is not rigid and, for a given person, can vary with his psychological state and his conscious and unconscious desires; the Rorschach tests are based on this variability. Even if this interpretation of perception of form has any chance of being correct, it does not lend itself to the construction of a mathematical model, and so we return to the following, more abstract definition: if E is a topological space, and G a group (or pseudogroup) operating on E, then a G-form is defined to be an equivalence class of closed sets of E modulo the action of

G. If, for example, E is a differential manifold and G a Lie group operating differentiably on E, there exists in general a continuum of equivalence classes, thus of G-forms, each G-form being parameterized by a system of real parameters. But when G is an infinite-dimensional group (or pseudogroup) there may only be a finite (or, at most, enumerable) number of G-forms with the property of structural stability.

C. Structural stability

A G-form A is called *structurally stable* if any form B sufficiently close to A in E is G-equivalent to A. Alternatively, a G-equivalence class F defines a structurally stable form if and only if the set of points in E of this equivalence class is an open set in E. The concept of structural stability seems to me to be a key idea in the interpretation of phenomena of all branches of science (except perhaps quantum mechanics) for reasons to be given later. Meanwhile we note only that forms that are subjectively identifiable and are represented in our language by a substantive are necessarily structurally stable forms; any given object is always under the disturbing influence of its environment, and these influences, however slight, will have some effect on its form, but because of its permanence they cannot take it out of its G-equivalence class. Therefore there is an open set in E consisting of the structurally stable forms; and the unstable forms, which can be changed by an arbitrary small perturbation, belong to its complement, which is closed. These unstable forms do not merit the name of forms and are strictly *nonforms*..

D. Nonform forms

We can distinguish two major classes of unstable forms, each at an end of a continuous spectrum: first, forms that are nonforms by reason of their very complicated internal structure—chaotic and not amenable to analysis; and, second, those that consist of a number of identifiable objects but whose composition seems contradictory or unusual, for example, the chimera and other monsters. Unstable forms of this second type are, in the space Y considered earlier, the bifurcation forms whose representative point lies on a threshold between two or more basins of attraction, and their appearance continually oscillates between the adjacent attractors. The effect is to upset or disquiet the observer—this is the technique known to and exploited by surrealist painters (see Photograph 1). On the other hand, the first type will be represented in Y by points that are in the closure of an infinite number of basins of attraction (in the static model to be defined in Chapter 4 these forms will be in strata of infinite codimension of the bifurcation set).

E. Geometrical forms

This is the place to consider the geometrical forms like lines, squares, and triangles. Theoretically the equivalence group of these forms is a finite-dimensional Lie group, and they form strata of infinite codimension in the Hausdorff space of spatial forms; from a Gestalt point of view, however, there is lit'le doubt that they in fact form an open set, since they do not stop being recognizable under small deformations. This is certainly a sign that the blossoming of geometrical thought in man is an extreme case of threshold stabilization, defined in Chapters 5 and 9

2.2. STRUCTURAL STABILITY AND SCIENTIFIC OBSERVATION

A. The conditions of scientific experiment

Let us return to the fundamental idea, structural stability; as we have said, the Universe is not chaos, and we can observe the recurrence of typical forms to which we give names. We must, however, take account of the conditions of scientific observation itself. The experimenter cannot observe all the universe at once but, for his experiments and observations, is compelled to isolate a subsystem S which is relatively independent of the rest of the universe. In practice he isolates and observes S in a box B whose geometric characteristics and the nature of whose attached measuring devices he specifies. Then he sets S in a certain state a which is defined by the *preparation procedure*, that is, the mode in which the box B is filled, described as precisely as possible. Having set up state a, the experimenter observes or tests the contents of box B some time after the preparation of a. Each experimenter hopes that, if another experimenter performs this experiment at another time and place with box B' obtained from B by the action of an element of the Galilean group G and with the same process of preparation of a, he will observe, to within experimental error, the same phenomena; without this hope, all would be in vain. However, no matter what precautions are taken to isolate S, the experimenter cannot remove entirely the interaction between S and the outside world, and *the conditions of a preparation procedure cannot be described and realized with perfect accuracy*—and these initial differences cannot but perturb the evolution of the system. Therefore approximately equal results (i.e., equivalent under G)

can be expected only after implicitly assuming that the evolution of S from state a is qualitatively stable, at least in respect to perturbations of the initial state and interaction with the outside world. In this way *the hypothesis of structural stability of isolated scientific processes is implicit in all scientific observation.*

We must, however, make precise what is meant by saying that the phenomena in boxes B and B' are "the same". Suppose, for the moment, that the experimenter can describe the local state of S at all points (x,t) of $B \times T$, for example, by the use of local probes. Then we can describe the qualitative agreement of events in B and B' as follows: there exists a ϵ-homeomorphism h of the product $B \times T$, $h:(x,t) \rightarrow (x,t)$, such that the composition $g \circ h$ transforms the process in B into the process in B'. From this point of view the local state of a system can also be described as a germ of the structurally stable evolution of the process.

B. The quantum objection

This description is classical and is not in the spirit of quantum mechanics. It would be rejected out of hand by a modern physicist, who would not fail to object on two grounds:

1. All experiment, all measurement, disturbs irreversibly the evolution of the process.
2. There is no local stability at the level of the single process, but only the statistical stability of a large number of events.

I am not sure that these difficulties do not appear in a large number of processes considered as classical; in biology, for example, it is often necessary for an observer to kill an animal subject in order to make his observation. Although this in practice introduces no difficulties into the task of describing the process, for he can repeat the experiment with another animal and allow it to continue to a later stage before intervening, a difficulty of the second type may appear, because he can never be sure that the two animals have the same genetical inheritance (genotype) and this unknown factor can have a most disastrous effect on the stability of the process. Personally I believe that the problem of preparing two electrons in precisely the same state is exactly the same problem confronting the biologist of breeding two ducks with the same hereditary characteristics. In quantum mechanics every system carries the record of every previous interaction it has experienced—in particular, that which created it, and in general it is impossible to reveal or evaluate this record. To conclude, the process described here must, like all models, be an idealization valid for many macroscopic processes, but I should not be surprised if it worked for phenomena of quantum mechanics, even though the uncertainty principle would not then permit the use of traditional spatial coordinates in the base space.

C. Isomorphic processes

We now need to define what we call *isomorphic processes*, that is, processes that can be transformed into each other by homeomorphism of the box B or the product $B \times T$. Observe first that a state a of the system S is, in general, not isolated in the set of states of the system, for all preparation depends on data given in the form of finite or infinite sets of real parameters c_i (e.g., the description of the metric form of box B could require an infinite number of parameters, or a function space). If some of these parameters are superfluous because they have no observable effect on the system, they can be neglected by passing to a quotient space (in geometric language). Finally the states of S appear as a topological space of which the connected components are finite or infinite-dimensional differential manifolds. We must give the case of infinite-dimensional manifolds an interpretation in terms of function spaces—for example, in present-day quantum mechanics they are the spaces L^2 of square summable functions with the structure of Hilbert space. More generally, they may be spaces C^m of m-times continuously differentiable maps with the appropriate topology. If the source manifold of the maps is compact, this manifold will have locally the linear structure of a Banach space.

The topology of the space of states can be recovered by introducing some measurements; suppose that, by using probes introduced into B, the experimenter can measure some characteristic local parameters. He will then obtain a function $g(x_i, t)$ that is characteristic of state a. It will not matter if this measurement destroys the normal evolution of the process, for state a at time $t = 0$ can be reproduced indefinitely. The functions $g(x, t)$ thus obtained will not be independent and could be calculated, in principle, from a finite number among them. In this way we can parameterize the states close to a by a function space or a finite product of function spaces.

D. The nature of empirical functions

This raises a difficult question on which physicists differ even today: the functions $g(x, t)$ are defined only on a finite mesh of points, for one can introduce into B only a finite number of probes and make a finite number of observations (life is finite, and measurements take a finite time . . .)—this being so, what hypotheses must be made about the mathematical nature of these functions, which are known only by their values at a finite number of points? Are they analytic, or even polynomials, or are they merely differentiable? I think that the reply is easy—all depends on what we want to do. If we wish to test a precise quantitative model, we must make numerical tests and calculations, and as only analytic functions are calculable (in fact, only polynomials of small degree), in this case the functions $g(x, t)$ will be analytic and will even be given by explicit formulae. If, on the other hand, we are interested only in the structural stability of the process, it is much more natural to suppose that the $g(x, t)$ are only m-times differentiable (m small: 2 or 3, in practice). Thus the local or global state of our system will be parameterized by a point of a function space, a Banach manifold, a situation analogous to that which defines the preparation procedure.

Here we are most interested in morphogenesis, the process characterized by the formation and evolution of structures and of forms represented in B. These structures will be characterized by a given closed set K in $B \times T$ such that at each point of K the process changes its appearance. Such a closed set K will be called a set of *catastrophe points* of the process. It is natural to say that a point (x, t) of $B \times T$ is a catastrophe if at least one of the functions $g(x, t)$ (or one of its first or second derivatives) has a point of discontinuity. We suppose that K is not locally dense, for then we should be reduced to considering the process in the neighborhood of some points (x_0, t_0) as chaotic and turbulent, when the idea of structural stability loses most of its relevance. Thus for two different realizations of the same state a in boxes B and B' the closed catastrophe sets K and K' will correspond under an ϵ-homeomorphism of $B \times T$, and so will be of the same topological type and have the same form in the two boxes.

E. Regular points of a process

In the open set $B \times T - K$ complementary to K the process will be called *regular*. The functions $g(x, t)$ will be continuously differentiable in a neighborhood of each regular point. If the local state is represented by a point of function space $L(X, Y)$, the evolution will be defined by a map $F: B \times T \times X \to Y$; now let us suppose that to each point (x, t) is associated a map $s(x, t): X \to Y$ defined by $F(x, t, m) = y$, $m \in X, y \in Y$. On the assumption that the local evolution of the process is defined by a differential operator of the type

$$\frac{ds(x, t)}{dt} = A(s, x, t), \tag{2.1}$$

such that the local problem defined by this partial differential equation is well posed, there will exist a unique solution of the Cauchy problem depending continuously on the initial data. For the isomorphism (which identifies two different realizations of the same structurally stable process) defined by a map

$$B \times T \xrightarrow{g \circ h} B' \times T',$$

where g is a displacement, and h a homeomorphism, we make the following restrictions:

1. The map h is a local diffeomorphism at each regular point of $B \times T$.
2. If (x', t') is the image of a regular point (x, t) under $g \circ h$, then (x', t') is a regular point of $B' \times T$ such that in a neighborhood of this point the process will be described by the same function space $L(X, Y)$ as at (x, t).
3. The law of evolution is defined by a differential equation of the form

$$\frac{ds'(x', t, m)}{dt} = B(s', x', t).$$

4. This differential operator is the transform of equation (2.1) under the diffeomorphism $g \circ h$ acting on variables (x, t), extended to a diffeomorphism of the form $(x, t, m, y) \rightarrow (x', t', m, y)$ affecting also the internal variables $m \in X$ and $y \in Y$.

2.3. STRUCTURAL STABILITY AND MODELS

In Chapter 1 we defined two types of models, *formal* and *continuous*. Formal models, whose kinematic is a formal system, have the following advantages: their description is simple, being axiomatic or combinational, and deduction within these models is formalized and theoretically could be mechanized; also the formal model is compatible with some indeterminacy of phenomena since deduction is an indeterminate operation. But they do have defects: some questions may be *undecidable* within the system (e.g., to know whether a proposition is or is not the consequence of a set of propositions); moreover, no dynamic is possible for them.

Continuous models, on the other hand, admit a dynamic; moreover, the use of differential models provides strict determinism; and even qualitatively indeterminate phenomena may be described by structurally unstable dynamical systems. However, these models too have inconveniences: they are difficult to describe, and if explicit differential equations are required, only a small number of sufficiently simple geometrical or algebraic objects can be used, conflicting in general with the a priori need for structural stability when dealing with a process that is empirically stable. Imposing this condition strictly leads not to a unique dynamical system, but to an open set of topologically equivalent dynamics, and we then reintroduce into the model a discrete factor making it analogous to a formal system.

We might be tempted to conclude from the foregoing that formal systems are superior, being the only ones that can be conveniently described, but this would be going too far. If, as Paul Valéry said, "Il n'y a pas de géométrie sans langage," it is no less true (as some logicians have hinted[2]) that there is no intelligible language without a geometry, an underlying dynamic whose structurally stable states are formalized by the language. As soon as a formal model is intelligible, it admits a *semantic realization*, that is, the mind can attach a meaning to each of the symbols of the system; thus in the model of E. C. Zeeman [1] the set of physiological processes occurring in the brain when we associate each symbol with its meaning forms such a dynamic. According to our model of language, the structurally stable attractors of this dynamic give birth to the symbols of the corresponding formal language.

These considerations demonstrate the importance of structural stability in mathematics, and we are now going to review briefly the many branches of mathematics where this problem occurs. We shall see that the problem is difficult and, despite its importance, had only recently been tackled and is still only very partially resolved.

NOTES

1. Note that the pseudogroup of equivalences of the shape of an animal has similar properties to the pseudogroup associated with the shape of a letter, for example, when handwritten. This coincidence is probably not accidental.

2. "The comparison described above suggests the very interesting problem of setting up a theory of proofs which are 'graspable' (intelligible) and not merely valid and, in particular, of intelligible combinatorial proofs. The corresponding geometric problem would be to find a theory of 'feasible' constructions which only involve points 'close' to the starting point and which are stable for 'small' changes in the data (this clearly requires the discovery of the metric appropriate to geometric intuition)." G. Kreisel and J. L. Krivine, *Elements of Mathematical Logic (Model Theory)*, North-Holland, 1967, p. 217.

REFERENCE

[1] E. C. Zeeman, Topology of the brain, *Mathematics and Computer Science in Biology and Medicine*, Medical Research Council, 1965.

CHAPTER 3

STRUCTURAL STABILITY IN MATHEMATICS

3.1. THE GENERAL PROBLEM

A. Continuous families and bifurcation

The circumstances in which structural stability enters a theory can be described, in what must necessarily be imprecise terms, as follows. We are given a continuous family of geometrical objects E_s; each object of the family is parameterized by a point of a space of parameters S; in practice S is Euclidean space or a finite or infinite-dimensional differential manifold. If E_s is the object corresponding to a given point s in S, it may happen that, for any point t sufficiently close to s in S, the corresponding object E_t has the same form as E_s (in a sense to be made precise in each specific case); in this case E_s is called a structurally stable or *generic*[1] object of the family, and the set of points s in S for which E_s is structurally stable forms an open subset of S, the set of generic points. The complement K of this open set is called the set of *bifurcation* points. The question, "Is K nowhere dense?" is what is usually called the problem of structural stability, and in most theories the object is to specify the topological structure of K and its singularities. We shall study the main theories in which this type of situation arises, proceeding, as far as possible, from the least to the most complicated, and from the best to the least known.

B. Algebraic geometry

Given a system of polynomial equations

$$P_j(x_i, s_k) = 0, \tag{3.1}$$

where the x_i and the s_k are coordinates in Euclidean spaces $\mathbf{R}^n$ and S, respectively, then for fixed $s = (s_k)$ the system defines an algebraic set E_s in $\mathbf{R}^n$, the set of solutions of the equation. The field of coefficients may be either the real or the complex numbers. The problem of topological stability of E_s has a positive answer: there exists in S a proper algebraic subset K (therefore nowhere dense) such that, if s and t are two points in the same component of S-K, the corresponding sets E_s and E_t are homeomorphic. This homeomorphism can be continuously deformed to the identity by an isotopy of the ambient space $\mathbf{R}^n$.

Remarks. We can improve this result by replacing K, in the case of real coefficients, by a smaller semialgebraic set K', or in the complex case by a constructible set K'. Let G be the graph in $\mathbf{R}^n \times S$ defined by equations (3.1), and $p:\mathbf{R}^n \times S \to S$ be the projection onto the second factor. We then stratify the map $p:G \to S$ and decompose G and S into a disjoint union of embedded differential manifolds (the strata) in such a way that the image under p of a stratum of G is a stratum of S, and the rank of this map is constant on each stratum. The patching of the strata along their boundaries must satisfy certain conditions on their tangent planes (see[7]).* Then we can show that the projection $p:p^{-1}(T) \to T$ is a fibration for any stratum T of S.

Next, considering algebraic (instead of topological) isomorphisms between the algebraic sets E_s gives an equivalence relation that is too fine for the structurally stable sets E_s (now called *rigid sets*) to be everywhere dense, because the algebraic structure of an algebraic set depends in general on continuous parameters (*moduli*; see [3]) and so can be deformed into a neighboring nonequivalent algebraic set. Also we can try to define the maximal continuous family containing the deformations of a given algebraic structure (e.g., modules, algebras, or Lie algebras), and results have recently been obtained in this direction. Two theories seem to dominate by virtue of the extent of their possible applications.

1. The deformation of G-structures, containing in particular the theory of Pfaffian systems and hence the general theory of systems of partial differential equations (see [5]).

2. The classification of the actions of a given group G on a given manifold M, which in principle includes the theory of ordinary differential systems (actions of $\mathbf{R}$ on M) and probably determines the algebraic theory of singularities of differentiable maps (actions of invertible jets $L^r(n) \times L^r(p)$ on $J^r(n,p)$; see [6]).

*Numbers in brackets refer to items in the reference list at the end of the chapter.

C. Geometric analysis

The results of Section 3.1. B cannot be extended to the situation of analytic functions $F_j(x_i, s_k)$ without some care. For complex scalars the theorem remains true provided that $p: G \to S$ is a proper map (i.e., the counterimage of a compact set is compact). The bifurcation set K is a constructible set contained in a proper analytic subset of S. For real scalars the situation is similar, but the nature of K is not yet fully known.[2] Because of the well-known fact that the projection of an analytic set is not necessarily analytic or even semianalytic (Osgood's example), an intrinsic characterization of these sets (which are very probably stratified) has not yet appeared in the literature.

The preceding remark on algebraic isomorphisms holds also for analytic isomorphisms: the structure of an analytic set depends, in general, on continuous moduli, forming a finite- or infinite-dimensional space (see [9] and [10]).

D. Differential topology

Let X and Y be two differential manifolds, X compact; then the function space $L(X, Y)$ of C^m-differentiable maps from X to Y, with the C^m-topology defined on a chart by

$$d(f,g) = \sum_{|k|=0}^{m} \left| \frac{\partial^{|k|}}{\partial x^k}(f-g) \right|$$

is an infinite-dimensional manifold modeled locally on a Banach space (i.e., a Banach manifold). This function space $L(X, Y)$ plays the part of the parameter space S; each point F in $S = L(X, Y)$ is a differentiable map $X \to Y$, and one tries to classify the form or topological type of these maps. Recall that two maps $f, g: X \to Y$ have the same type if there exist two homeomorphisms $h: X \to X$, and $k: Y \to Y$ such that the diagram

$$\begin{array}{ccc} X & \xrightarrow{f} & Y \\ {\scriptstyle h}\downarrow & & \downarrow{\scriptstyle k} \\ X & \xrightarrow{g} & Y \end{array}$$

is commutative.

Then there is, in $L(X, Y)$, a subspace K of codimension one (the closed bifurcation set) such that any two maps f and g of $L(X, Y)$ belonging to the same connected component of L-K have the same topological type. Moreover K has a stratified structure: K contains regular points that form an everywhere-dense open set in K, and this open set is a regular hypersurface of L; the complement in K of this open set is closed in K, and is a set of codimension at least two, and thus its regular points form a submanifold of codimension two, and so on. Furthermore, a neighborhood of a point of such a stratum of K has a model given by algebraic equations or inequalities. The local study of this situation, in the neighborhood of a stratum of the bifurcation set of codimension k, has a very important role in our model, which will be developed in more detail, together with the idea of the *universal unfolding* of a singularity.

Remarks. If F is the canonical map

$$F: X \times L(X,Y) \to Y \times L(X,Y) \to L(X,Y)$$

defined by

$$(x,f) \to (y = f(x), f) \to f,$$

then there exists a closed subset H of infinite codimension in $L(X,Y)$ such that the restriction of F to the counterimage under F of $L(X,Y)-H$ is a stratified map. The stratification of $L(X,Y)-H$ is then defined by the bifurcation subset K, which contains H as a closed subset. The maps in $L(X,Y)-K$ are structurally stable or generic. The maps in $L(X,Y)-H$ are stratified, and "almost generic" in the sense that any g in $L-H$ can be embedded in a q-parameter family in such a way that the map $X \times \mathbf{R}^q \to Y \times \mathbf{R}^q$ so defined is generic with respect to perturbations, leaving the parameter space $\mathbf{R}^q$ invariant. (See also the idea of the universal unfolding of a singularity at the end of this chapter.)

The proof proceeds as follows. We study first the local problem: let $g: \mathbf{R}^m \to \mathbf{R}^n$ be the germ of a differentiable map, sending the origin O in $\mathbf{R}^m$ to the origin O_1 in $\mathbf{R}^n$, and let z be a jet of order r of such a germ, that is, the terms of order $\leqslant r$ in the expansion of this map. Then there exists a larger integer $r+p$ (where p depends only on m, n, and r) such that extending the given jet to order $r+p$ and adjoining monomials of the form $\Sigma_\omega a_\omega x_\omega$ of degree $|\omega|$ with $r+1 \leqslant |\omega| \leqslant r+p$ gives, for almost all choices of coefficients a_ω, a jet of order $r+p$ which determines the local topological type of the germ of any map realizing this jet. The set C of points of the space of coefficients a_ω which do not have this property (i.e., the corresponding jet of order $r+p$ does not determine the topological type of the germ) is contained in a proper algebraic subset, the *bifurcation set*. Forming all possible subsets of bifurcation in the jet space $J^r(m, n)$ gives the *canonical stratification* of the jet space. A map $f: X \to Y$ whose rth derivative [a section of the fibration of r-jets $J^r(X, Y)$], for sufficiently large r (depending only on the dimensions m, n of X, Y), is transversal to the canonical stratification of $J^r(X, Y)$ is called a correct map and is locally structurally stable. For this reason the correct maps are everywhere dense; it is sufficient for global topological stability that, if $X_1, \ldots, X_k$ are the strata of the set of critical points of f, their images $f(X_1), \ldots, f(X_k)$ cut each other transversally. This condition comes from a slight generalization of the lemma on transversality.

The theory of the differential stability of differentiable maps (where, in the commutative diagram on p. 23, h and k are diffeomorphisms) is the subject of recent remarkable work by Mather [11]. It appears that, under certain conditions on dimension (unimportant in practice: e.g., $n < 7$), the *differentiable structurally stable maps are everywhere dense*. The same theory also resolves the problem of the formal stability of maps defined locally by formal power series.

E. Differential equations

Historically, the idea of structural stability was first introduced into mathematics by Andronov and Pontryagin [14] in the qualitative study of differential systems, defined by the following:

1. A configuration space that is a finite- or infinite-dimensional differential manifold M.
2. A vector field X on M.

The integration of this field gives rise, at least locally, to a one-parameter group of diffeomorphisms $h_t: M \to M$, which is the *flow* associated with the differential system. A dynamical system on M is defined in the same way. The set of vector fields X on M (a vector space of sections of the fiber space of tangent vectors to M) can be given the topology defined by the metric

$$d(X_0, X_1) = \sup_M \left\{ |X_0 - X_1| + \sum \left| \frac{\partial X_0}{\partial x} - \frac{\partial X_1}{\partial x} \right| \right\} \qquad \text{(the } C^1\text{-topology)},$$

and when M is compact it is a Banach space B. A differential system X on M is called structurally stable if any vector field X' sufficiently close to X in the C^1-topology has the following property: there is a homeomorphism h of M itself which transforms each trajectory of the field X' into a trajectory of X; in addition, we might impose the condition that this homeomorphism be small [i.e., the distance from x to $h(x)$ be small]. Equivalently, a dynamical system X on M is structurally stable if a sufficiently small (in the C^1-norm) perturbation of the field X does not alter the qualitative nature of the system. As it is not required that the homeomorphism h commute with time, the perturbed system may have a completely different structure from the original system after sufficient time has passed. Easy (but too lengthy to be given here) geometric conditions show that in general this is the only structural stability one can expect of a differential system. The problem of stability is then: Is the collection of structurally stable vector fields everywhere dense in the space B of all vector fields? A partial answer has been obtained recently: yes, when dimension $M \leq 2$ (Peixoto [18]); no, when dimension $M \geq 4$ (Smale [22]), and dimension $M = 3$ (Smale and Williams [24]).

Despite this negative answer one must not think that the problem of structural stability has no interest in dynamics, for, even when the dimension is greater than four, the function space B contains at least one relatively important open set where the structurally stable fields are everywhere dense. This set contains in particular the nonrecurrent gradient fields and also their generalization, known in the literature as *Morse-Smale systems* [17]. Note also that any system may be C^0-approximated by such a Morse-Smale system (see [21] and [25]).

Another important class of dynamical systems is that of conservative Hamiltonians, whose definition we recall here. Given an n-dimensional differential manifold M^n—the configuration space—one forms the fiber space of cotangent vectors $T^*(M^n)$, and in this space there is a canonical differential 1-form β [$\beta = \Sigma_i p_i\, dq_i$ in canonical coordinates (p_i, q_i)] whose exterior differential $\alpha = d\beta = \Sigma_i dp_i \wedge dq_i$ defines a symplectic structure on $T^*(M^n)$; a Hamiltonian dynamic on M is defined by giving a real-valued function, the Hamiltonian $H: T^*(M^n) \to \mathbf{R}$; the field defining the evolution dynamic is the symplectic gradient of the Hamiltonian, that is, the field X defined by Hamilton's equations $i(X) \cdot \alpha = dH$ (where i denotes the interior product).

The set of Hamiltonian dynamics on a given configuration space M^n is therefore naturally parameterized by the function space of Hamiltonians $L(T^*(M^n), \mathbf{R})$ with the C^∞-topology. However, we cannot infer from the correspondence of two Hamiltonians H and H' by a diffeomorphism $h: T^*(M^n) \to T^*(M^n)$ (i.e., the differential structural stability of these functions) the equivalence of their corresponding Hamiltonian dynamics X and X' because the diffeomorphism h does not leave the symplectic structure (or the 2-form α) invariant in the strict sense unless it is a canonical transformation of the phase space $T^*(M^n)$. To my knowledge this question of the structural stability of Hamiltonians has not been considered in the literature: the general attitude is that it is a doubtful problem and, apart from certain trivial cases (e.g., $n = 1$), a positive answer cannot be expected. Nevertheless, several globally structurally stable configurations are known and are of great practical importance, notably the trajectories of central type called *vague attractors* and studied by Kolmogoroff, Moser, and others, which probably correspond to "stationary states" in quantum mechanics. We return to this question in Chapters 4 and 5.

Remarks. This problem, to characterize the geometric structure of "generic" differential systems on a given space, is probably one of the most difficult and least known of contemporary analysis. Stephen Smale has recently put forward some very important ideas which I will describe here. A differential system (M,X) is said to satisfy *Axiom A* if the tangent space at each point $m \in M$ has a decomposition into vector subspaces of the form $T_m(M) = 1 + U_c + U_e$ (where 1 denotes the one-dimensional space defined by the field X), which is invariant under f_t [where f_t is the (local) one-parameter group generated by integrating x] and such that f_t is contracting ($|f_t| < 1$) on the factor U_c and expanding on U_e in some Riemannian metric. Let Ω be the set of nonwandering points of X. Smale postulates that Axiom A is satisfied in a neighborhood of Ω, and for such fields X he then proves that, when M is compact, the set Ω can be decomposed into a finite union of indecomposable closed sets Ω_i. Next, associate with each Ω_i the sets $\alpha^{-1}(\Omega_i)$ of trajectories leaving Ω_i, and $\omega^{-1}(\Omega_i)$ of trajectories finishing in Ω_i, and say that Ω_i *precedes* Ω_j if $\alpha^{-1}(\Omega_i) \cap \omega^{-1}(\Omega_j) \neq \varnothing$. This defines a partial order on the Ω_i, and the extremal Ω_i are the attractors of X; almost all trajectories of X finish in a neighborhood of one of these attractors. This gives a generalization of the structure on a manifold defined by Morse theory; the Ω_i correspond to the critical points and the sets $\alpha^{-1}(\Omega_i)$, $\omega^{-1}(\Omega_j)$ to the gradient cells associated with the critical points. If this ordering has no cycles, and only the coarse structure of the dynamic is considered, these systems have the structure of a gradient field;[3] in particular, there will be only a finite number of final attractor states. However, little is known of the topological structure of these Ω_i; they seem a priori to be very rigid and structurally stable, and examples are known in which the corresponding field is of conservative Hamiltonian type (e.g., geodesic flows on manifolds with negative curvature; see [15]). The arrangement of the basins of attraction of different attractors is in general very complicated and is not structurally stable. Moreover, we now know that these fields, described by Smale, are not dense in the space B of all fields (Abraham and Smale [13]).

Smale's counterexample in dimension four leaves no hope for structural stability in general; the bifurcation set K in the function space of vector fields B has a complicated topological structure even at points where it is not dense. We must observe, however, that only the stability of final states is of practical importance, and in this case it is not known whether dynamical systems with nonstable attractors can be locally dense in B. (A recent example due to Newhouse shows that flows having an infinite number of attractors may form a locally dense set.) It is very important in this respect to have more detailed information regarding the bifurcation set in B, and a classification of generic forms of bifurcations for gradient systems of codimension less than four will be given in Chapter 5. It would be most useful to know what remains of this classification for more general systems—to determine, for example, how a many-dimensional attractor (like an Anosov U-system) can disappear through deformation. Concerning this, we might say that the creation or destruction at time t_0 of a many-dimensional attractor needs, in general, an infinite number of changes of the topological type of the system at times t_i tending to t_0; this will correspond morphogenetically to a *generalized catastrophe* in the sense of Chapter 6. As we are far from understanding these transformations of attractors, the models that will here be presented, particularly in embryology, are bound to be imprecise.

We should also add that, as long as a system has one attractor of Hamiltonian type, the deformations of this attractor will be homeomorphisms and will depend on continuous moduli which allow local energy and entropy to be defined relative to this attractor. As a general rule the evolution of a dynamic in time takes place through successive Hamiltonian-like stages separated by short catastrophic transitional periods.

F. Functional analysis and partial differential equations

The structural stability of functional operators has never been the subject of a systematic study. In the case of partial differential equations the equivalent problem is almost exactly that of the well-posed problem, of which many classical examples are known, such as Dirichlet's problem for elliptic operators and Cauchy's problem for parabolic and hyperbolic operators. Only the linear theory has a satisfactory development; as soon as one leaves this theory, almost nothing is known. Probably the theorem of Cauchy-Kowalewska on the local existence and uniqueness of solutions of a system of analytic equations holds more generally, particularly when for every solution, the variational equation is of hyperbolic type; it is not unreasonable to expect that there is structural stability with respect to the initial data as well as perturbations of the differential system (although the use of a finer topology than the C^m-topology for the initial data may yet prove to be necessary). Another problem, important in morphogenesis, is the determination of structurally stable singularities of solutions of a system of partial differential equations; here the conceptually easiest variational problem is Plateau's problem to find the surface of minimum area bounded by a given curve. Although solutions can be demonstrated physically using films of soapy water, mathematical analysis is still unable to catalogue the structurally stable singularities of solutions (e.g., it is doubtful that there exists in $\mathbf{R}^3$ more than the triple edges and quadruple points imposed by Gibbs' phase rule).

The preceding remark concerning the Cauchy-Kowalewska theorem raises the question as to what extent analytic continuation is a structurally stable operation. Does there exist on the set of holomorphic functions on a compact set K a topology under which the domain of holomorphy of a function (or its complementary closed set) is a continuous function of the given function? It is obvious that such a topology must be much finer than the usual C^0-Banach topology. For this reason we are forced to the conclusion that structural stability and computability are, to a certain extent, contradictory demands, because all quantitative and effectively computable models must necessarily appeal to analytic functions, while a function that is differentiable, and no more, rarely lends itself to explicit evaluation.[4] Perhaps more comment is necessary on such an opinion, for it contradicts the well-established view that quantitative models in science are the best. Take, for example, a classical quantitative theory like celestial mechanics; even in this case, recent work on the three-body problem seems to indicate that the trajectories giving rise to unstable or catastrophic situations (like collision or enlargement to infinity of an orbit) are densely and intricately distributed throughout the stable Keplerian orbits. Thus on a large scale the evolution of a given planetary system is not structurally stable.

There seems to be a time scale in all natural processes beyond which structural stability and calculability become incompatible. In planetary mechanics this scale is of such an extent that the incompatibility is not evident, whereas in quantum mechanics it is so short that the incompatibility is immediately felt, and today the physicist sacrifices structural stability for computability. I hope that he will not have cause to regret this choice.

3.2. ALGEBRA AND MORPHOGENESIS

A. An example of bifurcation

We return in more detail to the formal mechanism which seems to me to control all morphogenesis. Let me explain in an elementary fashion, using an analogy between the development of an embryo and a Taylor series with indeterminate coefficients. We can roughly describe the overall development of the embryo in the following way: a totipotent egg divides, in the course of time, into masses of cells which acquire special and (in principle) irreversible functions, but there always exists in the animal a germinal line of totipotent cells which will result in the formation of reproducing cells (gametes) in the adult individual. On the other hand, consider a differentiable function of two real variables, zero at the origin, and its expansions up to, for example, the third order:

First order: $f = a_1x + b_1y;$

Second order: $f = a_1x + b_1y + a_2x^2 + 2b_2xy + c_2y^2;$

Third order: $f = a_1x + b_1y + a_2x^2 + 2b_2xy + c_2y^2 + a_3x^3 + 3b_3x^2y + 3c_3xy^2 + d_3y^3.$

The successive spaces of coefficients $J^1(a_1, b_1)$, $J^2(a_1, b_1, a_2, b_2, c_2)$, $J^3(a_1, \ldots, c_3, d_3), \ldots$ project canonically:

$$J^1 \xleftarrow{p_2} J^2 \xleftarrow{p_3} J^3 \leftarrow \cdots.$$

To the first order (in J^1) the topological classification of functions f is easy: either not both coefficients a_1, b_1 are zero, or $a_1 = b_1 = 0$. In the first case one of the derivatives $\partial f/\partial x$, $\partial f/\partial y$ is nonzero at the origin, and, by using the classical theorem on implicit functions, f can be changed locally into a linear function by changing to curvilinear coordinates in the Oxy-plane, so that the form of the function, its destiny, is irreversibly fixed no matter what the higher coefficients are. In the second case, when $a_1 = b_1 = 0$, nothing can be said about the topological nature of the function—we say that the origin $a_1 = b_1 = 0$ represents, in J^1, the bifurcation set K^1, the analogue of the germinal line in embryology. To the second order the points of J^2 projecting by p_2 outside K^1 have their destiny already decided; the points projecting to $O = K^1 \subset J^1$ have the following classification:

1. If the discriminant of $a_2x^2 + 2b_2xy + c_2y^2$ is nonzero, then (by a classical theorem of Morse) the addition of higher-order terms does not alter the topological type of the function, which remains that of a quadratic form.

2. If $b_2^2 - a_2c_2 = 0$, the topological type is not yet decided. Thus the surface with equation $b_2^2 - a_2c_2 = 0$ defines the bifurcation set K^2 in $J^2 \cap p_2^{-1}(O)$. The rank of the quadratic form is then used to classify K^2: only the zero form can be considered as totipotent, and a form of rank one, for example, $f = x^2$, is not topologically determined; adjoining a term of third order not divisible by x, (e.g., $x^2 + y^3$) gives a form whose topological type cannot be altered by addition of terms of higher order; adjoining a term of third order divisible by x gives a form whose topological type is again not determined—for example, $f = x^2 + 2xy^2$ and $g = x^2 + 2xy^2 + y^4$ are different types, for $g = (x + y^2)^2 = 0$ defines a double curve, whereas $f = 0$ does not.

Thus the general nature of these phenomena is as follows. Corresponding to each order k is an algebraic bifurcation subset K_k, and under the canonical map $p_k: J^k \to J^{k-1}$ we have $p_k(K_k) \subset K_{k-1}$. The sequence of bifurcation sets gives the formal analogue of the germinal line in embryology. Of course we deal here with a formal analogy, and it is out of the question to study it in detail. (In fact, the development of an embryo is much more complex, and later we shall see which interpretation seems adapted to the experimental facts of regulation and induction in embryology.)

There is an inconvenient feature in the algebraic model described above: the space of coefficients is not given once and for all, but appears in a sequence of Euclidean spaces $J^1 \leftarrow J^2 \leftarrow \ldots \leftarrow J^k \leftarrow \ldots$, each of which projects onto the preceding one (a projective limit). It is possible, however, to restrict attention once and for all to a given finite-dimensional space and to consider what we will call a *singularity of finite codimension* and its universal unfolding.

B. The universal unfolding of a singularity of finite codimension

Let f and g be two local maps of Euclidean spaces $\mathbf{R}^n \to \mathbf{R}^p$, which map the origin O on $\mathbf{R}^n$ to the origin O_1 in $\mathbf{R}^p$. If they have the same partial derivatives up to and including order r, we say (in the terminology introduced by Ehresmann) that these maps have the same *local jet* of order r at O. The set of these jets forms a vector space, written $J^r(n, p)$, whose coordinates are the coefficients of the Taylor expansion to order r of the functions defining the map at O. The example above showed that there is an algebraic bifurcation subset K^r in $J^r(n, p)$ with the following property: if the jets of order r of two maps f and g are the same and belong to the complement $J^r(n,p) - K^r$, then maps f and g have locally the same topological type, that is, there exist homeomorphisms h and h_1 of the range and image spaces such that $h(O) = O$, $h_1(O_1) = O_1$ and the diagram

$$\begin{array}{ccc} \mathbf{R}^n & \xrightarrow{f} & \mathbf{R}^p \\ {\scriptstyle h}\downarrow & & \downarrow{\scriptstyle h_1} \\ \mathbf{R}^n & \xrightarrow{g} & \mathbf{R}^p \end{array}$$

commutes. In this case we say that the jet of f is *determinate*. We can also have the case of the structurally stable jet, where any sufficient small deformation of the map f is topologically equivalent to f (in this case the homeomorphisms h and h_1 in the diagram above can displace the origin)—for example, a critical point of a function with a nondegenerate quadratic part is structurally stable. When is the jet at O of a map f not structurally stable? There are two cases, qualitatively very different.

1. Arbitrarily small perturbations (in the C^m-topology) of f give rise to an infinite number of different topological types [e.g., the flat function $y = \exp(-1/x^2)$ or the constant function $y = 0$ can be approximated near the origin by functions with an arbitrarily large number of maxima and minima]. In this case f is said to be of *infinite codimension*, and the jet of f cannot be determinate.

2. When these perturbations can give rise (locally) to only a finite number of topological types, f is of *finite codimension* and the jet is determinate. The codimension is q, where we can embed the map in a family of deformations parameterized by q variables so that the map F: $\mathbf{R}^n \times \mathbf{R}^q \to \mathbf{R}^p \times \mathbf{R}^q$ thus defined is structurally stable with respect to perturbations leaving the parameter space invariant. In particular a structurally stable jet defines a singularity of zero codimension. This family of deformations has a universal role with respect to C^k-deformations of f in the sense that any perturbation, parameterized by m variables, is induced from the universal family by a map $\mathbf{R}^m \to \mathbf{R}^q$. This universal family will be called the *universal unfolding* of the organizing center f. The mathematical theory is too complicated to be given here, and we will restrict attention to an example (in fact, the easiest example possible).

C. An example: The universal unfolding of $y = x^3$

Every deformation of $y = x^3$ is a differentiable function $y = f(x, v)$ depending on m parameters, denoted globally by v, such that $f(x, 0) = x^3$. We write

$$F(x, y, v) = f(x, v) - y.$$

Then, at the origin $x = y = v = 0$ of (x, y, v)-space, F satisfies

$$F(0,0,0) = \frac{\partial}{\partial x} F(0,0,0) = \frac{\partial^2}{\partial x^2} F(0,0,0) = 0, \qquad \frac{\partial^3}{\partial x^3} F(0,0,0) \neq 0.$$

We can therefore apply Weierstrass' preparation theorem (generalized by Malgrange to differentiable functions) and write

$$F(x, y, v) = h(x, y, v)[x^3 - 3a(y, v)x^2 + b(y, v)x + c(y, v)],$$

where the functions h, a, b, and c are differentiable and satisfy

$$h(0,0,0) \neq 0, \qquad a(0,0) = b(0,0) = c(0,0) = 0$$

and

$$\frac{\partial c}{\partial y}(0,0) = \frac{\partial}{\partial y}\left[\frac{F(x, y, v)}{h(x, y, v)}\right]_{x=y=v=0} = -\frac{1}{h(0,0,0)} \neq 0.$$

Under the transformation

$$x_1 = x - a(f, v), \qquad y_1 = -c(y, v) + a^3(f, v)$$

the graph of $F = 0$ is transformed into the graph of a polynomial of the third degree:

$$y_1 = x_1^3 + A(y, v)x_1, \qquad A = b - a^2.$$

As $\partial A / \partial y = 0$ at the origin, we can neglect, to the first approximation, the variation of the coefficient A_3 as a function of y and obtain the result that every deformation of $y = x^3$ can be transformed by changing to curvilinear coordinates $x \to x_1$, $y \to y_1$ into a polynomial $y_1 = x_1^3 + A(v)x_1$, which thus forms the universal unfolding of the singularity of $y = x^3$. As this family depends on only one parameter, A, the singularity has codimension one.

D. The general theory of the universal unfolding

To conclude we give the formal definition of the universal unfolding of a singularity of finite codimension. Let $y = f(x)$ be the germ of a differentiable map $\mathbf{R}^n \to \mathbf{R}^p$ whose jet at the origin has a singularity of codimension q; then the universal unfolding of the singularity f is a family

$$y = f(x) + \Sigma u_j g_j(x) = F(x, u),$$

parameterized by q variables u, with the following property: for every deformation $G(x, v) = y$, $G(x, 0) = f(x)$ of f there exist homeomorphisms

$$(x, v) \xrightarrow{h} (x_1, v) \quad \text{and} \quad (y, v) \xrightarrow{h} 1(y_1, v)$$

and a map $v \xrightarrow{k} u$ such that the diagram below

$$\begin{array}{ccc} (x, v) & \xrightarrow{G} & (y, v) \\ {\scriptstyle (h,k)}\downarrow & & \downarrow{\scriptstyle (h_1,k)} \\ (x_1, u) & \xrightarrow{F} & (y_1, u) \end{array}$$

commutes (all deformations have the same topological type as a map of the universal unfolding).

The theory of universal unfoldings can be interpreted in terms of functional analysis as follows. There is a bifurcation set Y with a stratified structure in the space of C^∞-maps $L(\mathbf{R}^n, \mathbf{R}^p)$. A singularity f of codimension q is a point in a stratum Y of codimension q in $L(\mathbf{R}^n, \mathbf{R}^p)$. The universal unfolding of the singularity is nothing more than a fragment of the transversal q-plane at f of the stratum of f. Therefore the universal family is not unique, but is defined up to a stratified isomorphism.

Remarks. As we said earlier, an arbitrary jet cannot be the organizing center of a finite universal unfolding unless this jet is determinate; but we can show that, if a jet z in $J^r(n,p)$ is not determinate, there exists an integer m (depending only on r, n, and p) such that almost all extensions of z of order $r+m$ are determinate jets, where "almost all" means, as usual in this theory, the complement of a nowhere-dense algebraic set (of bifurcation). For example, in $J^2(2,1)$ of local jets $\mathbf{R}^2 \to \mathbf{R}^1$, the jet $f = x^2$ is not determinate, but $f = x^2 + y^3$ is determinate and its universal unfolding is $F = x^2 + y^3 + uy$.

In the preceding theory we considered only the local singularities of germs of differentiable maps. Very probably the same theory extends to global maps of one manifold into another. We can associate with each point f of a stratum of codimension q a transversal q-plane which defines a universal unfolding. However, the strata of the bifurcation set Y which are defined by nonlocal singularities correspond in fact to maps for which the images of disjoint critical strata do not cut in general position. Very probably these singularities have a relatively trivial character, like that of the singularities of the barycentric subdivision of the simplices associated with Gibbs' phase rule; moreover, I do not think that the global theory is of much more interest (or difficulty) than the local theory.

E. The case of a function

Let f be a function having an isolated critical point at the origin O. In the algebraic sense, "isolated" means that the ideal generated by the first partial derivatives, $J(\partial f/\partial x_i)$, contains a power of the maximal ideal, $\mathfrak{m}$; that is, any monomial in the coordinates $x_{i_1}^{\alpha_1},\ldots,x_{i_k}^{\alpha_k}$ of sufficiently high degree can be expressed as

$$x_{i_1}^{\alpha_1} \cdot x_{i_2}^{\alpha_2} \cdot \ldots \cdot x_{i_k}^{\alpha_k} = \sum_i \lambda_i f_{x_i},$$

a linear combination of f_{x_i}, $\lambda_i \in C^\infty(0)$. Then the quotient algebra $\mathbf{R}[x_i]/J$ has finite dimension. Let $g_1, g_2, \ldots, g_k$ be a basis for this algebra.

According to Mather's theory of structurally stable mappings in the C^∞ sense, the map defined by

$$U_i = u_i, \qquad Y = f + \sum_i u_k g_k$$

is structurally stable. In all the cases that we shall consider in Chapter 5, this formula will provide the universal unfolding of the singularity of f at 0.

NOTES

1. The adjective "generic" is used in mathematics is so many different senses that to restrict its usage within the framework of a formal theory is probably unreasonable. Smale [23] has made a welcome suggestion that it should be reserved for properties of a topological space E and should never be applied to points of the space, a property P of E being generic if the set of points possessing that property is dense in E. Meanwhile, if a mathematician happens to use the term, he should specify its local meaning and adhere to it until further notice.

2. In fact, these sets have now been defined and shown to be stratified by Hironaka, who calls them subanalytic spaces (Ecole d'été de Cargèse, Corsica, August 1972).

3. The set separating the basins of attraction of different attractors of a Morse-Smale system do, however, have a much more complicated topology than the separatrices of gradients. This happens because the existence of structurally stable *heteroclinic points* (in the terminology of Poincaré [19]), which result in a very irregular boundary formation between adjacent basins. Moreover, stable configurations acting as attractors can be set up on these surfaces, and these can be regarded as the origins of the stable transitional regimes which often appear in the neighborhood of the shock wave separating the domains of two stable regimes in competition.

4. It might be argued that the importance accorded by analysis to the complex field and the theory of analytic functions during the last century has had an unfortunate effect on the orientation of mathematics. By allowing the construction of a beautiful (even too beautiful) theory which was in perfect harmony with the equally successful quantification of physical theories, it has led to a neglect of the real and qualitative nature of things. Now, well past the middle of the twentieth century, it has taken the blossoming of topology to return mathematics to the direct study of geometrical objects, a study which, however, has barely been begun; compare the present neglected state of real algebraic geometry with the degree of sophistication and formal perfection of complex algebraic geometry. In the case of any natural phenomenon governed by an algebraic equation it is of paramount importance to know whether this equation has solutions, *real* roots, and precisely this question is suppressed when complex scalars are used. As examples of situations in which this idea of reality plays an essential qualitative role we have the following: the characteristic values of a linear differential system, the index of critical points of a function, and the elliptic or hyperbolic character of a differential operator.

REFERENCES

The following list gives only the most recent or most complete or most typical article on each subject.

ALGEBRAIC GEOMETRY

Topological theory. The theory of the stratified sets is summarized in [1] and given more fully in [2].

1. R. Thom, La stabilité topologique des applications polynomiales, *L'enseignement mathématique* **8** (1962), 24–33.
2. R. Thom, Ensembles et morphismes stratifiés, *Bull. Amer. Math. Soc.* **75**, (1969), 240–284.

Theory of moduli. A large part of contemporary algebraic geometry deals with this. See, for example:

3. D. Mumford, Geometric invariant theory, *Ergebnisse der Mathematik*, Vol. 34, Springer, 1965.

Deformation of algebraic structures.

For Lie algebras:

4. A. Niejenhuis and R. W. Richardson, Jr., Cohomology and deformations in graded Lie algebras, *Bull. Amer. Math. Soc.* 72 (1966), 1–29.

For G-structures:

5. V. Guillemin and S. Sternberg, Deformation theory of pseudogroup structures, *Mem. Amer. Math. Soc.* No. 64, 1966.

For group actions:

6. R. Palais and S. Stewart, Deformation of compact differentiable transformation groups, *Amer. J. Math.* 82 (1960), 935–937.

GEOMETRIC ANALYSIS

Differential and topological theory.

7. H. Whitney, Local properties of analytic varieties, in *Differential and Combinatorial Topology, A Symposium in Honor of Marston Morse* (ed. S. S. Cairns), Princeton University Press, 1965, pp. 205–244.
8. S. Łojasiewicz, *Ensembles semi-analytiques*, Cours à la Faculté des Sciences d'Orsay, IHES Juillet, 1965.
   Also see [2].

Deformation of analytic singularities.

9. H. Grauert and H. Kerner, Deformationen von Singularitäten komplexer Räume, *Math. Annal.* 153 (1964), 236–260.
10. A. Douady, Le problème des modules pour les sous-espaces analytiques compacts, *Ann. Inst. Fourier* 16 (1966), 61–95.

DIFFERENTIAL TOPOLOGY

11. J. Mather, Stability of C^∞-mappings, I: *Ann. Math.* (2), 87 (1968), 89–104; II: *ibid.* (2), 89 (1969), 254–291; III: *Inst. Hautes Études Sci. Publ. Math.* 35 (1968), 279–308; IV: *ibid.* 37 (1969), 233–248; V: *Advan. Math.* 4 (1970), 301–336; VI: *Proceedings of the Liverpool Singularities Symposium* I, *Lecture Notes in Mathematics* No. 192, Springer, 1971, pp. 207–253.
    Also see [2].

GLOBAL THEORY OF DIFFERENTIAL EQUATIONS

12. R. Abraham and J. Marsden, *Foundations of Mechanics*, Benjamin, 1967.
13. R. Abraham and S. Smale, Nongenericity of Ω-stability, *Proceedings of the Symposium on Pure Mathematics* (*Global Analysis*), Vol. XIV, American Mathematical Society, 1970, pp. 5–8.
14. A. Andronov and L. Pontryagin, Systèmes grossiers, *C. R.* (*Dokl.*) *Acad. Sci. USSR* 14 (1937), 247–251.
15. D. Anosov, Roughness of geodesic flows on compact Riemannian manifolds of negative curvature (English translation), *Soviet Math. Dokl.* 3 (1962), 1068–1070.
16. V. I. Arnold and A. Avez, *Ergodic Problems of Classical Mechanics*, Benjamin, 1968.

17. J. Palis and S. Smale, Structural stability theorems, *Proceedings of the Symposium on Pure Mathematics* (*Global Analysis*), Vol. XIV, American Mathematical Society 1970, pp. 223–231.
18. M. M. Peixoto, Structural stability on two-dimensional manifolds, *Topology* **1** (1962), 101–110.
19. H. Poincaré, *Leçons sur la méchanique céleste*, Dover, 1957.
20. J. Robbin, Topological conjugacy and structural stability for discrete dynamical systems, *Bull. Amer. Math. Soc.* **78** (1972), 923–952.
21. M. Shub, Structurally stable diffeomorphisms are dense, *Bull. Amer. Math. Soc.* **78** (1972), 817–818.
22. S. Smale, Structurally stable systems are not dense, *Amer. J. Math.* **88** (1966), 491–496.
23. S. Smale, Differentiable dynamical systems, *Bull. Amer. Math. Soc.* **73** (1967), 747–817.
24. R. F. Williams, The "DA" maps of Smale and structural stability, *Proceedings of the Symposium on Pure Mathematics* (*Global Analysis*), Vol. XIV, American Mathematical Society, 1970, pp. 329–334.
25. E. C. Zeeman, Morse inequalities of Smale diffeomorphisms and flows (to appear).

CHAPTER 4

KINEMATIC OF FORMS; CATASTROPHES

4.1. SPATIAL PROCESSES

A. The morphology of a process

Suppose that a natural process, of any kind whatsoever, takes place in a box B; we then consider $B \times T$ (where T is the time axis) as the domain on which the process is defined. Also suppose that the observer has at his disposal probes or other means to allow him to investigate the neighborhood of each point x of $B \times T$. As a first classification of points of $B \times T$ we have the following: if the observer can see nothing remarkable in the neighborhood of a point x of $B \times T$, that is, if x does not differ in kind from its neighboring points, then x is a *regular point* of the process. By definition the regular points form an open set in $B \times T$, and the complementary closed set K of points in $B \times T$ is the set of *catastrophe points*, the points with some discontinuity in every neighborhood; "something happens" in every ball with center c when $c \in K$. The set K and the description of the singularities at each of its points constitute the *morphology* of the process.

This distinction between regular and catastrophic points is obviously somewhat arbitrary because it depends on the fineness of the observation used. One might object, not without reason, that each point is catastrophic to sufficiently sensitive observational techniques. This is why the distinction is an idealization, to be made precise by a mathematical model, and to this end we summarize some ideas of qualitative dynamics.

B. Attractors

Lᶜt (M, X) be a dynamical system defined by the vector field X on the manifold M. An *attractor* F of the system is a closed set invariant under X

and satisfying three conditions:

1. There exists an open invariant neighborhood U of F, called the basin of the attractor F, such that every trajectory starting from a point of U has F as its ω-limit set.

2. Every trajectory whose α-limit set contains a point of F is contained in F.

3. F is *indecomposable*, that is, almost every trajectory of X in F is dense in F.

An attractor F of a field X is called ***structurally stable*** if, for every field X_1 sufficiently close (in the C^1-topology) to X, there are an attractor F_1 and a homeomorphism h of a neighborhood of F onto a neighborhood of F_1, throwing trajectories of X onto trajectories of X_1. In addition we may require that h be close to the identity.

It has not been proved that any given field X on M has attractors, still less structurally stable attractors. However, according to some recent ideas of Smale (see Section 3.1.E), we might conjecture that when M is compact almost all fields X have a finite number of isolated, topologically stable attractors,[1] but unfortunately the topological structure of these attractors is still little understood. However, we shall be considering only attractors of simple types, like isolated points or simple closed curves, whose structural stability is immediate.

In Hamiltonian dynamics there are, strictly speaking, no attractors because the field X leaves a certain measure (the Liouville measure) invariant; hence no open set U can be transformed by X into a set U' contained strictly in U. There are certain closed invariant sets of positive measure that can be considered as ***vague attractors*** (like the set of satellite trajectories of a closed central trajectory studied by Kolmogorov, Moser, and others); a closed trajectory g and an infinite spiral on a torus around g may be thought to define the same thermodynamical state of the system. It is legitimate to consider these vague attractors because they have a property of stability under sufficiently small perturbations of the Hamiltonian, this leading not to topological structural stability but to a global stability under Hamiltonian perturbations.

C. The partition into basins

The basins associated with different attractors can be arranged in many topologically different ways. In the simplest case, that of gradient dynamics, the different basins are separated by piecewise regularly embedded hypersurfaces. For example, on a contour map the basins attached to different rivers are separated by watersheds, which are pieces of crest lines, and these separating lines descend to saddle points, where they meet like ordinary points, but rise to summits, where they may have flat cusp points. In all other cases, and particularly when the dynamic (M, X) has recurrence, the mutual arrangement of two basins can be very complicated. It can happen that the two basins interpenetrate each other in a configuration that is topologically very complicated yet structurally stable; for example, in two dimensions the curve separating the two basins can spiral around a closed trajectory. It is possible in such a case to speak of a situation of struggle or competition between two attractors, and when starting from a point close to the α-limit of the separatrix the final position can be sensitive to very slight variations in the initial position. In this case the phenomenon is practically indeterminate, and the probability relative to each attractor can be measured only by the local density of each basin.

These considerations apply even more to vague attractors of Hamiltonian dynamics, where we can scarcely talk of basins and the dominance of an atrractor is never certain. This hazy and fluctuating aspect of the situation in Hamiltonian dynamics is reminiscent of the probabilism of quantum mechanics.

4.2. MATHEMATICAL MODELS FOR REGULAR PROCESSES

We deal, in what follows, with two types of models, *static* and *metabolic*.

A. Static models

Here we are given two differential manifolds: U (the space of *internal parameters*) and V (the image manifold: in practice, the real or complex numbers), and we suppose that the process can be described locally about each regular point x by a field of maps of the form $G: W \times U \longrightarrow V$, where W is a neighborhood of x in $B \times T$. The local state of the process is described at each regular point by a map $g_x: u \longrightarrow v = G(x, u)$, with which we can, in theory, express all the observables of the system.

B. Metabolic Models

Here we are given, at each regular point x, the following:

1. A manifold M (the space of *internal parameters*).
2. A vector field X on M, depending on x.
3. An attractor $c(x)$ of the dynamic $(M, X(x))$.

We suppose that $X(x)$ depends differentiably on x. Only one factor determines the local state, namely, the neighborhood of the field $X(x)$ in a neighborhood of the attractor $c(x)$; the rest of the field $X(x)$ has only a virtual role. In particular, the observables of the system are determined by the attractor $c(x)$.

So far we have made no hypothesis concerning the differentiability of g and X; we now suppose that they are m-times differentiable, where m is sufficiently large. A catastrophe point is characterized by the fact that, at this point, the field g_x or $c(x)$ or one of its derivatives of order $\leqslant m$ has a discontinuity. In practice, in examples taken from thermodynamics, m is at most 2. It is a priori evident that a discontinuity in a derivative of high order has every chance of escaping detection.

C. Evolution of fields

We suppose that the evolution of the static or metabolic field around a regular point $x = (y, t), y \in B$, is determined by a differential equation of the form

$$\frac{dg}{dt} = H(g, y, t) \qquad \text{(static case)}$$

or

$$\frac{dX}{dt} = H_{c(x)}(x, y, t) \qquad \text{(metabolic case)}$$

where the operators H lead to locally well-posed problems; that is, there is existence and uniqueness in the Cauchy problem relative to this evolution operator, and the solution depends continuously on the initial data throughout a neighborhood of these data $g(y, 0)$. Note that the evolution equation in the metabolic model depends on the attractor $c(x)$.

D. Equivalence of models

It is important to note that there is no reason for a model to be unique. Consider the easiest case, that of a static field on $B \times T$ with values in a function space $L(U, V)$. Given two other manifolds U_1, V_1, and maps $k: U_1 \to U$, $j: V \to V_1$, then $g_x \in L(U, V)$ induces a field g_{1x}, defined by $g_{1x} = j \circ g \circ k$. Of course there is no reason for the evolution operator of g_1, induced from the evolution operator of g by a diffeomorphism $h: X \to X$ of base spaces, to lead to a well-posed problem except in the obvious case in which the maps j and k are diffemorphisms. We have, therefore, two equivalence relations between static models: first, the one denoted by D, for which j and k are diffeomorphisms; and second, E, when j and k are not necessarily bijective maps. It is possible for local models to be isomorphic at points x and x' modulo E without being isomorphic modulo

D; this type of change of behavior is, in principle, completely unobservable but cannot be excluded a priori. It would occur, for example, if a degree of freedom of the internal space U came into play at time t_0, having been fixed up to this moment. In what follows we will allow the possibility of the occurrence of such *silent catastrophes*. Finally, the mathematical problem of whether there exists in each equivalence class modulo E a universal model, the simplest model that induces every other representative by suitably chosen maps j and k, has never been considered.

E. Isomorphic processes

We say that two processes of the same type over open sets W, W_1 in space-time are isomorphic if there is a homeomorphism $h: W \longrightarrow W_1$ such that (1) h maps the catastrophe set $K \subset W$ onto the catastrophe set $K_1 \subset W_1$, and (2) h, restricted to the regular points of W, is a diffeomorphism onto the regular points of W_1. In most cases, the isomorphism extends to an isomorphism of the corresponding local fields: at regular points, x and $x_1 = h(x)$, the corresponding local fields, whether static or metabolic, are equivalent modulo D in the sense of Section 4.2.D.

The metric structure of the base space will enter into the definitions (of ordinary and essential catastrophes) which follow, and this introduces an arbitrary a priori restriction. It should be noted, however, that the theory applies essentially to structurally stable natural processes, and, as we have seen in Chapter 2, this implies a comparison between boxes that are homologous under a Galilean transformation (in the nonrelativistic case, the only one considered here); the introduction of the Euclidean metric is then quite justified, depending, as it does, on the implicit hypothesis that the elementary underlying dynamics satisfy Galilean invariance.

4.3. CATASTROPHES

A. Ordinary catastrophe points

A point y of $B \times T$ is an *ordinary catastrophe point* if the intersection of the catastrophe set K and the ball $b_r(y)$ of center y and sufficiently small radius r has, as model, a nonempty embedded semianalytic[2] polyhedron without interior point, and also the restrictions of the processes to the open balls $b_r(y)$ are all isomorphic for sufficiently small r.

The catastrophe points z sufficiently close to y [contained in the ball $b_r(y)$] also satisfy the same conditions as y; the set K is semianalytic and nonvoid in a neighborhood of z, and so there exists a canonical structure on $K \cap b_r(z)$ which establishes an isomorphism of the local processes. Thus the ordinary catastrophe points form an open set in the set K of catastrophe points.

An ordinary catastrophe point can easily be isolated; however, in all models of morphogenetic fields that we shall consider such an isolated catastrophe can be structurally stable only in the very trivial case in which the dimension of the space-time is one.

B. Essential catastrophe points

A catastrophe point that is not ordinary will be called *essential.* The essential catastrophe points M_1 form a closed subset K of catastrophe points.

The possibility that the set M_1 has a nonempty interior cannot be excluded, as we shall see from examples, because a natural process on a domain of space-time can very well become locally chaotic in a necessary and structurally stable fashion. Faced with such a situation, the observer will ordinarily try to change from the original model to a *reduced model* expressing in some way the average theormodynamic state of the original model. It should be noted that the global stability of the universe depends on the law of compensation: when the catastrophes are frequent and close together, each of them, taken individually, will not have a serious effect, and frequently each is so small that even their totality may be unobservable. When this situation persists in time, the observer is justified in neglecting these very small catastrophes and averaging out only the factors accessible to observation. We shall later describe this operation, called *reduction of a morphogenetic field,* which in general leads to the replacement of a metabolic model by a static model.

An essential catastrophe y cannot, by definition, be isolated in the set of catastrophes because the process restricted to any ball $B_r(y)$ changes its isomorphism class as r tends to zero; this implies that the boundary of $B_r(y)$ meets new catastrophes for arbitrarily small r. We must distinguish two types of essential catastrophe points: those belonging to the closure of the set of ordinary catastrophes, and those having some neighborhood containing only essential catastrophes. In models that we shall study, the points of the first type will lie on hypersurfaces of which one side contains only ordinary catastrophes and the other side essential catastrophes, or no catastrophes at all; the points of the second type form an open set (chaotic processes) or have a complicated topological structure like that of the Cantor set.

4.4. MORPHOGENETIC FIELDS ASSOCIATED WITH LOCAL CATASTROPHES

A. Static models

We come now to the fundamental construction which seems to me to determine the morphology of a process from its dynamic. Consider first the case of a static model; let W be the base space (a domain of space-time), the function space $L(U, V)$ the fiber, and $s: W \rightarrow L(U, V)$ the section field. We suppose that there is a closed subset H' of the bifurcation set H in $L(U, V)$ for which the following essential hypothesis holds: *the local process changes its phenomenological appearance at a point only when a map $g: U \rightarrow V$ of the internal dynamic changes its topological type or form.* Then each point $x \in W$ for which $s(x) = g$ is structurally stable is a regular point of the process, and each catastrophe point y is such that $s(y)$ belongs to the bifurcation set H of $L(U, V)$. In fact, in the only known case of a static model (where V is the real line $\mathbf{R}$; this case is considered in Chapter 5) the stable regimes, the attractors of the local dynamic, correspond to the minima of $g: U \rightarrow \mathbf{R}$, and here the catastrophe points are counterimages of strata H' of H under s, where H' is the set of functions $g: U \rightarrow \mathbf{R}$ with at least two equal minima (Maxwell's convention). These strata H' often have as free boundary the strata $H_1 \subset H$, where two minima coincide or where a minimum decomposes through bifurcation into two distinct minima. We shall describe the simplest case of this phenomenon, called the Riemann-Hugoniot catastrophe, in Chapter 5. This collection of ideas may be summarized in terms of a railway metaphor: branching (*bifurcation*) produces catastrophe.

The set $H' \subset H$ defined in this way by Maxwell's convention is invariant under the action of diffeomorphisms of the internal manifolds U and V, and so the definition of the catastrophe points of the process is independent of the chosen representative in the D isomorphism class of the static fields. Usually we will suppose that the section s is a transversal map of W on H'; this will determine the topological nature of the catastrophe set in a neighborhood of each point. We shall give, in Chapter 5, a description of all these types when the dimension of the space W is less than four. This situation gives only ordinary catastrophes, *and the structurally stable existence of essential catastrophes in our space-time shows that the static model is inadequate or the transversality hypothesis is incorrect.*

The static model does account, however, for the important phenomenon of the propagation of wave fronts and the theoretical determination of all the stable singularities of wave fronts.

B. Stable singularities of wave fronts

Consider a linear rth-order differential operator acting on real functions $f(x, t)$, $x \in \mathbf{R}^n$, $t \in \mathbf{R}$, defined by

$$\frac{\partial^r}{\partial t^r} f = \sum_{|\omega| = r} a_\omega D_\omega f + \sum_{|\omega'| < r} b_{\omega'} D_{\omega'} f,$$

where the coefficients a_ω depend only on x. The terms of maximum order $\Sigma a_\omega D_\omega f$ form the *principal part* of the operator. When one of the $(r - 1)$th derivatives of the solution f has a localized discontinuity at $t = 0$ on a hypersurface with equation $S(x, 0) = 0$, the support of this discontinuity is spread in $\mathbf{R}^n$ like the hypersurfaces $S(x, t) = 0$, where $S(x, t)$ is the solution of *Jacobi's equation*:

$$\frac{\partial S}{\partial t} + H\left(x_i, \frac{\partial S}{\partial x_i}\right) = 0,$$

for which $S(x, 0)$ is the given function, and where

$$H^r = \sum_{|\omega| = r} a_\omega p^\omega = G$$

and, as usual, we write $\partial S / \partial x_i = p_i$.

It is well known that the variation of the hypersurface is obtained by forming the *characteristic system* of Jacobi's equation:

$$\dot{x}_j = -\frac{\partial H}{\partial p_j}, \quad \dot{p}_i = \frac{\partial H}{\partial x_i},$$

which defines the *bicharacteristics* of the given equation. Because

$$\dot{x}_i = -\frac{1}{r} G^{(1-r)/r} \frac{\partial G}{\partial p_i},$$

we see immediately that a unique bicharacteristic $g(p)$ passes through each covector (x_i, p_i), and $g(p)$ is invariant under the dilation $(p_i) \rightarrow \lambda(p_i)$ because of the homogeneity. Consider the set of all the covectors originating from a given point x_1 as the internal space U, and let p be a covector in U, with $g(p)$ the associated bicharacteristic; then in general this bicharacteristic will pierce the initial wave front $S(x, 0)$ at a point $x_0(p)$. The action $\int_{x_0}^{x_1} \Sigma_i p_i \, dx_i$ evaluated between x_0 and x_1 along this bicharacteristic is equivalent to $\int_{x_0}^{x_1} H\,dt$ and is extremum for the covector $p_1 \in U(x_1)$, which is tangent to the wave front at x_1. In this way we are led to associate with each point x_1 a differentiable function on the normalized space of covectors with origin x_1, and *to take the extremum of this function at each point,* giving what is now known as a *Lagrangian manifold.* This is just another way of expressing Huyghens' construction describing a wave front at time t as the envelope of the spherical waves centered on the wave front at time $t=0$ (see Figure 4.1).

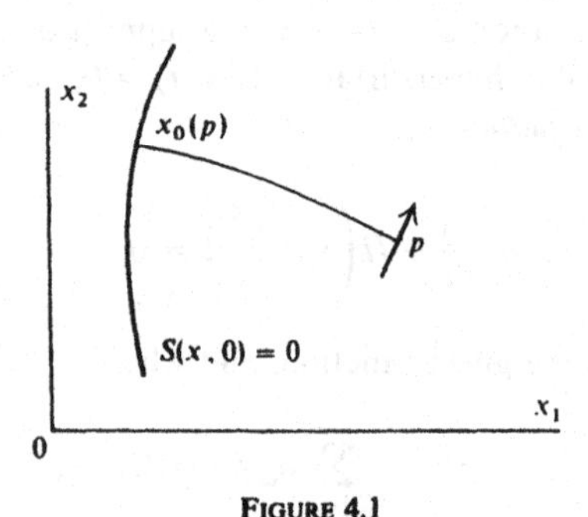

FIGURE 4.1

In our case of three-dimensional space, the normalized covectors with given origin form a sphere S^2, and so we can obtain singularities of corank two (see Section 5.2.C) for the function S on this space U; inverting the construction shows immediately that this gives all the singularities. Among these, only those with codimension less than four will be structurally stable, and they will be studied in Chapter, 5.

C. Metabolic models

The theory for a metabolic model with fiber dynamic (M, X) over W is as follows. In the function space F of vector fields on X there is a bifurcation subset H consisting of systems in the neighborhood of which the configuration of attractors of the field X is not structurally stable. This set H contains a subset H' of those fields X for which the attractor $c(x)$ associated with the section s is not structurally stable, and the catastrophe points of the process then are the counterimage of H' by the section $s: W \to F$. Although probably the set H' is still stratified, it is not locally closed in F, in contrast with the static model. This means that the set $K = s^{-1}H'$ may have a complicated topological structure containing, in particular, essential catastrophe points.

4.5. PRELIMINARY CLASSIFICATION OF CATASTROPHES

A. The domain of existence and basin of an attractor

Consider a natural process defined over domain W of space-time and described by a metabolic model with fiber (M, X). If the local dynamic (M, X) at $y \in W$ is in a bounded state defined by a structurally stable attractor $c(y)$, the point y is a regular point of the process, and all the regular points that can be joined to y by paths lying in the set of regular points form a domain W called the *domain of existence* of the attractor, an open subset of space-time. This must not be confused with the *basin* of the attractor, an open subset of the manifold M of internal states.

In every neighborhood W of an ordinary catastrophe point $y \in K$ there are at least two attractors c and c_1 of (M, X) whose domains of existence meet W. It is natural to suppose that such a competition is possible only when the basins of c and c_1 have a common frontier in M. This is the geometrical interpretation of the Heraclitean maxim that all morphogenesis is the result of a struggle.

B. Conflict and bifurcation catastrophes

If each of the attractors involved in a catastrophe enters into competition in the neighborhood of the given point only with one or several other attractors, this is called a *conflict catastrophe*. The behavior of the shock waves separating the different domains of existence is difficult to specify in general. In most cases each attractor c_i has an associated potential V_i, and the shock wave separating the domain c_i from that of c_j is given by the equation $V_i(y) = V_j(y)$. This is called *Maxwell's convention*, and under general conditions it leads to Gibbs' phase rule.

When an attractor enters into competition with itself (after continuous variation) it is called a *bifurcation catastrophe*; the simplest example is the Riemann-Hugoniot catastrophe described in Chapter 5. Here again Maxwell's convention specifies the topological structure of local shock waves; however, the convention applies only when the domain W can be mapped into the universal unfolding of the singularity of the local dynamics by a map of maximum rank, and this means that W must be sufficiently polarized. On the other hand, the only existing universal model known occurs in the case of gradient dynamics, and thus in principle well-defined local models leading to ordinary catastrophes arise only from gradient dynamics with previously polarized domains. The other cases (of recurrent or insufficiently polarized domains) can give rise to essential catastrophe points leading to the appearance of generalized catastrophes (see Chapter 6).

4.6. THERMODYNAMICAL COUPLING

A. Microcanonical entropy

Suppose that the following are given:

1. A configuration space, an n-dimensional differential manifold V with local coordinates q_i.
2. The phase space, the fiber space $T^*(V)$ of covectors of V, with local coordinates q_i, p_i.
3. A time-independent Hamiltonian, that is, a real-valued function $H: T^*(V) \rightarrow \mathbf{R}$.

Let α be the canonical 2-form of $T^*(V)$, defined locally by $\alpha = \Sigma dp_i \wedge dq_i$. Then the evolution field X in $T'(V)$ is given by Hamilton's equation $i(X)\cdot\alpha = dH$, where i is the interior product. Now it is known that the field X has H as a first integral (when V is compact, it is generally the only integral), and leaves the 2-form α invariant. The nth exterior power of α defines a measure invariant under X, the Liouville measure α^n. Write $m(x)$ for the Liouville volume of $\{p \in T^*(V);\, H(p) < x\}$; then the derivative $a(x) = dm/dx$ represents the $(2m-1)$-volume of the energy hypersurface $H = x$. The *microcanonical entropy* of the system is the function $S(x) = \log a(x)$, and the *formal temperature* of the system is $T = (dS/dx)^{-1}$; geometrically the temperature is the reciprocal of the average over the energy hypersurface of the mean curvature of this hypersurface.

The introduction of the logarithm $\log a(x)$ can be motivated as follows. Consider two conservative Hamiltonian systems S_1 and S_2 with configuration spaces V_1 and V_2 and Hamiltonians H_1 and H_2, and suppose that these systems are thermodynamically coupled so that they evolve at almost all instants as if systems S_1 and S_2 had no interaction, except that at moments of very brief duration they exchange energy by random catastrophic processes, with the effect of making the probability that the system is in a domain D of the energy hypersurface proportional to the

Liouville measure of D (the *ergodic* hypothesis). Now the hypersurface with total energy c and equation $H_1(x_1) + H_2(x_2) = c$ has its Liouville volume A, given by

$$A(c) = \int a_1(c - t)a_2(t)\, dt;$$

hence the most probable value of the parameter t is that for which the product of the volumes of the hypersurfaces $H_1 = c - t$ and $H_2 = t$ is maximum. This gives

$$\frac{1}{a_1}\frac{da_1}{dt} = \frac{1}{a_2}\frac{da_2}{dt},$$

that is, the two systems S_1 and S_2 have the same formal temperature $T_1 = T_2$, or $dS_1/dH_1 = dS_2/dH_2$.

B. The interaction of two systems

If two systems S_1 and S_2, with total energy c, are thermodynamically coupled, the equilibrium regime will occur at the value of t for which the total entropy $S_1(c - t) + S_2(t)$ is maximum. If there is only one sufficiently sharp maximum, the joint system will evolve toward this state and stay there, neglecting fluctuations. On the other hand, if there are several maxima or if the maximum is very flat, we cannot say anything a priori about the evolution of the system, for it could take on one of these maxima or could fluctuate without a well-determined limit.

The importance of thermodynamical coupling comes from the fact that, if two dynamical systems over two contiguous open sets U_1 and U_2 are put into spatial contact, we can suppose (without any special hypothesis concerning the nature of the interaction) that the two systems are thermodynamically coupled. If the functions $S(H)$ have, for the two systems, convex increasing graphs (the usual case), the addition of such a function to the function obtained by the transformation $t \rightarrow c - t$ gives a function with a single sharp maximum (Figure 4.2), and the two systems will adopt the unique equilibrium regime. But other cases are also possible when dealing with unusual dynamics of other systems than large numbers of identical particles.

Consider the following case: if each of the graphs of $S(H)$ has a sharp maximum at value a, the sum of the entropies $S_1(t) + S_2(c - t)$ will have two different sharp maxima, near $t = a$ and $t = c - a$, when $c \neq 2a$. In such a case we would expect the following situation: over each open set U_1, U_2 each system will keep, more or less, its own regime, and on the common frontier of these two open sets the two regimes will be separated by a surface shock wave or a narrow fluctuating boundary zone; the two systems will almost refuse to interact and each will evolve independently (see Figure 4.3).

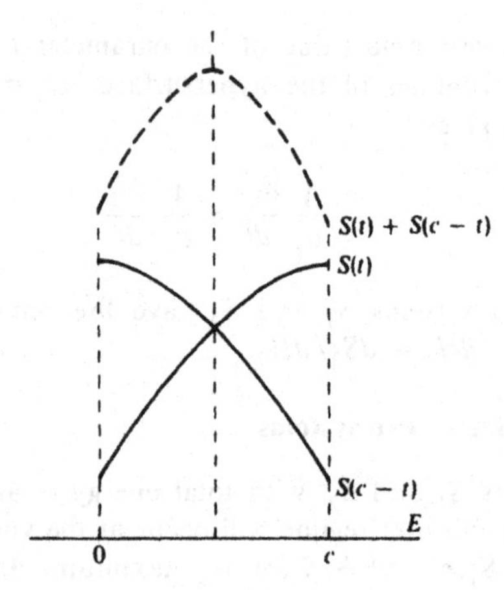

FIGURE 4.2

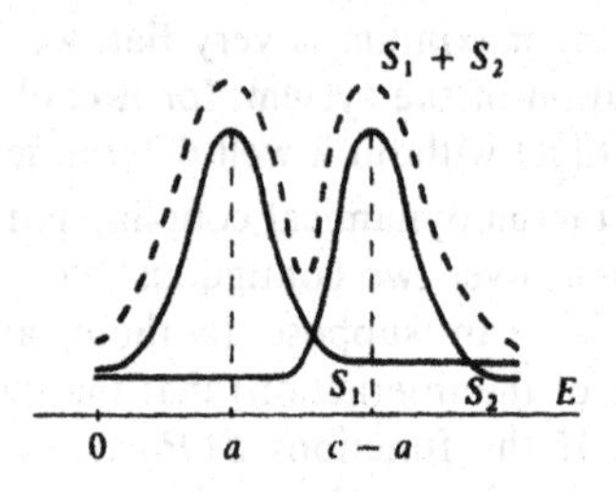

FIGURE 4.3

C. The approach to the equilibrium state with thermodynamical interaction

Let us return to the situation of two classical Hamiltonian systems coupled thermodynamically; if we write A_1 and A_2 for the volumes of the energy hypersurfaces $H_1 = E_1$ and $H_2 = E_2$, respectively, the entropies of the two systems are defined by $S_1 = \log A_1$, $S_2 = \log A_2$ and the temperatures by $T_1^{-1} = dS_1/dE_1$, $T_2^{-1} = dS_2/dE_2$; T_1 and T_2 are then scalar functions defined on the energy axes with the dimensions of energy. Furthermore, if the H_i depend on external parameters (e.g., a point in a space G_i) and there are other first integrals than energy which add vectorially like energy and form spaces P_i, we can define the functions T_i on the product $G_i \times P_i$. When equilibrium occurs at the maximum of entropy on the hypersurface $H_1 + H_2 = E$, nothing is known, in general, of the approach to the equilibrium state; this equilibrium position (supposed unique) is given by $T_1(E_1) = T_2(E - E_1)$, and it is reasonable to suppose that the evolution of the energy E_1 of the first system is defined by $dE_1/dt = k(T_2 - T_1)$ (k a scalar constant), that is, the direction of the approach to the equilibrium state in the space $E_1 \times G_1 \times P_1$ is given by the gradient of the entropy.

D. Polarized dynamics

Suppose now that there is a field of *polarized dynamics* S_1 on U_1, by which we mean that at each point x of U_1 there is a local dynamic characterized by a local function $s(h, x)$, where each function $s(h, x)$ has a sharp maximum at $h = m$, m varying strongly with x (we shall consider later the conditions to be imposed on this function field in order to realize local thermodonamical equilibrium), and that on U_2 is a dynamic characterized by a function $S_2(H)$ with a very flat plateau-like maximum. Then the local linking between S_1 and S_2 will give rise in the total entropy to a function with a sharp maximum m' close to m (see Figure 4.4) depending on the point x considered in U_2—that is to say, the polarization of the field dynamic S_1 will diffuse throughout U_2, and there will be a spatical extension of this polarization. If the dynamic S_2 had had only one sharp maximum, the regime S_2 could not have adopted the regime S_1 and there would not have been any extension of the polarization. This gives a model to explain the phenomenon of competence of a tissue in embryology, and in this model the loss of competence by aging of a tissue can be described as a change in the function $S(H)$, whose maximum, initially flat, becomes sharper and sharper. We shall see later how to interpret this phenomenon.

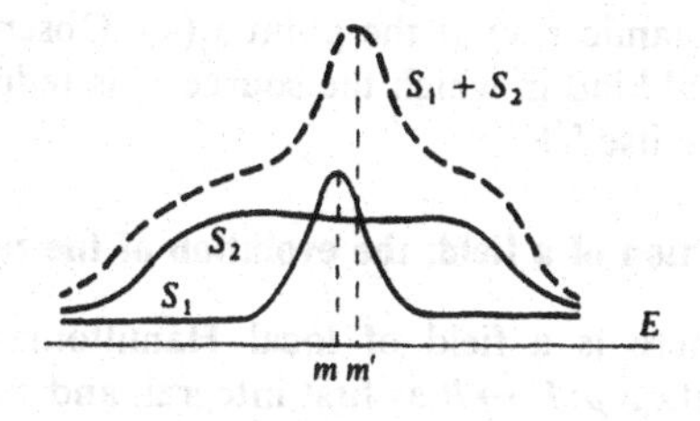

FIGURE 4.4

E. The pseudogroup of local equivalences of a field

The theory of fields of local dynamics that we shall use differs from classical thermodynamics in an essential point: usually one considers a discrete set A of dynamical systems in interaction without any topology (other than the discrete topology) on set A; in our theory of dynamical fields we take for set A a topological space U (most frequently a domain of space-time $\mathbf{R}^4$), and the interaction between systems E_x is not significant unless the associated points x of U are neighboring. Moreover, there will exist a pseudogroup G of transformations acting both on U and on the total configuration space of the system and leaving the total dynamic invariant. When G acts transitively on U, we call the field *homogeneous* on U; otherwise, *polarized.* We shall show later how to define a manifold P over each point x of U parameterized by the approximate first integrals of the fiber dynamic; we will have then a reduced field defined by a section of a fiber space on U with fiber P, and the equation of evolution of the field, defined by the condition of local maximum entropy, will be invariant under G.

4.7. THE REDUCED FIELD

A. The definition of the reduced field

Suppose a metabolic field on an open set W in $\mathbf{R}^n$ whose space of internal states is a manifold F with Hamiltonian fiber dynamic $H(F)$ is given; the metabolic field is then defined by a section $s:x \longrightarrow H(F)$, which defines a vector field on F, and by a particular point $s_1(x)$ in F. Let E be the energy, the first integral of the field $H(F)(x)$. The *reduced field* of the metabolic field (s,s_1) is the static field defined by $g:W \longrightarrow E\ (s_1)$; hence it ignores the fine internal structure of the fiber dynamics $H(F)$ and considers only the energy E at the base point $s_1(x)$. More generally, if the fiber dynamic $H(F)$ has a system of *global first integrals* defined by a differentiable fibration $p:F \longrightarrow P$, the reduced field of (s,s_1) is the static field defined by $g(x)=p[s_1(x)]$, where $p[S_1(x)]$ denotes the set of values of the first integrals of the dynamic $s(x)$ at the point $s_1(x)$. Observe that these static fields are of a trivial kind in which the source U is reduced to a point and the fiber $L(U,V)$ is itself V.

B. The self-interaction of a field; the evolution of the reduced field

Suppose that there is a field of local Hamiltonian dynamics on W admitting the fibration $p:F \longrightarrow P$ as first integral, and suppose also that the local dynamic over an open subset U of W is in the vague attractor $c(x)$;

then this attractor is, in some average sense, contained in a fiber of the fibration $F \to P$. Frequently there is a continuum of attractors of type $c(x)$ contained in a fiber, and this continuum is a symplectic manifold on which we can define a Liouville measure A, and then an entropy S relative to c by the formula $S = \log A$. Using S, we define the temperature T by $T^{-1} = dS/dE$.

Further suppose that a pseudogroup of local dynamical equivalences G is associated with the attractor c, where G operates both on W as a subgroup of displacements and on the continuum of attractors in the fiber F; when G and its isotropy group (the subgroup of G leaving one point fixed) are transitive on U, the corresponding dynamic is homogeneous and isotropic in U. Let us then look for conditions for local equilibrium. In equilibrium the local system above a disc D in U must exchange energy in equilibrium with the rest of the system, which we suppose invariant (thermostatic); by an earlier assumption (Section 4.6.C) the exchange of energy is proportional to the temperature difference, and so $dE = 0$ if and only if $\int_D (T - T_0)$ is zero on the disc D (where T_0 denotes the temperature at the center of D). Therefore at equilibrium the temperature is a harmonic function, $\Delta T = 0$, and when the system is not in equilibrium it will evolve according to the equation $dE/dt = k/\Delta T(E)$. When the local dynamic is that of a linear oscillator, T and E will be identified, up to a scalar factor, and this will give the heat equation. For quantum fields, E can be a complex energy, and the equation will then be the Schrödinger equation.[3]

It is quite conceivable that some natural processes approach equilibrium through a superposition of these two methods with a reversible component of Hamiltonian character and an irreversible gradient component.

We must note that the validity of evolution equations of the form $dE/dt = \Delta T$ depends on the possibility of defining a local temperature associated with a well-defined attractor c of the local dynamic. In circumstances where this attractor ceases to be defined and is captured by other attractors, new catastrophes with morphogenetic effects appear in W and the evolution has, strictly speaking, no more meaning.

NOTES

1. This statement is valid in the C^0-topology. See Section 3.1.E and [21] and [25].

2. Recall that a semianalytic set is the union of a finite number of sets, each defined by a finite number of analytic equations or inequalities; then these sets are triangulable in analytically embedded simplices (Łojasiewicz). The fact that sections of semianalytic sets by concentric spheres of sufficiently small radius are all homeomorphic can be established by the isotopy theorem for transversal sections of stratified sets.

3. *Quantum fields.* As an additional remark, let us consider what happens during the coupling of two quantum systems. A quantum system is a Hamiltonian system admitting first integrals in even numbers which underlie a manifold P with a symplectic structure; in particular, the energy E is complex, the argument of E is the phase of the system, and $|E|$ is the scalar energy. Above each point of P is a phase manifold whose Liouville volume A allows the definition of *entropy*, $S = \log A$. Here again there are first integrals defined by P for the product dynamic of two coupled systems of this type, and an addition map $P_1 \times P_2 \longrightarrow P$, which defines global first integrals of the joint system; in particular, the energies can be added (as complex numbers). Here (as opposed to the previous case) we suppose that the catastrophes by which the systems interact are reversible in time (this will be so if, for example, these catastrophes are generated by one-parameter subgroups of the total phase space, leaving invariant the global symplectic structure of the space). Then the volume $A(P)$ of the fiber above the point (p_1, p_2) is left invariant by the interaction. Therefore the entropy S is left invariant, and there is no approach to equilibrium in the kernel of the addition map $P_1 \times P_2 \longrightarrow P$ through constant S; the field of evolution will be itself a Hamiltonian field defined (in direction) by $dp/dt = i \cdot \text{grad } S$ ($i \cdot$grad being the symplectic gradient). The coupled system will move around its equilibrium state, never reaching it.

CHAPTER 5

ELEMENTARY CATASTROPHES ON $\mathbf{R}^4$ ASSOCIATED WITH CONFLICTS OF REGIMES

Εἰδέναι χρὴ τὸν πόλεμον ἐόντα ξυνόν, καὶ δίκην ἔριν· καὶ γινόμενα πάντα κατ' ἔριν καὶ χρεών.

It should be known that war is universal, that strife is justice, and all things come into existence by strife and necessity.

HERACLITUS

5.1. FIELDS OF GRADIENT DYNAMICS AND THE ASSOCIATED STATIC MODEL

A. The competition between local regimes

Let W be an open subset of space-time $\mathbf{R}^4$ with a field (M, X) of local dynamics over it, where the compact fiber manifold M is the space of internal coordinates, and, for each $x \in W$, $X(x)$ is a vector field on M depending differentiably on x. Throughout this chapter (except in Section 5.6) we make the hypothesis that for each $x \in W$ the field $X(x)$ is a gradient field on M:

$$X(m; x) = -\operatorname{grad} V(m; x),$$

where $V(m; x)$ is a potential function on M depending differentially on x, and the gradient is taken with respect to a fixed Riemannian metric on M.

The metabolic field (M, X) can be replaced by a static field whose fiber is the function space $C^\infty(M, \mathbf{R})$ of differentiable functions on M, where $V(m; x)$ is the value of this field at x.

The stable local regimes at a point x of W are defined by the structurally stable attractors of X, that is, by the minima of the potential $V(m; x)$. At a regular point x of the process there will, in general, be several such minima of which only one can dominate; how should this be chosen from all the theoretically possible regimes? Here we adopt an arbitrary, but convenient and simple convention, namely Maxwell's convention.

B. Maxwell's convention

The convention is that, when several stable attractors c_i are in competition at a point x of W, the chosen state is one, c_{i_0} of minimum potential, that is,

$$V(c_{i_0}) \leqslant V(c_i) \qquad i \neq i_0.$$

A similar rule was used by Maxwell to deal with the indeterminancy in v in van der Waals' equation $F(p, v)=0$ in an interval of p where there are three real roots for v describing a mixture of liquid and gaseous states: I will not discuss the physical justification of this convention. Let me say immediately that we are not concerned here with the quantitative validity of the convention; our only interest is in describing the topological structure of shock waves (catastrophe points) in the neighborhood of a singularity of the potential V where a controlling attractor undergoes bifurcation. To this end $V(c)$ might well be replaced by any other function expressing roughly the stability of the attractor, (e.g., the volume or the diameter of the basin of attraction in M or, possibly better, the relative height of the lowest saddle on the boundary of the basin), and the structure of the catastrophe points would scarcely be altered. In general, however, the convention has two drawbacks.

1. It leads to geometrically clear-cut separating shock waves, whereas in nature these transitional regions tend to thicken and fluctuate under the effect of viscosity and diffusion. Frequently, after some time has elapsed, stable transitional regimes are formed whose competition gives rise to a submorphology, for example, von Kármán's vortex street in the theory of wakes [1], or the formation of structured membranes in biology.
2. It takes no account of the very common phenomenon of *delay* in the formation of a new regime: suppose that an attractor c_1 controls a point x of W and then a new attractor c_2 with much lower potential appears at x as x moves in W. According to Maxwell's convention, the state must then immediately change to the new regime c_2, but in fact the new regime will

generally appear much later and often in the characteristically ramified form of the generalized catastrophe described in Chapter 6.

5.2. THE ALGEBRAIC STUDY OF POINT SINGULARITIES OF A POTENTIAL FUNCTION

A. The catastrophe set

According to Maxwell's convention, a point x of W is catastrophic in only two cases: either the absolute minimum of the potential $V(m; x)$ is achieved at at least two points c and c' *(conflict points)*, or this minimum, achieved at a unique point m_0, stops being stable (*bifurcation points*). Let K denote the set of functions V in $C^\infty(M, \mathbf{R})$, having either of these two properties (i.e., the absolute minimum of V is achieved at least at two, distinct or coincident, points of M). If the given field is $s: W \longrightarrow C^\infty(M, \mathbf{R})$ [so that $V(m; x) = s(x)$], the set of catastrophe points of the field is precisely the counterimage $s^{-1}(K)$. The set of strata of finite codimension in $C^\infty(M, \mathbf{R})$ forms an everywhere-dense subset of the set K; and if the natural hypothesis that the field s is transversal to K is added, $s^{-1}(K)$ will have only strata with codimension less than four, each of them giving rise to a specific type of local catastrophe in W. Let us consider first the bifurcation points.

B. Bifurcation strata

Definition. Let V be a real differentiable function of n variables $x_1, x_2, \ldots, x_n$; then the origin O is a ***strictly isolated singular point*** of V if the ideal J generated by the partial derivatives V_{x_i} in the algebra of germs of C^∞-functions at O contains a power of the maximal ideal defining O, or, equivalently, if every monomial M in the x_i of sufficiently high degree can be written as a linear combination

$$M = \sum_i g_i V_{x_i}$$

with C^∞-coefficients g_i.

Lemma. Let O be a singular point of V that is *not* strictly isolated. Then the jet of infinite order of V at O belongs to a pro-set A of infinite codimension in the space $J(n, 1)$ of jets of functions.

Proof. To say that the singular point O of V is not strictly isolated is equivalent to saying that there is a curve of singular points passing through

O, in a formal sense at least. Write A_1 for the set defined by $V_{x_i} = 0$ in $J^1(n, 1)$; describing the V for which j^1V cuts A_1 nontransversally at O, we get in $J^2(n, 1)$ the condition

$$\text{Hessian } V = \det\left(V_{x_i x_j}\right) = 0$$

defining a set A_2 projecting into A_1. Similarly, by induction, we can define sets $A_k \subset J^k(n, 1)$ by writing that $j^{k-1}V(x)$ is nontraversal to A_{k-1} at O, and codimension $A_k >$ codimension A_{k-1}. Then the sequence

$$A_1 \leftarrow A_2 \leftarrow \cdots \leftarrow A_{k-1} \leftarrow A_k \leftarrow \cdots$$

defines the pro-set A of the lemma.

Corollary. Every differentiable m-parameter family of functions $V(x_i)$ can be approximated arbitrarily closely in the C^s-topology by an m-parameter family of functions all of whose singular points are strictly isolated.

C. A study of isolated singular points; corank

Definition. Let M be an n-dimensional manifold and O $(x_1 = x_2 = \cdots = x_n = 0)$ a singular point of V. Then the Taylor series of V at O begins with a quadratic form $Q(x_i)$ of rank $n - k$, and the integer k is called the *corank* of the singularity O.

The corank can be defined intrinsically as follows. Let T be a local foliation at O of codimension s, defined by s coordinate functions $u_1, u_2, \ldots, u_s$, and say that T is *adapted to the singularity* O of V if the restriction of V to the leaf $T(O)$ containing O has a nondegenerate quadratic critical point at O. The corank of O is then the minimum codimension of an adapted foliation.

Let $(u_1, u_2, \ldots, u_s; y_1, y_2, \ldots, y_{n-s})$ be a coordinate system adapted to the foliation T. The singular point O is defined in the leaf $T(O)$ by the system of equations

$$\frac{\partial V}{\partial y_i} = 0, \quad j = 1, 2, \ldots, n - s.$$

This system is of maximum rank in $T(O)$ and hence is of maximum rank in each leaf $T(u)$ for sufficiently small u and defines there a unique point $c(u)$ that is a nondegenerate quadratic critical point of the restriction of V to the leaf $T(u)$. The point $c(u)$ depends differentiably on u, and $c(O) = O$.

By a classical theorem of Morse [2], there is a coordinate system Y_j,

$j = 1, 2, \ldots, n-s$, in the leaf $T(u)$ in which V can be written as

$$V = V_1(u) + \sum_{j=1}^{n-s} \pm Y_j^2, \tag{M}$$

where $V_1(u) = V[c(u)]$.

I now claim that $V_1(u)$ has a critical point at the origin $u = O (dV_1 = 0$ for $u = O)$, which is, of zero rank and hence the Taylor series of $V_1(u)$ begins with a term of the third, or higher, degree. Suppose, on the contrary, that the rank of this point is strictly positive; then we can find a direction in the plane $Y = 0$ (e.g., the u_1-axis) along which V (or V_1) is strictly of the second degree; $V = O(u_1^2)$. Then the dimension of an adapted foliation can be increased by one; in fact the foliation defined by $(u_2, u_3, \ldots, u_s)$ will be adapted.

It follows from this construction that the condition for a critical point O of a function V of n variable to be of corank k, $k \leqslant n$, is the vanishing coefficients of a quadratic form in k variables $u_1, u_2, \ldots, u_k$ in the quadratic term of the Taylor series of V. This imposes $\frac{1}{2}k(k+1)$ independent conditions.

Proposition. In the space $J^2(n, 1)$ of jets or order two, the jets of singular points of corank k form a linear manifold of codimension $\frac{1}{2}k(k+1)$ in the space of singular jets of order one.

Corollary. The only stable critical points of families of functions parameterized by less than four variables (and so of our field s of Section 5.2.A) are those of corank at most two, for the points of corank three have codimension $\frac{1}{2} \cdot 3 \cdot 4 = 6$.

D. The residual singularity

We now show an important property, that the singularity of $V_1(u)$, called the *residual singularity* of the singularity at O, is independent (in a sense that we shall explain) of the chosen adapted foliation T. Suppose that the adapted foliation T, defined by coordinate functions $u_1, u_2, \ldots, u_s$, is replaced by another adapted foliation T' with functions $v_1, v_2, \ldots, v_s$ (themselves functions of u_i and Y_j). The set of critical points of the restrictions of V to the leaves $T'(v)$ of T' is then given by the system

$$\left(\frac{\partial V}{\partial Y_i}\right)_{v = \text{const.}} = \left(\frac{\partial V}{\partial Y_i}\right)_{u = \text{const.}} + \sum_j \frac{\partial V}{\partial u_j} \cdot \frac{\partial u_j}{\partial Y_i} = 0,$$

and so, substituting equation (M),

$$\mp Y_i = \frac{1}{2}\sum_j \frac{\partial V}{\partial u_j}\cdot\frac{\partial u_i}{\partial Y_i}.$$

The new residual singularity is given by the function

$$V_2(v) = V[c'(v)].$$

As the set of critical points $c'(v)$ forms an s-manifold that can also be parameterized by u, the function V_2 can be written as a function of u:

$$V_2(u) = V_1(u) + \frac{1}{4}\sum_i \pm\left(\sum_j \frac{\partial V}{\partial u_j}\cdot\frac{\partial u_j}{\partial Y_i}\right)^2.$$

We now use the following property (suggested by a result of Tougeron [3]): If V is a function that is 2-flat at the origin O, then every function $V + g$, where g belongs to the square J^2 of the ideal J generated by the partial derivatives $\partial V/\partial x_i$, is equivalent to V, that is, there exists a local change of coordinates h: $x_i \longrightarrow x_i$ such that $V \circ h = V + g$. Therefore the residual singularity $V_1(u)$ is independent, up to differentiable equivalence, of the adapted foliation T.

This fact has important consequences: the bifurcation of a critical point of corank k depends only on k and the type of residual singularity, and not on the dimension $n \geqslant k$ of the space on which the function is defined, and so we can always suppose that this space has dimension k and the critical point under consideration has zero rank. This algebraic behavior has fundamental importance in our model, *as it shows that the type and dynamical origin of a catastrophe can be described even when all the internal parameters describing the system are not explicitly known.*

We are now going to describe the singularities of codimension less than four and will consider strata of conflict and bifurcation. We shall however, do little more than mention the transversal intersection of the strata of disjoint singularities, as these have no new morphology. Furthermore, when considering space-time as foliated by hyperplanes t = constant, it is of interest to describe the local evolution of certain singularities as functions of time, and this will give what we call *transition strata.*

5.3. CATASTROPHIES OF CORANK ONE

By the theory of Section 5.2 we may suppose that the manifold of internal states M is one dimensional with coordinate x.

A. Strata of codimensio.. zero

These are functions whose absolute minimum is generic and hence defined locally by a quadratic minimum $V = x^2$. The corresponding points in W are regular points of the process.

B. Strata of codimension one

There are two types of these strata.

*1. **Strata of fold type.*** Here the potential can be taken to be $V(x) = \frac{1}{3}x^3 + ux$ (see Section 3.2.C, where this potential is described as the universal unfolding of the singularity $V = \frac{1}{3}x^3$). On one side of the stratum U there are two regimes, one stable c_1 and the other unstable c' (a minimum of V and a critical point of V of index one annihilating each other by collision on U), the other critical points remaining unchanged (see Figure 5.1). When Maxwell's convention is applied strictly, such a stratum can have no morphogenetic effect, as the new attractor c_1 is such that $V(c_1)$ is greater than the absolute minimum $V(c)$. Therefore the shock wave of folded type does not, in principle, exist. If the possibility of delay is allowed, an attractor c_1 of greater potential than the potential of the lowest regime (we might speak of a metastable regime) can be destroyed by a fold of the form $s^{-1}(U)$. The corresponding zone of W can then adopt either a metastable regime of lower potential or the ground regime, depending on the topological arrangement of the basins of attraction (see Figure 5.2).

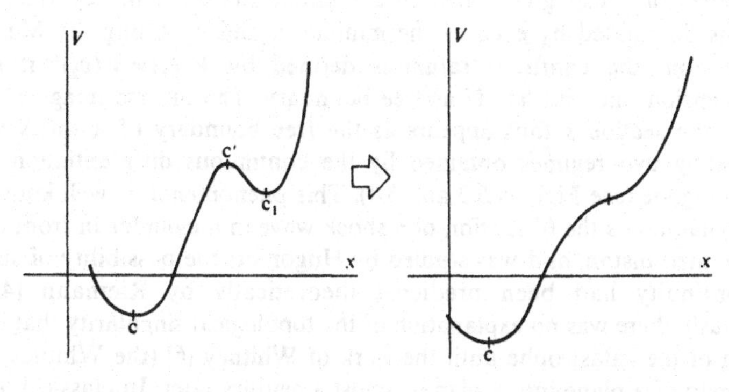

FIGURE 5.1 FIGURE 5.2

2. Conflict strata. This is the set of functions V having two equal absolute minima c_1 and c_2. On one side of the stratum Z, $V(c_1) < V(c_2)$, while $V(c_1) > V(c_2)$ on the other. The inverse image $s^{-1}(Z)$ gives rise to a shock wave separating regimes c_1 and c_2.

C. Strata of codimension two

There are four types of these strata.

1. The transversal intersection of two fold strata. These strata, like the fold strata, are of no morphogenetic interest.

2. Strata of cusp points and the Riemann-Hugoniot catastrophe. These strata are very important in morphogenesis, for they give birth to the simplest of catastrophes, the *Riemann-Hugoniot catastrophe*. Let c be the critical point of the potential V associated with a point O of such a stratum U. Writing (u, v) for coordinates in L transversal to U, and supposing that $V(c) = 0$, we can express the potential V locally as

$$V = \tfrac{1}{4}x^4 + \tfrac{1}{2}ux^2 + vx$$

with respect to a coordinate x in M (depending differentiably on u and v).

In a (u, v)-plane transversal to U, the semicubical parabola with equation $4u^3 + 27v^2 = 0$ defines the fold strata of K of codimension one meeting at U. Thus the stratum U appears as an edge of regression (*arête de rebroussement*) on the bifurcation hypersurface K. Outside this semicubical parabola there is only one minimum and so only one possible regime; inside, there are two minima defining regimes in competition. This shows immediately that there cannot be one continuous regime inside the parabola; the cusp gives birth to a conflict stratum and separates the regions dominated by each of the minima c_1 and c_2. Using the Maxwell convention, the conflict stratum is defined by $V(c_1) = V(c_2)$; it is of codimension one and has U as free boundary. The inverse image $s^{-1}(U)$ under the section s thus appears as the free boundary of a shock wave separating two regimes obtained by the continuous differentiation of a single regime (see Figures 5.3 and 5.4). This phenomenon is well known in gas dynamics as the formation of a shock wave in a cylinder in front of an accelerated piston, and was studied by Hugoniot; the possibility of such a discontinuity had been predicted theoretically by Riemann [4, 5]. Curiously there was no explanation of the topological singularity that is the origin of the catastrophe until the work of Whitney [6] (the Whitney cusp of a map of a plane into a plane) almost a century later. In classical optics the singularity gives birth to the structurally stable cuspidal structure of caustics (Photographs 2a, b, and c). In geomorphology it is the origin of a fault (Photographs 3a and b).

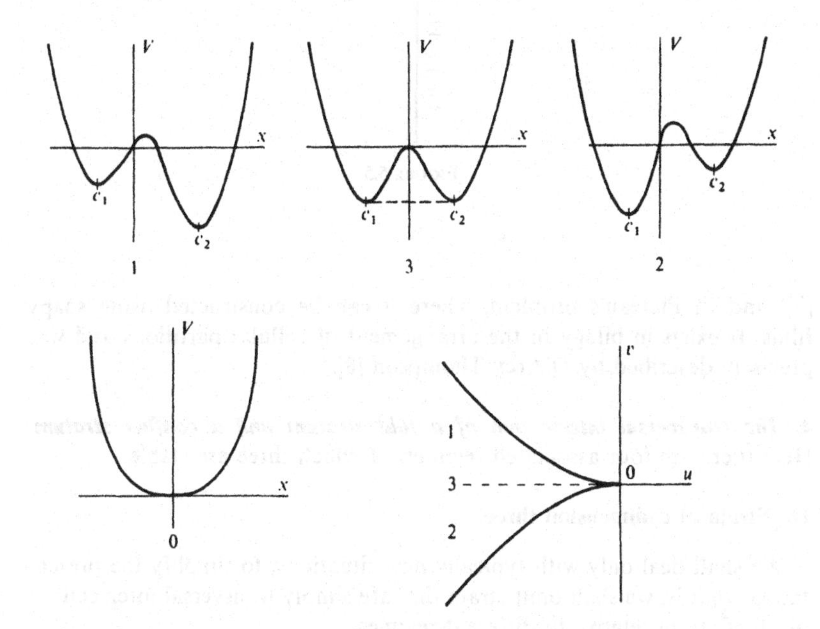

Figures 5.3 and 5.4. Figures 0, 1, 2, and 3 (Figure 5.3) represent the local potentials corresponding to the positions indicated on the universal unfolding (Figure 5.4).

3. Conflict strata. When three regimes c_1, c_2, and c_3 are in competition, the set of points for which the potential V has three equal (distinct and nondegenerate) absolute minima forms locally a stratum of codimension two. This stratum appears as a triple edge, the common boundary of three one-codimensional conflict strata defined by

$$V(c_1) = V(c_2) < V(c_3); \quad V(c_2) = V(c_3) < V(c_1);$$

$$V(c_3) = V(c_1) < V(c_2).$$

The inverse image of this stratum under s defines a triple edge, the common boundary of three strata of conflict (see Figure 5.5).

This singularity is known in gas dynamics (notably in Mach's reflection

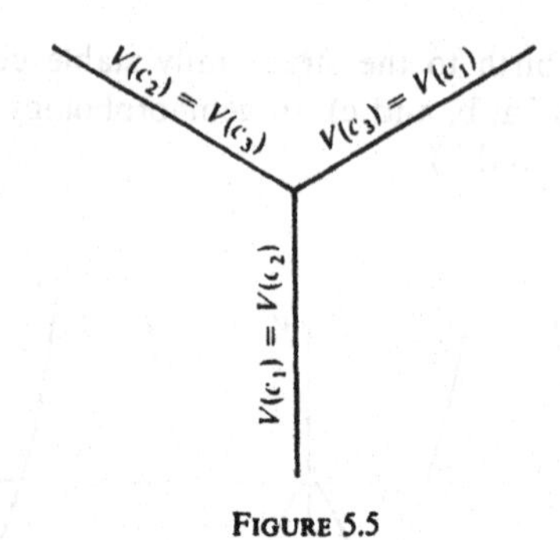

FIGURE 5.5

[7]) and in Plateau's problem, where it can be constructed using soapy films. It exists in bilogy in the arrangement of cellular partitions and was precisely described by D'Arcy Thompson [8].

4. The transversal intersection of a fold stratum and a conflict stratum. Here there are four associated regimes, of which three are stable.

D. Strata of codimension three

We shall deal only with typically new situations, to simplify the presentation; that is, we shall omit strata that are simply transversal intersections of other strata relative to different regimes.

1. The swallow's tail. This is a singularity of a map $\mathbf{R}^3 \to \mathbf{R}$ (or its suspension). If dim $M=1$ (with coordinate x) and (u,v,w) denotes coordinates transversal to the stratum U, the equations for the local model are as follows:

$$V = \tfrac{1}{5}x^5 + \tfrac{1}{3}ux^3 + \tfrac{1}{2}vx^2 + wx$$

$$\frac{\partial V}{\partial x} = x^4 + ux^2 + vx + w = 0. \qquad \text{(S)}$$

In $\mathbf{R}^3$, with coordinates (u,v,w), the discriminant of equation (S) defines a surface K, which can be investigated best by examining sections by the planes $u =$ constant. Looking for the points of $\mathbf{R}^3$ where the fourth-degree equation (S) has a triple root in x, we have

$$0=\frac{\partial^2 V}{\partial x^2}=4x^3+2ux+v,$$

$$0=\frac{\partial^3 V}{\partial x^3}=12x^2+2u,$$

giving a parametric representation of a curve C:

$$u=-6x^2; \qquad v=8x^3; \qquad w=-3x^4.$$

This curve is the edge of regression of the surface K; itself, it has a cusp at the origin. For negative u, the section of K by a plane $u=-a^2$ has two cusps, symmetric with respect to the v-axis; as this section is the envelope of a one-parameter family of lines, without a stationary line, it has no point of inflection, and the only possible form of this curve is a swallow's tail (see Figure 5.6). For positive u, there are no cusps and the section is a simple convex curve. Starting with $u=-a^2$ and then increasing u, the curvilinear triangle of the swallow's tail shrinks and coincides with the origin for $u=0$; the section has a point of infinite curvature at the origin. Note that the surface K has a line of selfintersection given algebraically by $v=0$. $w=u^2/4$. In the algebraic or analytic model this double line lies completely in K, and so for positive u the section of K would have an isolated double point but since we are only considering real values of x, such a singularity has no significance in our models of real situations (see Figure 5.7).

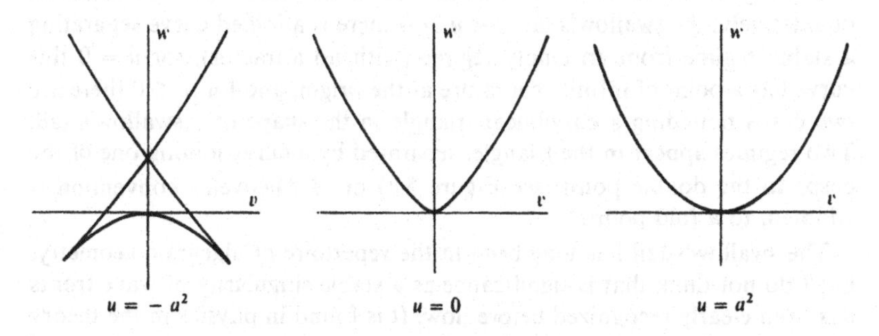

FIGURE 5.6. Sections of the swallow's tail (see also Figure 5.7).

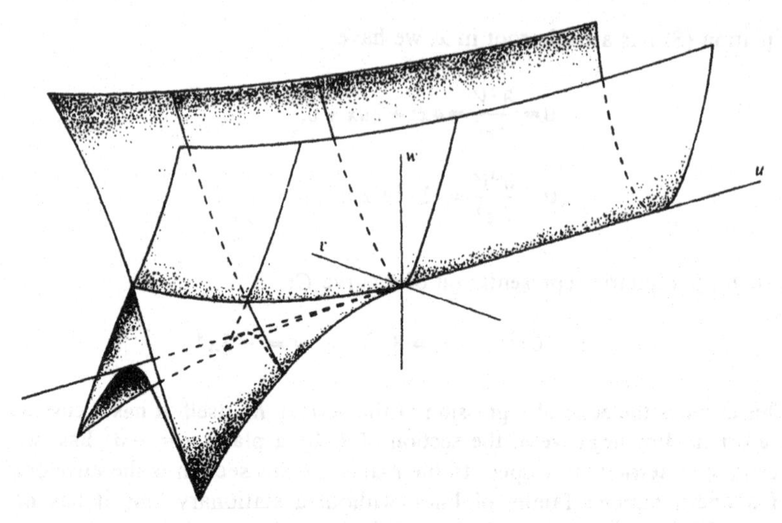

FIGURE 5.7. The swallow's tail.

Let us now investigate whether there can exist conflicts of regimes in the neighborhood of such a singularity; for this the equation must have four real roots (if not, V cannot have two minima). We must therefore consider the interior of the curvilinear triangle of the swallow's tail for $u < 0$. Since one of the cusps of this triangle defines a Riemann-Hugoniot catastrophe, there will be a shock wave in the triangle, the line of conflict between two regimes, which starts from the cusp and ends at the double point of the section[1] or in its neighborhood, on a fold line,. This, it is easy to see, is the only topological possibility for the division of the regimes. Following the variation of the sections as u decreases, we can describe, qualitatively, the swallow's tail: for $u > 0$ there is a folded curve separating a stable regime from an empty regime (with no attractor), for $u = 0$ this curve has a point of infinite curvature at the origin, and for $u < 0$ there are two cusps bounding a curvilinear triangle in the shape of a swallow's tail. Two regimes appear in the triangle, separated by a curve joining one of the cusps to the double point (see Figure 5.8) or, if Maxwell's convention is adopted, to a fold point.

The swallow's tail has long been in the repertoire of algegraic geometry, but I do not think that is significance as a stable singularity of wave fronts has been clearly recognized before now. It is found in physics in the theory of wave fronts of plasmas [9] and, more recently, as a singularity of Landau surfaces associated with certain Feynman diagrams [10]. It can also be realized easily in classical geometric optics as a singularity of caustics (Photograph 4). In regard to embryology, we propose in Chapter 9 to consider the extremities of the blastopore furrow as swallow's tails.

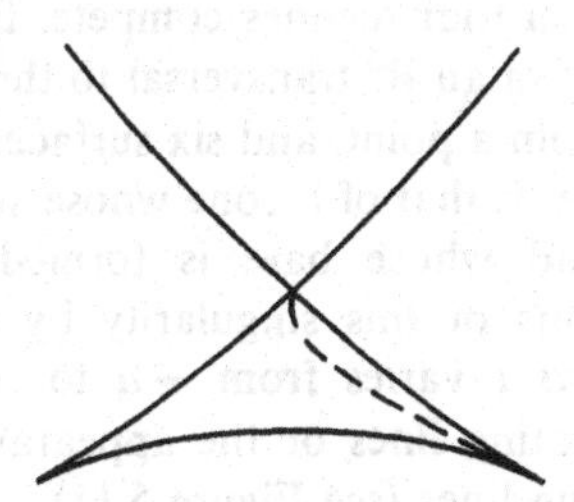

FIGURE 5.8. The shock wave associated with a swallow's tail.

2. Transition strata: transitions on cusp strata. In our space of three parameters the Riemann-Hugoniot stratum is a curve; the section by the plane w = constant can thus have singularities when this plane is a tangent to the curve (in fact, we are dealing here not with a new singularity but with one-parameter variations—in time, for example—of hyperplane sections of the cusp singularity). There are two types of contact.

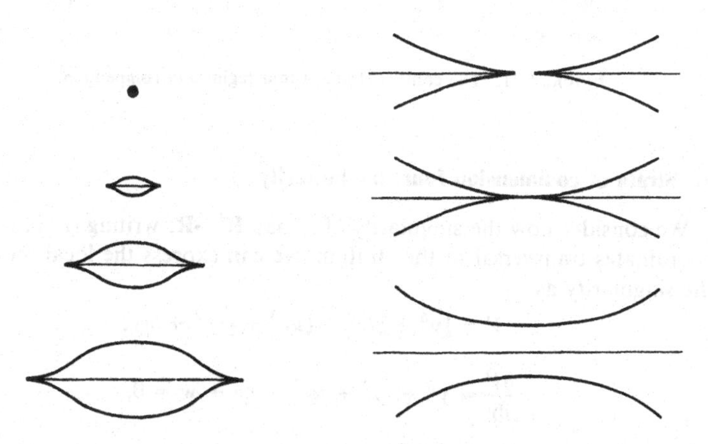

FIGURE 5.9. The lip singularity. FIGURE 5.10. The beak-to-beak singularity.

1. The *lip singularity*: two cusps appear at $t=0$ and then separate, remaining joined by a separatrix of the two regimes (see Figure 5.9).

2. the *beak-to-beak singularity*: two symmetric cusps join at their extremeties and the corresponding separating lines rejoin (see Figure 5.10).

3. Conflict strata. When four regimes compete, this defines a stratum of codimension three in K; in an $\mathbf{R}^3$ transversal to this stratum there are four triple edges emerging from a point, and six surfaces of conflict. The global topological configuration is that of a cone whose summit is the center of a regular tetrahedron, and whose base is formed by the edges of the tetrahedron. The sections of this singularity by the plane t = constant show an interchange, as t varies from $-a$ to $+a$, of the connections between the four separating lines or the appearance of a triangle at the vertex of three separating lines (see Figure 5.11).

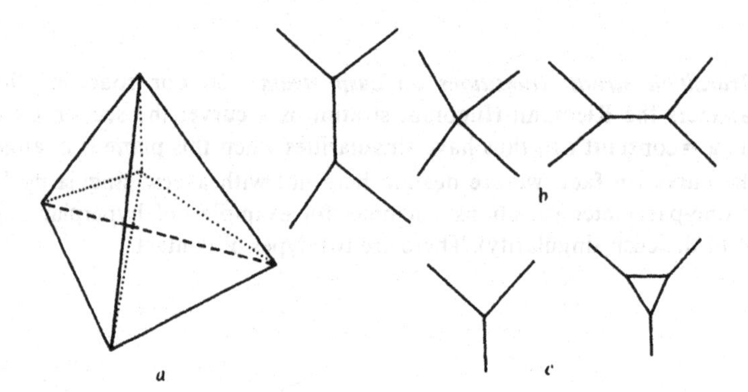

FIGURE 5.11. The conflict strata of four regimes in competition.

E. Strata of codimension four: the butterfly

We consider now the singularity of a map $\mathbf{R}^4 \to \mathbf{R}$; writing (u, v, w, t) for coordinates transversal to the stratum, we can express the local model for the singularity as

$$V = \tfrac{1}{6}y^6 + \tfrac{1}{4}ty^4 + \tfrac{1}{3}uy^3 + \tfrac{1}{2}vy^2 + wy,$$

$$\frac{\partial V}{\partial y} = y^5 + ty^3 + uy^2 + vy + w = 0, \qquad \text{(B)}$$

where y is the coordinate of M.

We can picture the hypersurface D^3 in space-time $\mathbf{R}^4(u, v, w, t)$ whose equation is the discriminant of equation (B), by finding the parametric representation of the surface A^2 of points where B has a triple root and the curve C^1 where B has a quadruple root. In order to avoid going into details that a patient reader can reconstruct for himself, I shall restrict myself to giving a qualitative description of D^3 by considering intersections with the hypersurface t=constant. For $t>0$ (Figure 5-12*a*) the section is a surface with an edge of regression (the section of A^2); this curve is convex and has no singularities. For $t=0$ the curve has a point of infinite curvature at the origin. For $t<0$ (Figure 5-12*b*) two cusps emerge from it in a configuration resembling a swallow's tail but spread into the third dimension; the surface, a section of D^3 by the plane $t = -\alpha^2$, has a double line W having two end points at the cusp points of C^1, and two cusps at points where the cusp line A cuts one of the sheets associated with the other branch of C^1. The plane of symmetry $v = w = 0$ cuts this surface D^3 in a curve with three cusps and without points of inflection, that is, a curve in the form of a butterfly on which there are three points of self-intersection where this plane meets the double curve W.

What conflicts of regime can be associated with this singularity? The equation B has five real roots in the curvilinear quadrilateral inside the butterfly curve, and so there can be conflict between three regimes (the minima of V). The only possibie configuration of the division between these three regimes is indicated in Figure 5-12*c* where three separating surfaces leave a triple point and terminate in two cusps (Riemann-Hugoniot) and a double point. The variation can be described qualitatively as follows: for $t > 0$, there is a surface shock wave whose boundary is a Riemann-Hugoniot curve; for $t < 0$ the boundary of the shock wave exfoliates and gives birth to a triple edge (see Figure 5-12*c*).

Although this singularity is structurally stable, it is difficult to observe directly because of its transient nature. On the other hand, it is easier to demonstrate the result of a space $\mathbf{R}^3$ passing across it; there is, as we have already seen, an exfoliation, a swelling of a shock wave with a free surface. Frequently the boundary of the shock wave proceeds, while the two extremities of the triple line join, leading to a cell joined to a vertical partition (Figure 5.13). There are many instances of structures of this type in biology, for example, the alveole of vein walls. In semantic interpretations of elementary catastrophes, the butterfly appears as the organizing center of the *pocket*.

Two symmetric exfoliated surfaces joined along their boundaries give a closed blister, a configuration corresponding to a catastrophe of a potential of codimension five. Although defined by an unstable singularity of a potential, this formation of blisters on a shock wave separating two

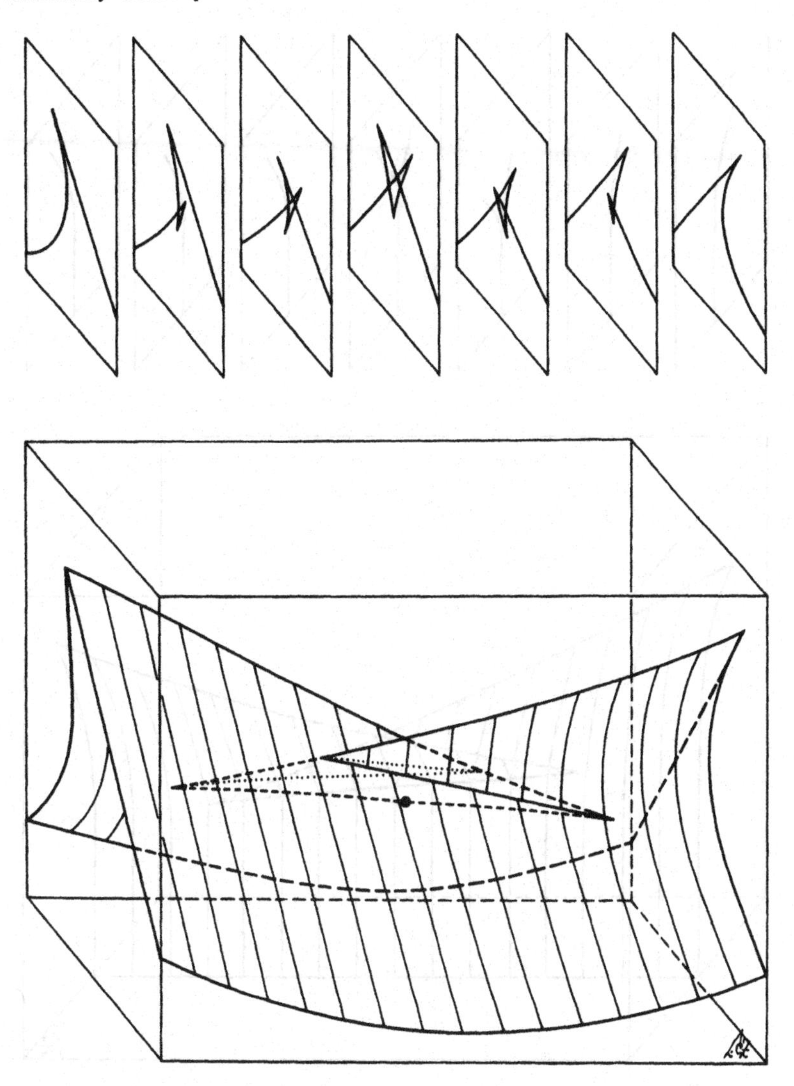

FIGURE 5.12b The butterfly: the bifurcation set when $t < 0$.

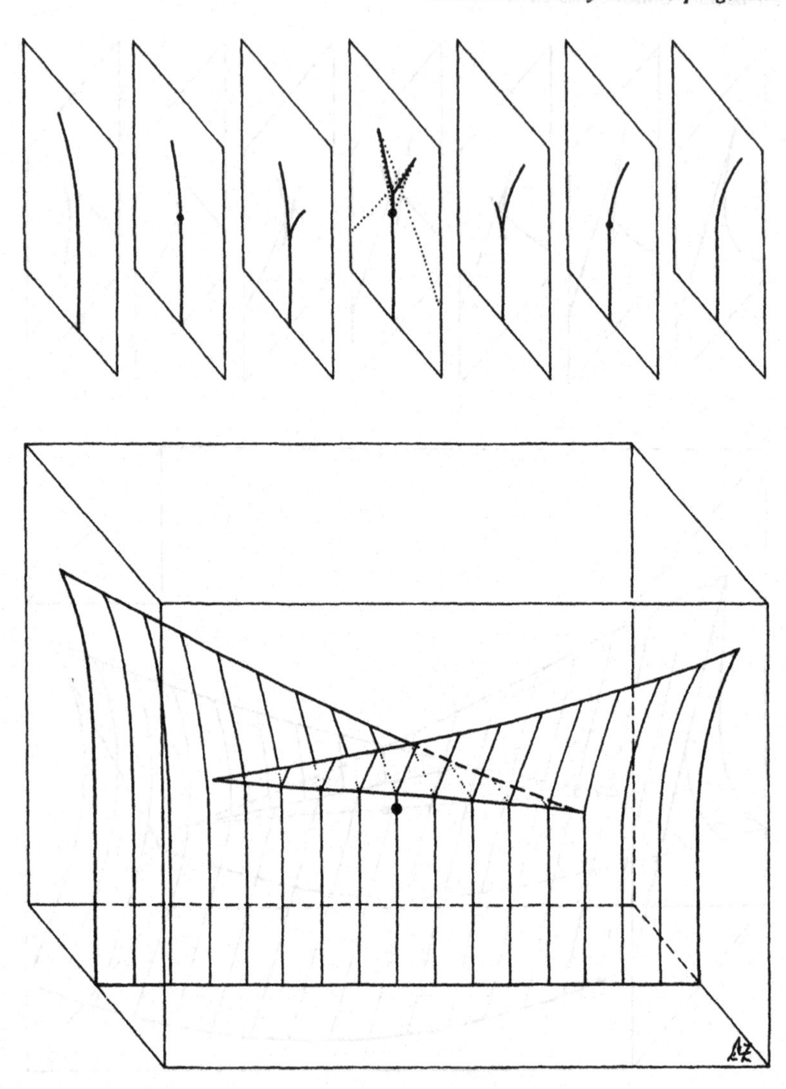

FIGURE 5.12*c* The butterfly: the shock wave (using Maxwell's convention) when $t < 0$.

regimes is often observed. Frequently the closed blister then evolves into an open blister, an alveolus, a phenomenon representing the capture of the intermediate regime of the blister by one of the bordering regimes on the side of the opening. Another illustration of this process, apart from the blister that erupts on our skin, is the formation of small domes on a beach at low tide (see Photograph 5).

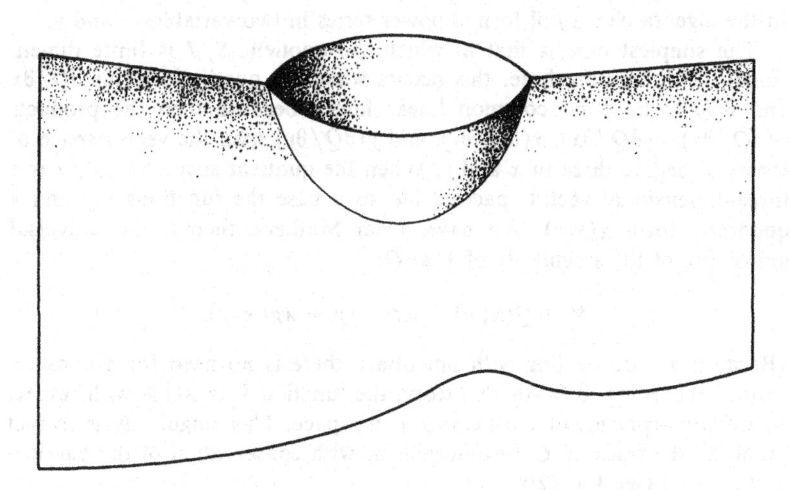

FIGURE 5.13. Evolution of the exfoliation into a blister.

5.4. ELEMENTARY CATASTROPHES OF CORANK TWO

A. Umbilics

So far we have considered only singularities of a potential $V: M \longrightarrow \mathbf{R}$ of corank one, where, if M is an n-dimensional manifold, the potential V at the singular point O has as a second differential a quadratic form of rank $n - 1$. Now it can happen that it has, structurally stably, rank $n - 2$; to say that a quadratic form in n variables depends only on $n - 2$ variables means that three coefficients must be zero (e.g., the coefficients a, b, and c in $ax^2 + 2bxy + cy^2$), and this singularity can be stable, occurring at isolated points in $\mathbf{R}^3$. We shall call these points *umbilics* for reasons that will appear later.

B. Classification of umbilics

Suppose that M is a two-dimensional manifold with local coordinates (x, y); the other $n - 2$ variables of the general theory will play no part in the description of the singularities. The Taylor series of the potential V at O will start with the cubic term $Q(x, y)$; and, according to the theory of Mather (see Section 3.2.E), to classify the singularities we must consider the ideal J generated by the formal partial derivatives $\partial V/\partial x$ and $\partial V/\partial y$ in the algebra $S(x, y)$ of formal power series in two variables x and y.

The simplest case is that in which the quotient S/J is finite dimensional, of dimension three: this occurs when the quadratic forms $\partial Q/\partial x$ and $\partial Q/\partial y$ have no common linear factor because then the products $x(\partial Q/\partial x)$, $y(\partial Q/\partial x)$, $x(\partial Q/\partial y)$, and $y(\partial Q/\partial y)$ span the vector space of forms of degree three in x and y. When the quotient space $S(x,y)/J$ is a three-dimensional vector space, it has as a base the functions x,y and a quadratic form $g(x,y)$. We have, from Mather's theory, the universal unifolding of the singularity of V at O:

$$V = Q(x, y) - ux - vy + wg(x, y).$$

(Because we are dealing with potentials, there is no need for a constant term.) According to Mather's theory, the function V is stable with respect to diffeomorphisms of the (x, y, u, v, w)-space. This singularity is in fact stable in the sense of C^∞-isomorphisms with conservation of the parameters u, v, w) (see Lu [12]).

When the quadratic form g has mixed signature, we call the singularity a *hyperbolic umbilic*; when g is positive (or negative) definite, we call it *elliptic*.

These types can be defined more intrinsically as follows. The involutive linear pencil defined by $aQ_x = bQ_y = 0$ does not depend on the coordinate system (x, y) and its double lines are distinct, because common linear factors of Q_x and Q_y define invariant lines of the involution, and conversely. When the double lines are real, we have a hyperbolic umbilic; taking these double lines as axes, we may write $Q_x = 3x^2$, $Q_y = 3y^2$ and so

$$Q = x^3 + y^3.$$

We can take $g = xy$ for the base of quadratic forms, and this gives

$$V = x^3 + y^3 + wxy - ux - vy, \qquad \text{(H)}$$

the hyperbolic umbilic and its universal unfolding.

When the double lines are complex conjugates, we can take them to be the isotropic lines $x^2 + y^2 = 0$; to obtain this we put

$$Q = x^3 - 3xy^2,$$

$$Q_x = 3(x^2 - y^2), Q_y = -6xy.$$

Taking $g = x^2 + y^2$ gives

$$V = x^3 - 3xy^2 + w(x^2 + y^2) - ux - vy, \tag{E}$$

the elliptic umbilic and its universal unfolding.

C. The morphology of umbilics

It is difficult to describe the hypersurfaces of $\mathbf{R}^5$ defined by the equations (H) and (E), and so we study first their apparent contours B on the universal unfolding space (u, v, w), and associate with each region, defined by the bifurcation set, the topological type of the corresponding function $V(x, y)$ on the space (x, y) of internal parameters. Later on, we shall give an interpretation of the configuration obtained by supposing that the regime associated with an attractor of grad V occupies all the domain that it controls.

1. The hyperbolic umbilic: the crest of the wave. Here we have

$$V = x^3 + y^3 + wxy - ux - vy. \tag{H}$$

Consider the set C of critical points of $V(x, y)$, when (u, v, w) are constant; this set is a three-dimensional manifold defined by $V_x = V_y = 0$, that is,

$$u = 3x^2 + wy, \qquad v = 3y^2 + wx.$$

The projection of C on (u, v, w)-space has as critical values the bifurcation set B. When $w = 0$, this gives the map $\mathbf{R}^2 \to \mathbf{R}^2$ defined by $u = 3x^2$, $v = 3y^2$, the map defined by the "four-folded handkerchief" (see Figure 5.14). The curve of critical values consists of the two positive half-axes Ou and Ov; adjoining fourth-order terms to V clearly gives a map of the type $u = 3x^2 + y^3$, $v = 3y^2 + x^3$, and then the curve B in the Ouv-plane consists of two arcs, each with a simple cusp at the origin (see Figure 5.15). This singularity can also occur as a singularity of caustics in geometric optics (see Photograph 6). When w has a small, nonzero value, the double point of C disappears, giving a critical curve with two arcs (see Figure 5.16), one a convex curve and, in its interior, the other with a simple cusp; the same configuration occurs regardless of the sign of w because changing x to $-x$, y to $-y$, and w to $-w$ does not change C. The degenerate situation $u = v = w = 0$ gives, for the internal parameters (x, y), the curve $x^3 + y^3 = 0$ with a triple point and one real branch; this curve decomposes into the line $x + y = 0$ and the imaginary circle $x^2 - xy + y^2 = 0$. Varying the parameters (u, v, w) gives three generic situations.

1. No critical point for V: here V has the topological type of the linear function $x+y$.
2. Two critical points for V: either a minimum and a saddle point (case 2a) or a maximum and a saddle point (case 2b).
3. Four critical points for V: here $B=0$ might be, for example, the intersection of $x+y=0$ with the real circle $x^2+y^2-h=0$, giving one maximum, one minimum, and two saddle points.

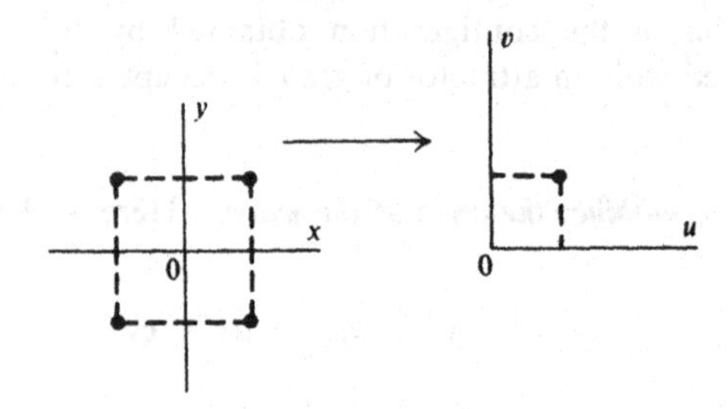

FIGURE 5.14. The four-folded handkerchief.

Case 3 corresponds to the internal region of Figures 5.15 and 5.16; the cusp of Figure 5.16 corresponds to Figure 5.16*C*, in which the real circle touches the line and so the three critical points coincide at the point of contact. The two cases 2a and 2b correspond to the zone between the two critical curves of Figure 5.16 or to the interior of the cusps of Figure 5.15, and case 1 corresponds to the outside zones of Figures 5.15 and 5.16.

Now let us study the morphogenetic field associated with a hyperbolic umbilic; as only one stable regime is possible [a minimum of $V(x, y)$] in

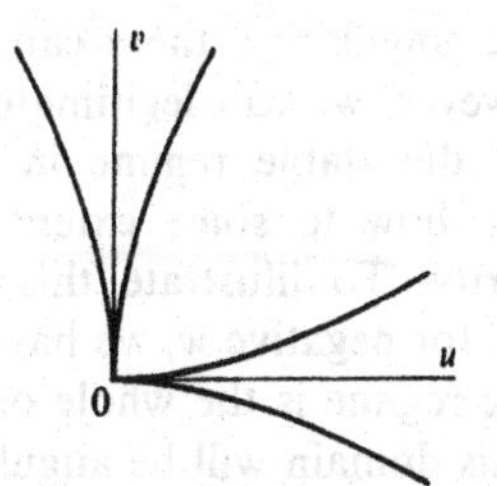

FIGURE 5.15. The four-folded handkerchief made generic.

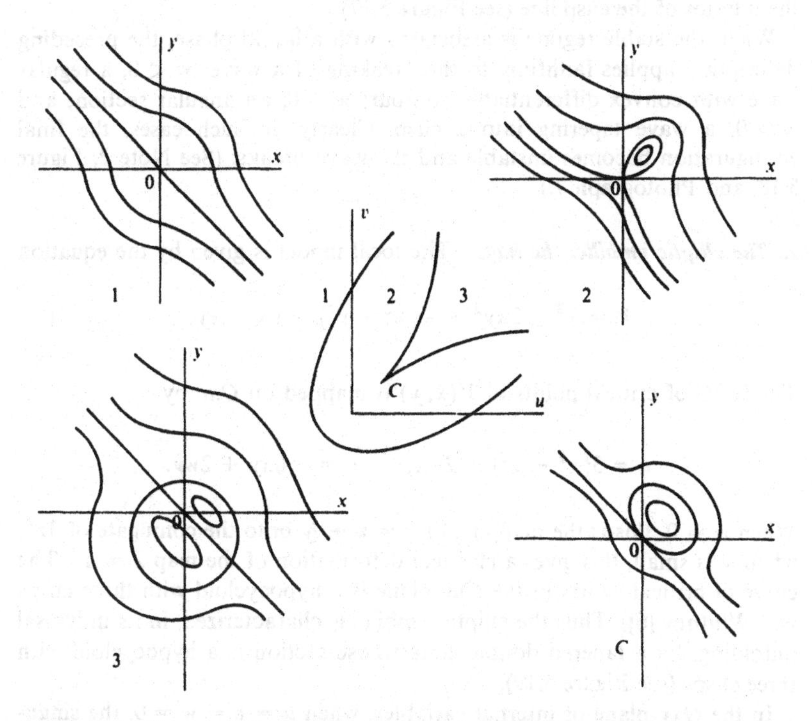

FIGURE 5.16. The universal unfolding of the hyperbolic umbilic (at center), surrounded by the local potentials in regions 1, 2, and 3 and at the cusp C.

the neighborhood of the singularity, there cannot, strictly speaking, be a conflict of regimes. However, we can legitimately suppose that the global form of the domain of the stable regime in the (u, v, w)-space of the universal unfolding will show to some extent where the local dynamic regime has the singularity. To illustrate this idea, we suppose that w represents time and that, for negative w, we have case 2a; then the domain of existence of the stable regime is the whole of the interior of the convex curve c. When $w = 0$, this domain will be angular at O, and for positive w we have case 2b (because the sign of V must be changed to restore the same equation) and so the domain of the stable regime will be reduced to the interior of the cusp line (see Figure 5.17).

When the stable regime is associated with a liquid phase, the preceding description applies faithfully to the breaking of a wave: $w < 0$, a regular wave with convex differentiable contour; $w = 0$, an angular section; and $w > 0$, a wave tapering into a cusp. Clearly, in such cases, the final configuration becomes unstable and the wave breaks. (See Note 2, Figure 5.18, and Photograph 7.)

2. *The elliptic umbilic: the hair.* The local model is given by the equation

$$V = x^3 - 3xy^2 + w(x^2 + y^2) - ux - vy. \tag{E}$$

The set C of critical points of $V(x, y)$ is mapped on Ouv by

$$u = 3(x^2 - y^2) + 2wx, \qquad v = -6xy + 2wy.$$

When $w = 0$, this is the mapping of $z = x + iy$ onto the conjugate of $3z^2$; when w is small, this gives a classical deformation of the map $Z = z^2$. The curve of critical values in the Ouv-plane is a hypocycloid with three cusps (see Whitney [6]). Thus the elliptic umbilic is characterized, in its universal unfolding, by a tapered double cone whose section is a hypocycloid with three cusps (see Figure 5.19).

In the Oxy-plane of internal variables, when $u = v = w = 0$, the singularity defines a curve with a triple point with real distinct tangents, $x^3 - 3xy^2 = 0$; this critical point is called an *Affensattelpunkt* (see Figure 5.20-0). A generic deformation of this situation transforms the triple point into three double points, with the triangle they define containing either a maximum or a minimum of V (Figures 5.20*a* and 5.20*b*, respectively). There is also, in particular for $w = 0$, a generic deformation for which V has only two saddle points. Consequently only one stable regime is possible in the neighborhood of the singularity. As the change from w negative to w positive (with $u = v = 0$) transforms case a into case b, it

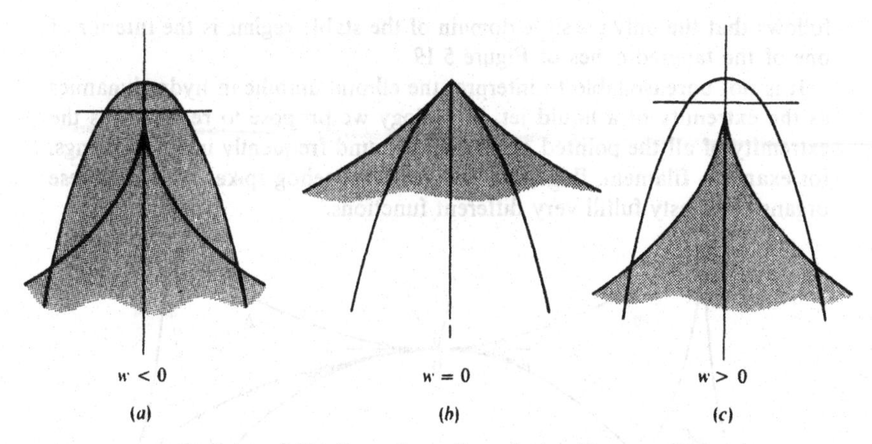

FIGURE 5.17. Successive sections of a breaking wave.

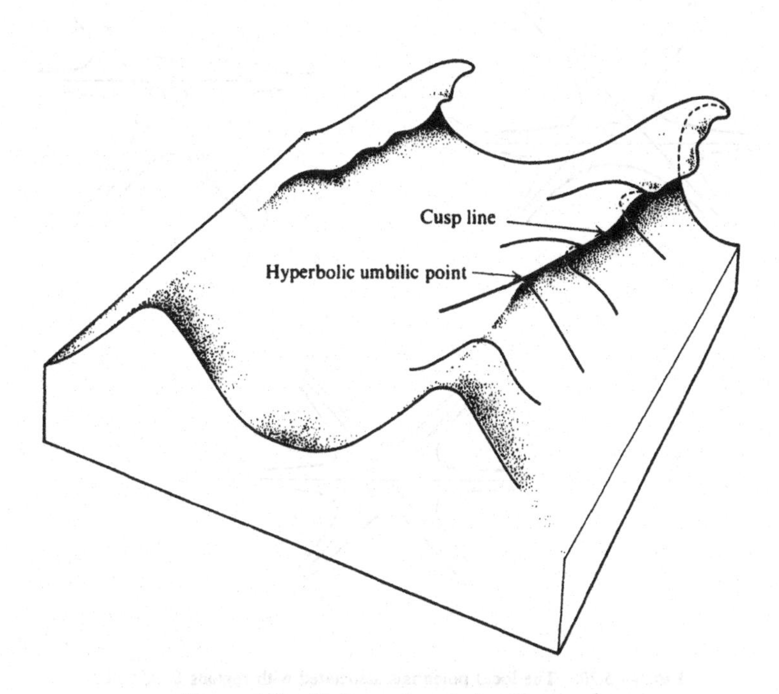

FIGURE 5.18. Bird's eye view of a breaking wave.

follows that the only possible domain of the stable regime is the interior of one of the tapered cones of Figure 5.19.

It is not unreasonable to interpret the elliptic umbilic in hydrodynamics as the extremity of a liquid jet. In biology we propose to regard it as the extremity of all the pointed organs to be found frequently in living beings, for example, filament, flagellum, hair, and hedgehog spike, although these organs obviously fulfill very different functions.

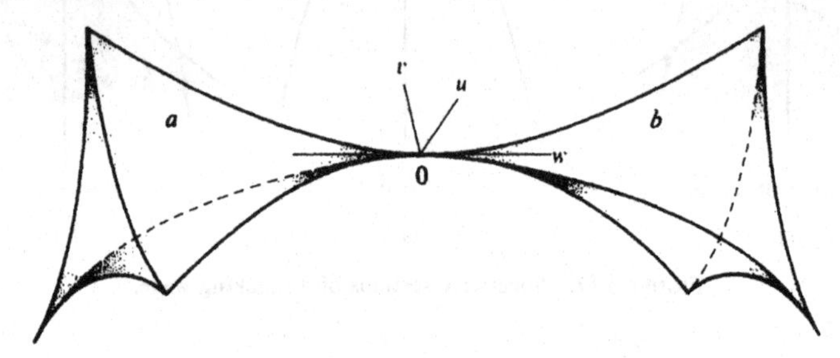

FIGURE 5.19. The universal unfolding of the elliptic umbilic.

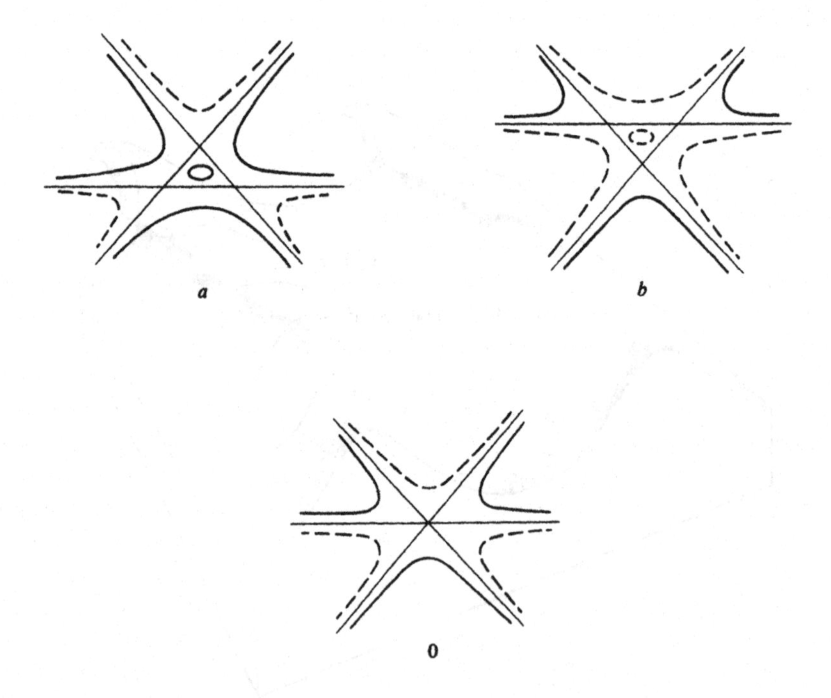

FIGURE 5.20. The local potentials associated with regions 0, *a*, and *b*.

3. A remark on the terminology. An *umbilic*, in classical differential topology, of a surface S in $\mathbf{R}^3$ is a point at which the two principal radii of curvature are equal; then for each umbilic M there is a unique center of curvature $q(M)$ on the normal to S at M. It is such a point that we here, with perhaps an abuse of terminology, call an umbilic. Consider the set of spheres with variable radius r centered on S. They form a manifold Q of dimension $2 + 3 = 5$ which maps naturally into $\mathbf{R}^3 \times \mathbf{R}$ (with parameter r); at the center of curvature $q(M)$ this map has a singularity identified locally with the umbilic that we have defined. The classical classification due to Picard has much in common with our classification: the umbilics through which three lines of curvature pass have at their centers of curvature an elliptic umbilic in our terminology, while the umbilics with only one line of curvature have at their centers of curvature our hyperbolic umbilic (see [13] and [14]). This follows because the normals tangent to the edges of regression of the focal surface meet the surface along curves whose tangents at the umbilic are the Darboux directions, which are therefore tangential to the lines of curvature passing through the umbilic.

In biology the umbilic, or navel, is in principle the singularity at which the successor organism separates from its parent. As we will see, this idea corresponds, in some sense, to the parabolic umbilic, which we now describe.

4. The parabolic umbilic: the mushroom. It is possible to have a structurally stable transition point in $\mathbf{R}^4$ between a hyperbolic and an elliptic umbilic because these singularities are of codimension three; we call such transition points *parabolic umbilics*.

Let $V(x,y) = Q(x,y) +$ terms of higher degree, where Q is a cubic form, be the local Taylor expansion of V about 0. Whenever the two quadratic forms Q_x and Q_y have a common linear factor, or, equivalently, whenever the involutive pencil $aQ_x + bQ_y = 0$ has coincident double rays, we say that the origin is a parabolic umbilic; if this linear factor has equation $x = 0$, this implies that $Q(x,y) = x^2y$. (This follows immediately from writing $Q_x = xA$, $Q_y = xB$, where A and B are linear factors, and using $Q_{xy} = Q_{yx}$; hence $x(A_y - B_x) = B$ and $Q_y = kx^2$.) However, it is necessary to observe that the third-order jet $V = x^2y$ is not determinate in the sense of Chapter 3, and to obtain a determinate jet we must stabilize it by adjoining higher-order terms; it can be shown that fourth-order terms are sufficient [12]. We shall begin by studying, qualitatively, the fourth-order jet

$$V(x,y) = x^2y + \frac{y^4}{4};$$

it has been shown by Lu that all other stabilizations are equivalent to this one. The gradient of V is given by the differential equation $dx/2xy = dy/(x^2 + y^2)$. Along the line $y = mx$ the limiting position of the field is given by

$$\lim_{x \to 0} \frac{x^2(1 + m^3x)}{2mx^2} = \frac{1}{2m}.$$

In addition to the vertical Oy, there are two directions, with slopes $\pm 1/\sqrt{2}$, at which the limiting position of the field is precisely this same direction. One can show that, if the origin is blown up, the corresponding points on the counterimage circle of O are ordinary quadratic points of the induced field, and so these two points define two separatrices which reappear by projection in the Oxy-plane. We see immediately that for negative y the branches of these separatrices c and c' are crest lines, while for positive y they are the boundary of an open set of trajectories leading toward the origin. Thus the singular point looks qualitatively like a saddle point, but, unlike an ordinary quadratic saddle point, it has two special features.

1. Whereas the crest line is regular at an ordinary saddle point, here it is angular at O.
2. Whereas the only gradient lines ending at the ordinary saddle point are the two thalweg lines (which together form a regular line transversal to the ridge line), here a whole sector of gradient lines (bounded by c and c', and with positive y) ends at O, while for negative y there is only one thalweg line, the negative y-axis.

The situation can be described intuitively as follows: There is a saddle point in a chain of mountains whose configuration is highly asymmetrical. Approaching in the direction of decreasing y, the traveler will arrive at the saddle point from an open, wide, gentle ramp; from the other direction he will reach it only after climbing a narrow gorge with steep sides (see Figure 5.21).

The essential point of this description is that, whereas in a classical quadratic situation the threshold point O will be reached with zero probability, here it will be reached from an open set of initial positions. This is the first example of the phenomenon of *threshold stabilization*, which we shall return to in biological morphogenesis. The parabolic umbilic also appears in all cases in which a basin flows directly into a lower basin, as when water flows from a jug into a bowl.

The parabolic umbilic is the most difficult of our singularities to describe because it is of codimension four. I am endebted to Chenciner [15] for a very detailed investigation of the discriminant variety, of which I

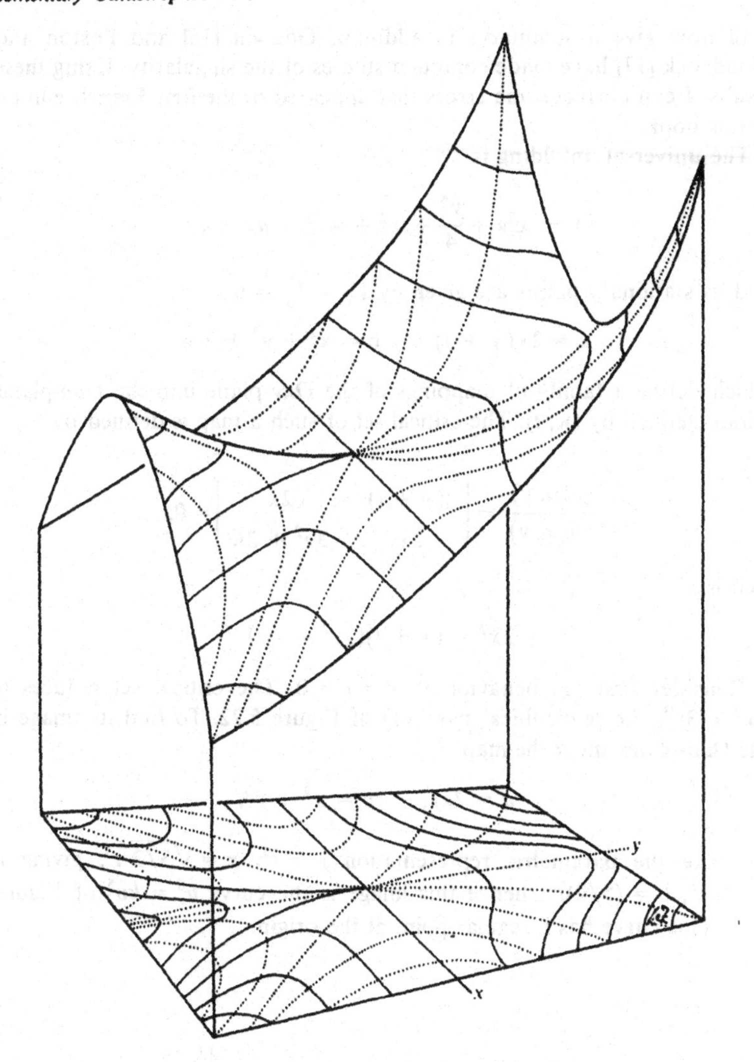

FIGURE 5.21. Graph and contours of the function $V = x^2y + y^4/4$.

shall now give a summary; in addition, Godwin [16] and Poston and Woodcock [17] have made computer studies of the singularity. Using these results, I can correct some errors that appeared in the first French edition of this book.

The universal unfolding is

$$V = x^2y + \frac{y^4}{4} + tx^2 + wy^2 - ux - vy,$$

and its stationary points are given by $V_x = V_y = 0$:

$$u = 2x(y + t), \qquad v = x^2 + y^3 + 2wy,$$

which define a family of mappings of the Oxy-plane into the Ouv-plane, parameterized by (w, t). The critical set of such a map is defined by

$$\frac{\partial(u, y)}{\partial(x, y)} = \begin{vmatrix} 2(y + t) & 2x \\ 2x & 3y^2 + 2w \end{vmatrix} = 0,$$

that is,

$$2x^2 = (y + t)(3y^2 + 2w).$$

Consider first the behavior at $w = t = 0$. The critical set reduces to $2x^2 = 3y^3$, the semicubical parabola of Figure 5.22. To find its image in the Ouv-plane under the map

$$u = 2xy, \qquad v = x^2 + y^3,$$

we take the parametric representation $y = t^2$, $x = \sqrt{3/2}\, t^3$, giving $u = \sqrt{6}\, t^5$, $v = (5/2)t^6$; hence this image is the curve $u^6 = kv^5$ of Figure 5.23. This curve has a regular point at the origin.

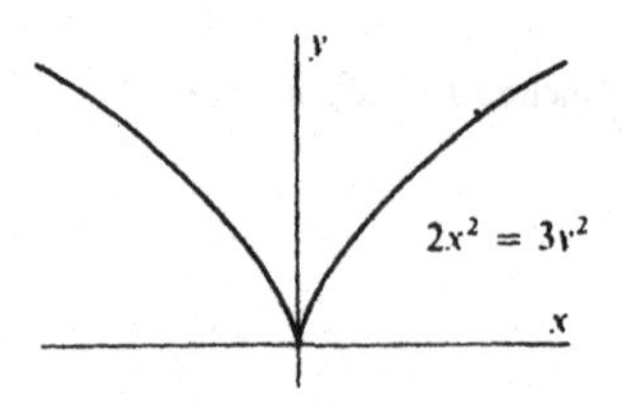

FIGURE 5.22

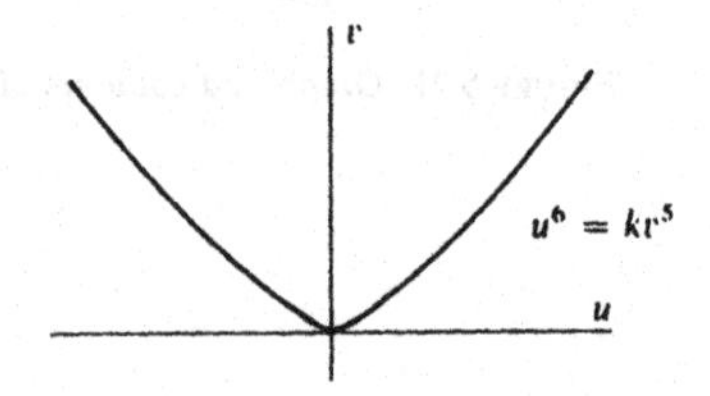

FIGURE 5.23

Next we investigate the image Z of the critical set in the $Ouvwt$-space in a neighborhood of the origin. To do this we consider the intersection of Z with a sphere whose radius is sufficiently small, so that the intersection with any smaller sphere has the same topological type.

Now we delete from this sphere the "polar" regions $t = w = 0$, retaining only a tubular neighborhood of the equator $u = v = 0$. Figure 5.24 is the equatorial section by the Otw-plane on which each of the sixteen marked points corresponds to a topological type of the section of Z by the plane normal to the equator at that point. Figure 5.25 gives the sequence of sixteen sections. (Also see Photographs 8 and 9.)

Qualitatively, at 1 there is a curve having a cusp pointing downward; then, at 2, a new point appears at the origin, where the lip formation begins; this grows (3), pierces the cusp (4), and crosses it (5) to form the phallic mushroom (5) characteristic of the parabolic umbilic. Next, at 6, the cusp meets the lower branch of the lip in a hyperbolic umbilic, and then the two branches cross (7) to form a curvilinear triangle piercing laterally a convex curve. The triangle shrinks, first touching the curve (8) and then shrinking inside it (9) to form a hypocycloid with three cusps, and finally vanishes (10), in an elliptic umbilic, reappearing immediately with the same orientation (11). Then its lower cusp meets the curve (12) and pierces it (13); at 14 the curve and the upper edge of the triangle touch in a beak-to-beak singularity which separates (15), producing two symmetric swallow tails which are reabsorbed in the curve (16) to lead to the original configuration.

Tracing the locus of the principal changes of topological type on the Otw-plane, we find lines of lip points (2), axial piercing (4), hyperbolic unbilics (6), double lateral piercing (8), elliptic umbilics (10), axial piercing (12), beak-to-beak singularity (14), and swallow tails (16). These have equations as follows.

Lines of umbilics (6 and 10): After making the substitution $x = a + X$, $y = b + Y$ in V, the second-order part is

$$X^2(b + t) + 2aXY + Y^2\left(\frac{3b}{2} + w\right),$$

which is zero when $t = -b$, $a = 0$, and $w + 3t^2/2 = 0$. This parabolic equation defines the locus of umbilics. The third-order part is

$$X^2Y + bY^3 = Y(X^2 + bY^2).$$

When b is negative and t is positive, this is an elliptic umbilic (10) with three real roots; when b is positive and t is negative, a hyperbolic umbilic (6).

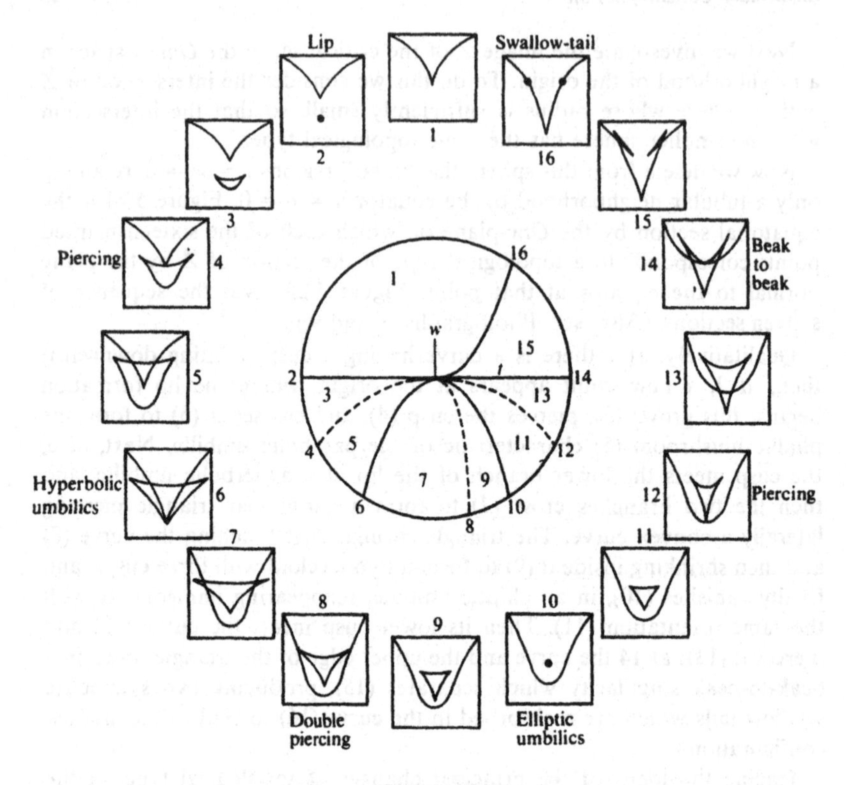

FIGURE 5.24

Lines of axial piercing (4 and 12): The counterimage of the axis of symmetry Ov, defined by $u=0$, is $2x(y+t)=0$; we therefore need the condition that the point $x=0$, $y=-t$ and one of the points $x=0$, $y=\pm\sqrt{(-2w/3)}$ both have the same image on Ov. Then, since $v = x^2+y^3+2wy$,

$$-t^3-2wt=\pm\left(\frac{-32w^3}{27}\right)^{1/2}.$$

Squaring this gives

$$\frac{-32w^3}{27}=4w^2t^2+4wt^4+t^6,$$

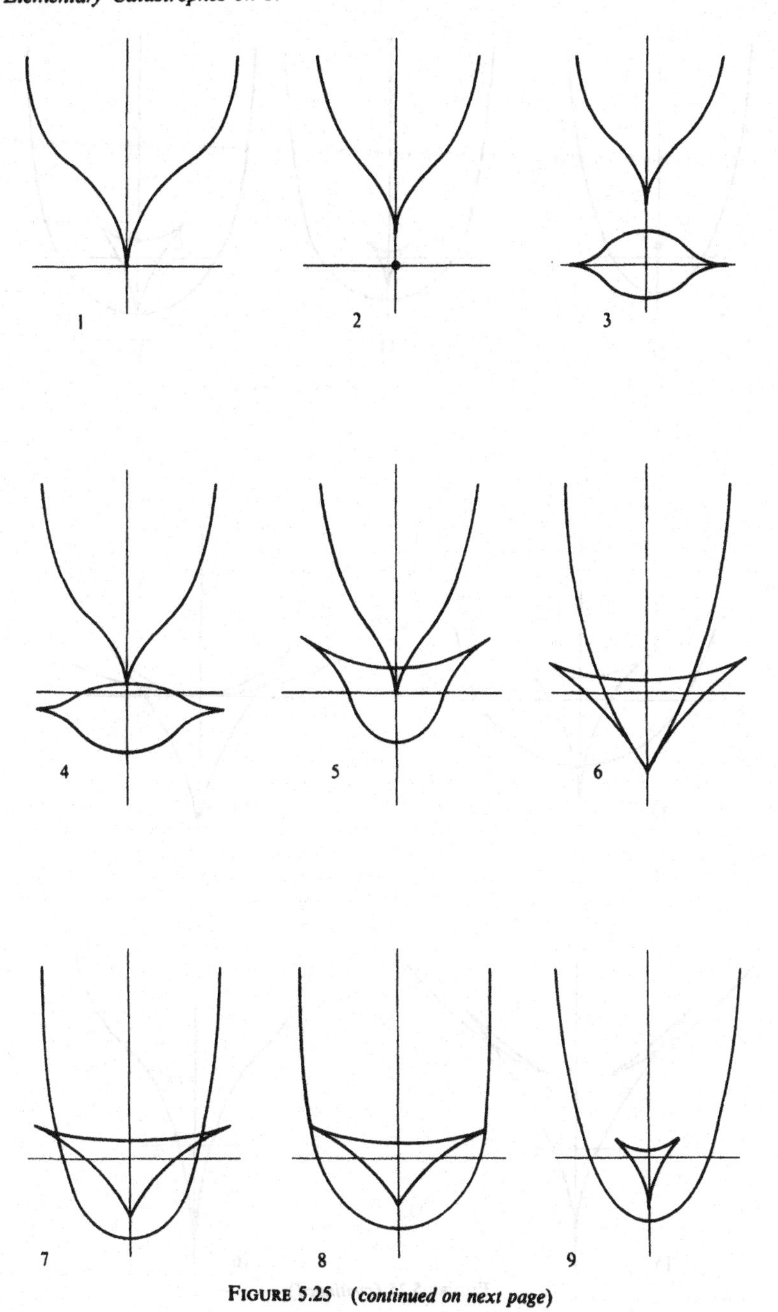

FIGURE 5.25 (*continued on next page*)

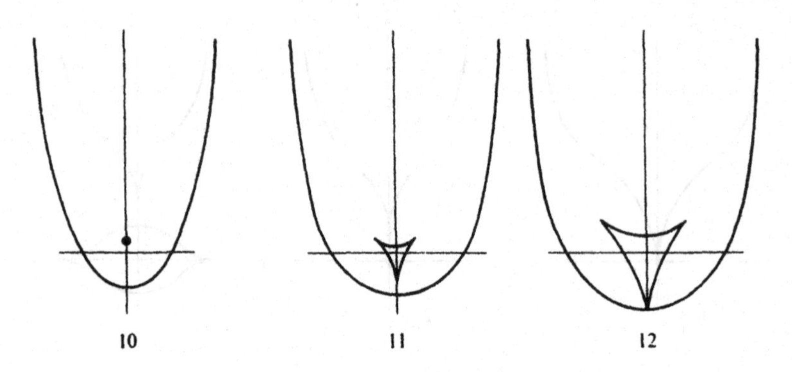

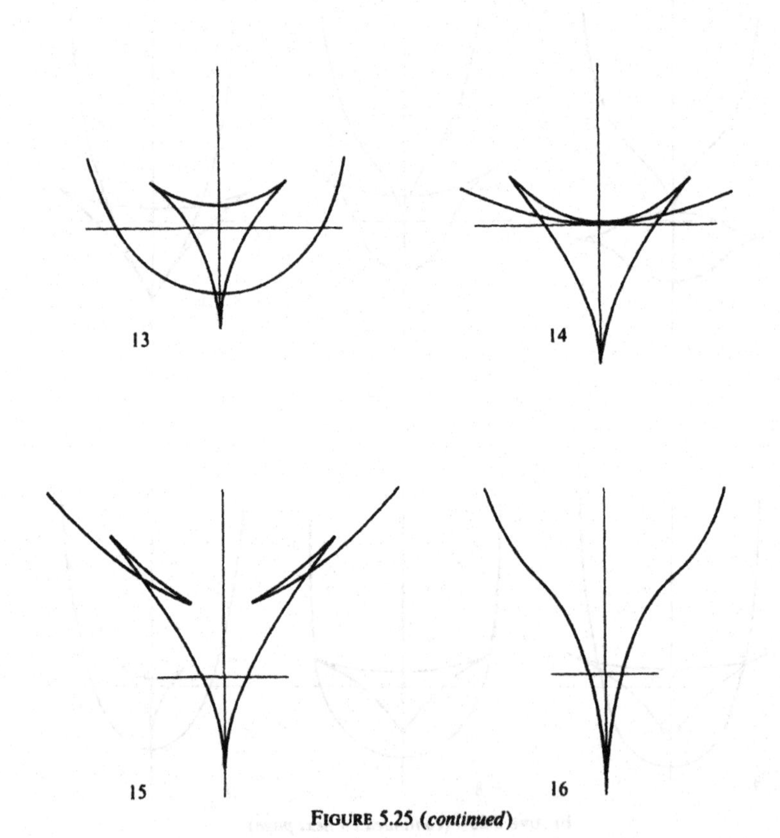

FIGURE 5.25 (*continued*)

or

$$\tfrac{32}{27}w^3 + 4w^2t^2 + 4wt^4 + t^6 = \tfrac{32}{27}\left(w + \tfrac{3}{2}t^2\right)^2\left(w + \tfrac{3}{8}t^2\right) = 0.$$

The double line $(w + \frac{3}{2}t^2)^2 = 0$ corresponds to the umbilics, and $w = -\frac{3}{8}t^2$ is the equation of the line of axial contact.

The line of swallow tails (16): We substitute $y + t = Y$; then the critical curve of the map f from (x,y) to (u,v) has the equation

$$x^2 = \left(\frac{Y}{2}\right)\left[3(Y-t)^2 + 2w\right] = P(Y).$$

A vector $(\delta x, \delta Y)$ will be tangential to the critical curve at (x, Y) if $2x\,\delta x = P'(Y)\,\delta Y$, that is, $\delta x / P'(Y) = \delta Y / 2x$. Also this vector $(\delta x, \delta Y)$ is in the kernel of $j^1(f)$ if $y\,\delta x + x\,\delta y = 0$. Combining these equation gives

$$0 = YP'(Y) + 2x^2 = YP'(Y) + 2P(Y),$$

and substituting the expressions for $P(Y)$ and $P'(Y)$ gives the equation for the cusps:

$$\frac{15Y^3}{2} - 12tY^2 + 3\left(w + \frac{3t^2}{2}\right)Y = 0.$$

Eliminating the trivial root $Y = 0$, we obtain a quadratic equation that has a double root (corresponding to coincident cusps) if $w = t^2/10$.

The line of double lateral piercing (8): With the same notation as in the preceding paragraph.

$$\begin{aligned} v &= x^2 + y^3 + 2wy = P(Y) + (Y-t)^3 + 2w(Y-t) \\ &= \frac{5Y^3}{2} - 6tY^2 + 3\left(w + \frac{3t^2}{2}\right)Y - 2wt - t^3 = Q(Y). \end{aligned}$$

The derivative

$$Q'(Y) = \frac{15Y^2}{2} - 12tY + 3\left(w + \frac{3t^2}{2}\right)$$

has two zeros, of which one, in a neighborhood of the origin, defines the real cusps of the image curve; let d denote this root. Then, writing the other (simple) root of $Q(Y) = Q(d)$ as a, $2d + a = 12t/5$, and this other root defines the simple points of the critical curve with ordinate $Q(d)$. It remains to work out the condition that these points have the same image as the cusps defined by d; for this, the condition that they have equal values of u is

$$(xY)(a) = (xY)(d),$$

or, in terms of squares,

$$x^2Y^2(a) = x^2Y^2(d).$$

Now

$$Y^2P(Y) = \frac{3Y^5}{2} - 3tY^4 + AY^3,$$

where $A = 3t^2/2 + w$. Hence, by substitution,

$$\frac{3}{2}(d^4 + d^3a + d^2a^2 + da^3 + a^4) - 3t(d^3 + d^2a + da^2 + a^3)$$
$$+ A(d^2 + da + a^2) = 0.$$

Noting that, if we substitute $A = kt^2$, this equation is homogeneous in t, a, and d, we see that there exists solutions of the form $d = ct$, $a = c't$ which could now be evaluated by using $2d + a = 12t/5$. Hence, we find the locus as a curve defined by a relation of the form $A = kt^2$, that is, $(3/2 - k)t^2 = w$. No other exponent γ in $w \sim t^\gamma$ is possible, as can be seen from the equation as t tends to zero.

D. A general remark on bifurcation catastrophes

A pure bifurcation singularity at an isolated point can occur only when the critical point to be unfolded is itself stable in the sense of being a (degenerate) minimum of the potential V. When the dimension of the space of internal variables is one, there are only two bifurcation catastrophes:

$V = x^4$, the cusp, and $V = x^6$, the butterfly.

When dealing with the syntactical interpretation of catastrophes, these two singularities play an important part in the syntactical organization of nuclear phrases. The cusp defines the morphologies of capture and emission at the origin of the ternary structure of subject-verb-object; the butterfly defines the source-message-receptor structure of communication. Singularities of odd degree ($V = x^3$, $V = x^5$) define transient processes (finishing, beginning, etc.) of lesser structural importance.

As far as umbilics are concerned, they can be associated with a locally stable bifurcation catastrophe only by considering them in the universal unfolding of the singularity $V = x^4 + y^4$, called the *double cusp* by Zeeman. This singularity has algebraic codimension eight and topological codimension seven, and so can be manifested generically on $\mathbf{R}^4$ only through associations of catastrophes. The unfolding of a double cusp can have four minima, and so corresponds to interactions involving the conflict of four regimes. Some of these interactions can be localized in the neighborhood of the parabolic umbilic and can be given a linguistic interpretation.

As an example, take

$$V = \frac{x^4 + y^4}{4} + x^2 y;$$

this has universal unfolding

$$V = \frac{x^4 + y^4}{4} + x^2 y + wx^2 + ty^2 - ux - vy.$$

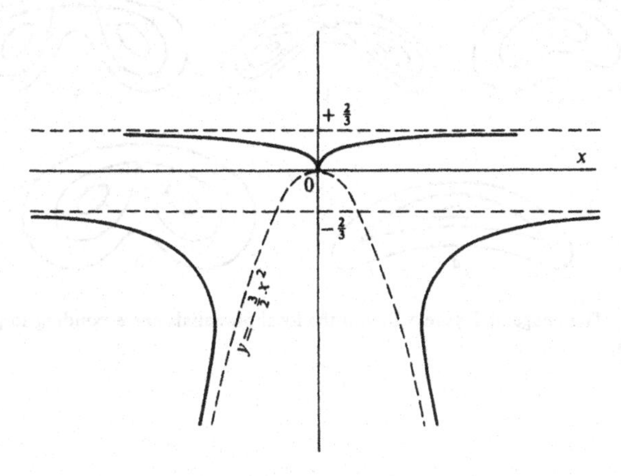

FIGURE 5.26

When $w = t = 0$, the critical set is $x^2(4 - 9y^2) = 6y^3$ (Figure 5.26), made up of three curves having asymptotes $y = \pm\frac{2}{3}$ and the parabola $y = -(3/2)x^2$. The image of this curve in the Ouv-plane is given in Figure 5.27, together with the typical potentials in regions 1, 3, 5, d, and O. Varying the parameter w will give rise to configurations of the type shown in Figure 5.28 in the neighborhood of the elliptic umbilic at the origin, and the image in the Ouv-plane will have the form of Figure 5.29.

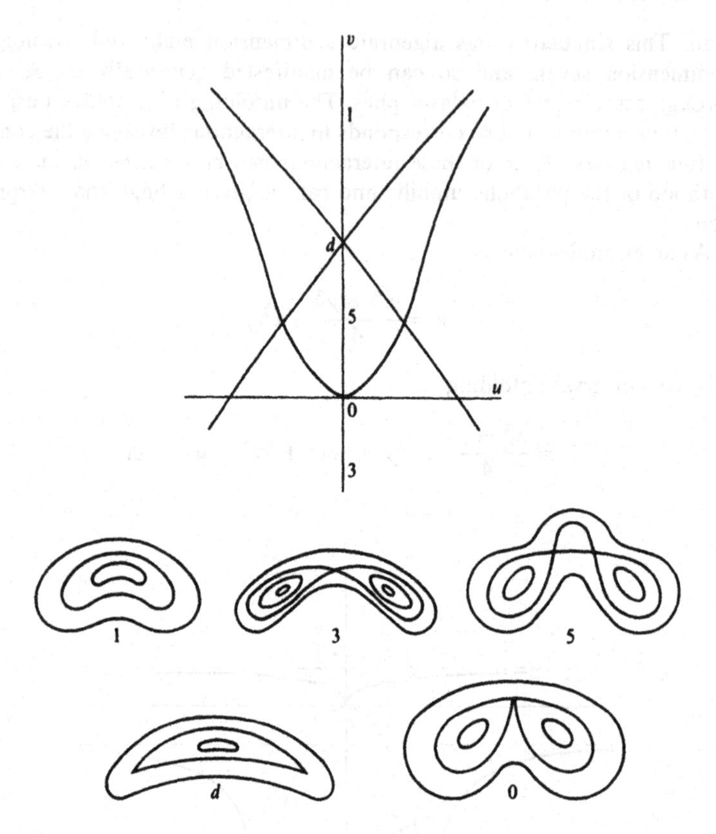

FIGURE 5.27. The image of Figure 5.21 with the local potentials corresponding to positions 1, 3, 5, *d*, and *o*.

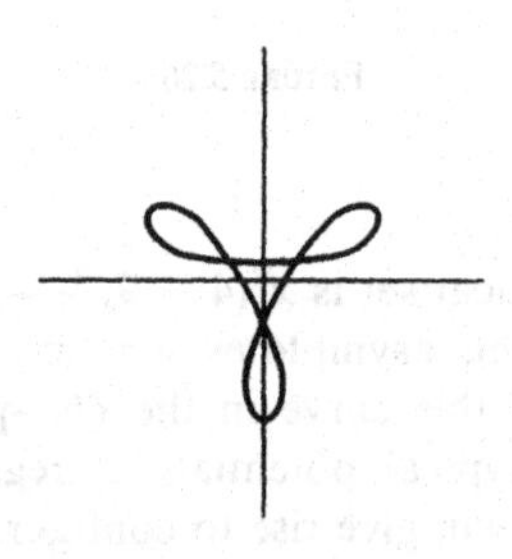

FIGURE 5.28

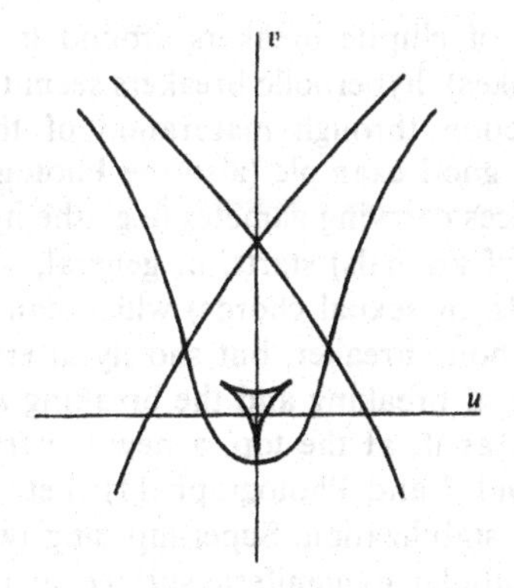

FIGURE 5.29

5.5. THE MORPHOLOGY OF BREAKERS

J'ai vu bondir dan l'air amer
Les figures les plus profondes...

PAUL VALÉRY, *Le vin perdu.*

There are two stable types of breakers in $\mathbf{R}^3$: the *hyperbolic breaker*, which occurs when the crest of a wave becomes angular, and then forms a cusp, and the wave breaks, and the *elliptic breaker*, which produces sharp, pointed spikes with theoretically triangular cross section; and the usual transition is in the sense elliptic ⟶ hyperbolic. Thus the parabolic umbilic has a material realization during the collision between a pointed object with triangular, hypocycloid cross section and a surface which it finally ruptures "hyperbolically" along a cusp with hypocycloid section. The spear and the bayonet are military realizations of this catastrophe.

The elliptic ⟶ hyperbolic transition, illustrated in the sequence of Figures 5.25-10, 9, 8, 7, 6, 5, suggests the germination of a mushroom, when we form the surface of revolution of these curves. The unstable cusp extremities, representing the edge of the mushroom cap, are the unstable breaking zones, corresponding biologically to zones of gametogenesis and sporulation. The transition of Figures 5-25-15 to 14, an evolution toward an elliptic state, is reminiscent of a jaw closing in a beak-to-beak singularity, with the swallow tails corresponding to the teeth; the closure of the neural plate in the embryology of vertebrae is probably an illustration of this.

Whereas examples of elliptic breakers around in biology (e.g., prickly surfaces of hair or spikes), hyperbolic breakers seem to be much more rare; however, the destruction through maturation of the gills of *Coprinus*[3] seems to constitute a good example (also see Photograph 10). In fact, the formation of all surfaces carrying gametes (e.g., the hymeniferic surfaces of fungi or the gonads of animals) starts, in general, with a state of folding (the formation of gills, or sexual chords) which can be interpreted as the beginning of a hyperbolic breaker, but shortly afterward the ridge of the fold stabilizes instead of breaking and the breaking zone is forced back to the faces of the gills, as if, at the top, a new hyperbolic umbilic appears (see Figures 5.25-6 and 7 and Photograph 11); here, without doubt, is an example of threshold stabilization. Superimposing two hyperbolic folds of this type can give a tubular hymeniferic surface, as in *Boletus*.

Although, in hydrodynamics, the hyperbolic umbilic is easy to observe, the elliptic umbilic cannot be realized by a liquid jet; this is so because of surface tension, which prevents the boundary of a liquid from having stable cusps. Consequently the hyperbolic $\longrightarrow$ elliptic transition is difficult, if not impossible, to realize. However, even though the elliptic umbilic is forbidden in hydrodynamics, it can still occur under certain conditions of initial symmetry. For example, if the crest of a wave reaches the breaking angle simultaneously along the length of the wave, there exists a highly unstable situation that leads, in the terminology of Chapter 6, to a generalized catastrophe. We see later that the situation can stabilize, forming a periodic structure that destroys the symmetry; in this case the wave, initially hyperbolic, will break out in elliptical spikes at regular intervals along the crest. However, these spikes acquire rotational symmetry, and a drop breaks off the end (Figure 5.24-4). The diadem of a splash gives a fine example of this phenomenon (Photograph 12). The general behavior is as if the wave in the *Owt*-plane which is attempting to realize the forbidden hyperbolic$\rightarrow$elliptic transition is broken into fragments, each of which flows toward the top of the circle of Figure 5.24, realizing the sequence of surfaces of revolution of the curves in Figures 5.25-10, 9, 8, 7, 6, 5, 4, 3. Similarly, the basidia on the faces of the gills of a fungus provide another example of a locally elliptic situation located on the crest of an initially hyperbolic wave that is stabilized at its summit (see Figure 5.25-7, Photograph 13, and Note 3).

Returning to the universal unfolding of the compact form of the parabolic umbilic,

$$V = x^2y + wy^2 + tx^2 + \frac{x^4 + y^4}{4} - ux - vy,$$

we can describe the composed transition hyperbolic $\longrightarrow$ elliptic$\rightarrow$hyperbolic with the aid of a very interesting loop g in the Owt-plane (Figure 5.30). Starting from a point a_0 close to the hyperbolic umbilic with $w > 0$, the loop rejoins the w-axis and passes through the parabolic umbilic $w = t = 0$. It then moves into the $w < 0$, $t < 0$ quadrant along the curve defining the cusps in Figure 5.30 b_1 and moves to b_2, where $u < 0$. Then the point crosses the Ot-axis at a_1, corresponding to an elliptic umbilic, and returns to its starting point, avoiding the origin.

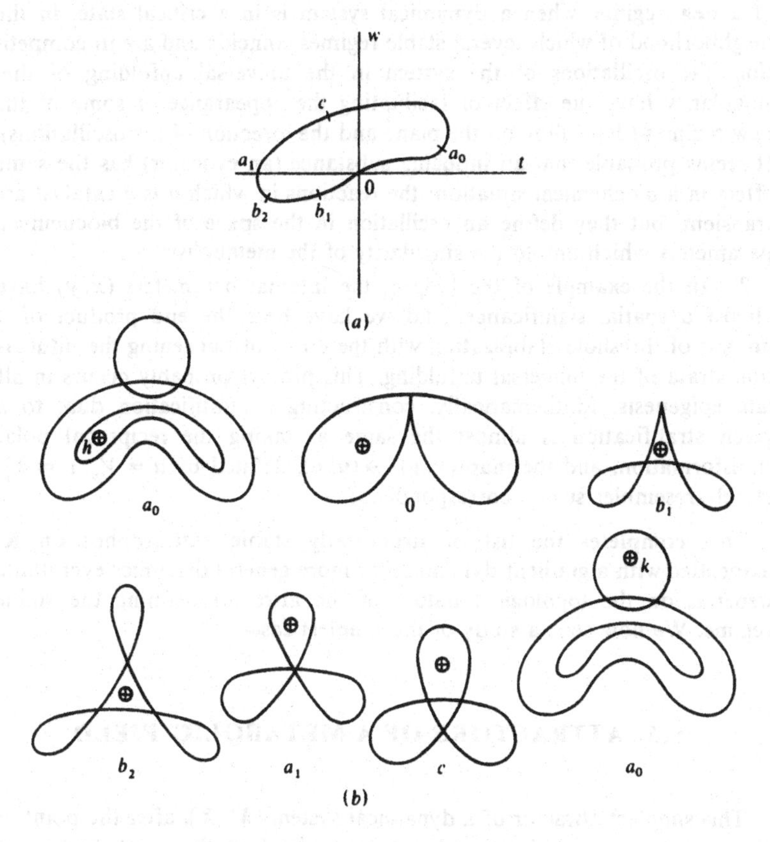

FIGURE 5.30. An ejection loop.

Following the corresponding variation in the (x, y)-space of internal parameters, we try to lift the loop at each point to an attractor (a minimum of V) of the corresponding figure. This is possible in the way shown in

Figure 5.30; the global effect of the loop is to transform the regime h at the initial point a_0 into another central regime k. Thus the total effect of the loop is the evocation (in the sense of Waddington) of a new regime; and, interpreting the (x, y)-variables as spatial variables, we cannot fail to observe the phallic nature of the process, which appears to be a kind of ejaculation, while the inverse loop is a process of capture. There are two conclusions to be drawn from this example.

1. In regard to the mechanism of embryological induction or evocation of a new regime, when a dynamical system is in a critical state, in the neighborhood of which several stable regimes coincide and are in competition, the oscillations of the system in the universal unfolding of the singularity have the effect of facilitating the appearance of some of the new regimes (depending on the plane and the direction of the oscillations). It seems probable that an inducing substance (an evocator) has the same effect in a biochemical situation; the reactions in which it is a catalyst are transient, but they define an oscillation in the space of the biochemical parameters which unfold the singularity of the metabolism.
2. In the example of the loop g, the internal parameters (x, y) have almost a spatial significance, and we have here the end product of a process of threshold stabilization with the effect of thickening the bifurcation strata of the universal unfolding. This process probably occurs in all late epigenesis. Mathematically, constructing a stratification dual to a given stratification is almost the same as taking the reciprocal polar transformation, and the map $(x, y) \rightarrow (u, v)$, defined by $u = V_x$, $v = V_y$, closely resembles such a correspondence.

This completes the list of structurally stable catastrophes on $\mathbf{R}^4$ associated with a gradient dynamic. For more general dynamics everything depends on the topological nature of the attractor defining the stable regime. We now start a study of the simplest cases.

5.6. ATTRACTORS OF A METABOLIC FIELD

This simplest attractor of a dynamical system (M, X), after the point, is the closed generic trajectory. Let us consider first the point, because a point attractor of a metabolic field of attractors differs in some cases from a minimum of potential. At such a point u, the field $X(u)$ is zero, and the matrix of coefficients of the linear part of the components of $X(u)$ in a local chart has all its characteristic values in the half-plane $\mathrm{Re}(z) < 0$; this is obviously a structurally stable situation. Such an attractor can disappear stably in two ways.

1. One and only one characteristic value of the Jacobian matrix becomes zero, as a function of t; this happens when the attractor meets an unstable point and both are destroyed by collision.

2. The real part of a characteristic value is zero for $t = 0$, and the imaginary part remains nonzero; the conjugate characteristic value undergoes the same evolution. This case is the subject of a classical study by Hopf [18], who called it *bifurcation (Abzweigung)*. Qualitatively, the following happens. The singular point becomes unstable but, in the plane of the associated characteristic vectors, is surrounded by a small invariant attracting cycle, and the evolution is as if the initial point attractor dilates into a two-dimensional disc in this plane, while the center of the disc becomes a repeller point of the field (see Figure 5.31). Finally there is always a stable regime, but the associated attractor is no longer a point but a closed trajectory. The new attractor is topologically more complicated than the initial point attractor; in general terms, the new regime contains more information than the initial regime (e.g., the period of the cycle is a new parameter).

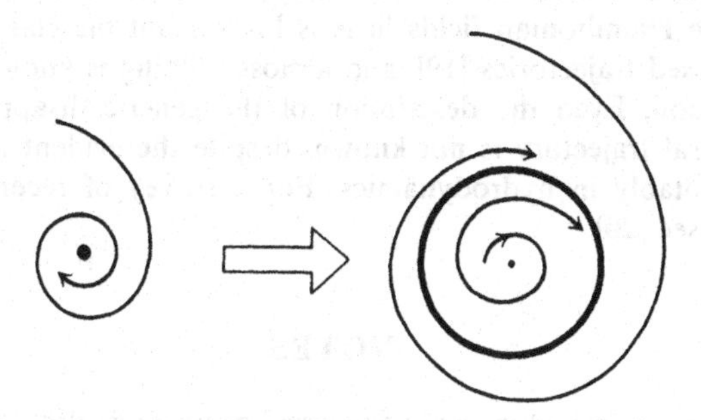

FIGURE 5.31. The Hopf bifurcation of a point attractor into a closed trajectory.

The simplest attractor, after the isolated point, is the attracting closed cycle. The local study of such a trajectory uses the classical technique of Poincaré-Floquet. We cut the trajectory locally by a germ of a traverse hypersurface F; then, if the trajectory leaving from a point m of F meets F again for the first time at m', the map $m \longrightarrow m'$ defines a local diffeomorphism h of F. If the trajectory is attracting, the Jacobian of this diffeomorphism has characteristic values less than one in modulus. Here again the situation is structurally stable, but the attractor can disappear by a variation of t in the two ways indicated above. In case 1 it is destroyed by collision with an unstable trajectory; in case 2 (which can occur only when the dimension of the ambient manifold is greater than three) there is a bifurcation and the trajectory changes into a closed unstable trajectory, the center of a small invariant attracting torus.

This is an example of an invariant attracting manifold of a field X on M. The mathematical study of attracting manifolds of a field is delicate; some are structurally stable when they are sufficiently attracting and the field in the interior of the manifold is strongly transversally dilating, whereas others degenerate easily into manifolds of lower dimension or closed trajectories. As was said in Chapter 3, little is known of the topological structure of structurally stable attractors of a field X, and so a systematic study of the transformation of attractors in the deformation of a field is, for the moment, out of reach. It seems to me, however, that this is the essential geometric phenomenon intervening in a large number of morphogenetic processes, such as changes of phase in physics and the phenomenon of induction in embryology. In the absence of this theory we restrict ourselves to a qualitative classification of these processes, described in the next chapter, where they are called generalized catastrophes.

Even the classification of the stable singularities of a two- or three-parameter family of closed trajectories is unknown; finally, in the case of conservative Hamiltonian fields little is known but the classification into types of closed trajectories [19], and almost nothing is known about their transformation. Even the description of the generic disappearance of a closed central trajectory is not known, despite the evident interest in the problem, notably in hydrodynamics. For a survey of recent results and references, see [20].

NOTES

1. This assertion holds when we take the function measuring the stability of the minimum to be the relative height of the lowest saddle. Under Maxwell's convention, it will end on the fold line.

2. "This expresses that when breaking is about to occur in the neighbourhood of the crest, the wave is in the form of a wedge on angle 120°. This agrees with observation of waves just before breaking and with a theoretical result obtained by Stokes." L. M. Milne-Thomson, *Theoretical Hydrodynamics*, Macmillan, 1962, p. 414.

3. *Coprinus*, with its remarkable mode of maturation, has an eccentric position among agarics. In most species, maturation starts from the edge of the gills and progresses to the surface of the hymenium; this process can easily be seen because the spores are very dark. At the same time the gill frequently undergoes a kind of self-destruction and liquefies after the last generation of basidia, being finally reduced to a blackish line. Translated from H. Romagnesi, *Nouvel atlas des champignons*, Bordas, 1965, Vol 1, p. 36.

4. *Reduced stability of elliptic states.* The elliptic quadratic forms $AX^2 + BXY + CY^2$, in the real projective plane with homogeneous coordinates (A, B, C), are those with imaginary roots, so that they satisfy $B^2 - 4AC < 0$, and thus form a concave subset of the plane in the sense that every projective line contains hyperbolic points where $B^2 - 4AC > 0$. If it is assumed that the point (A, B, C) moves as if controlled by the dynamic associated with the natural metric of the projective plane, this point will always adopt hyperbolic states after a sufficiently long interval of time. We have seen in the preceeding interpretation that elliptic states can be interpreted as states of tension, and hyperbolic states, of release; this explains why a state of tension, although necessary for life, must always be limited and followed by release. This perpetual elliptic-hyperbolic dialectic recalls the yin-yang opposition of Taoism, or the excitation-inhibition opposition of neurophysiology. Because of the very nature of the spatial transport of the male sexual act, the masculine sex has a more elliptic nature than the feminine sex; perhaps this can explain (a generalization, already to be found in Empedocles and roughly verified from *Escherichia coli* to man) why males are more hairy (in a generalized sense) than their mates and are biologically more fragile.

The important role that Freud attributes to sexual symbolism, particularly in dreams, is well known; it can be argued that the geometric-dynamic forms representing the sexual processes occur in so many objects of live and inert nature because they are only structurally stable forms in our space-time realizing their fundamental function as the union of gametes after spatial tranport. We might say that these forms existed before sexuality, which itself is only their genetically stabilized manifestation. Also see what is said in Chapter 9 on sexual chreodes.

REFERENCES

1. The vortex street in hydrodynamics.

   The original article: T. von Kármán, Über den Mechanismus des Widerstandes den ein bewegter Körper in einer Flüssigkeit erfährt, *Nachr. K. Ges. Wiss., Gottingen* (1912), 247–556.

   A quicker treatment: W. Kaufman, *Fluid Mechanics*, McGraw-Hill, 1960, p. 220.

2. M. Morse, The calculus of variation in the large, *Amer. Math. Soc. Colloquium Pub.* No. 13, 1964.
3. J. C. Tougeron, Une généralisation du théorème des fonctions implicites, *C. R. Acad. Sci. Paris*, Ser. A-B **262** (1966), A487-A489.
4. R. Courant and K. O. Friedrichs, *Supersonic Flow and Shock Waves*, Interscience Publishers, 1948, p. 115.
5. B. Riemann, uber die Fortpflanzung ebener Luftwellen von endlicher Schwingungsweite, *Abh. K. Ges. Wiss., Göttingen* **8** (1860), 43–65.
6. H. Whitney, On singularities of mappings of Euclidean spaces I. Mappings of the plane into the plane, *Ann. Math.* **2**, 62 (1955), 374–410.
7. For the Mach reflection, see [4], p. 334.
8. D'A. W. Thompson, *On Growth and Form*, abridged edition, Cambridge University Press, 1961, p. 100, Figure 38.
9. R. Courant and D. Hilbert, *Methods of Mathematical Physics*, Vol. II, Interscience Publishers, 1961, p. 617, Figure 54.
10. R. C. Hwa and V. L. Teplitz, *Homology and Feynman Integrals*, Benjamin, 1966, p. 140.
11. J. Mather, Stability of C^∞-mappings, I: *Ann. Math.* (2) **87** (1968), 89–104; II: *ibid.* (2) **89** (1969), 254–291; III: *Inst. Hautes Etudes Sci. Publ. Math.* **35** (1968), 279–308; IV: *ibid.*

37 (1969), 233–248; V: *Advan. Math.* **4** (1970), 301–336; VI: *Proceedings of the Liverpool Singularities Symposium* I, *Lecture Notes in Mathematics* No. 192, Springer, 1971, pp. 207–253.

12. Y. C. Lu, Sufficiency of jets in $J^r(2, 1)$ via decomposition, *Invent. Math.* **10** (1970), 119–127.
13. G. Darboux, *Leçons sur la Théorie Générale des Surfaces*, Vol. IV, Note VII, Gauthier-Villars, 1896.
14. I. R. Porteous, The normal singularities of a submanifold, *J. Diff. Geom.* **5** (1971), 543–564.
15. A. Chenciner, *Exposé au Séminaire I.H.E.S: Méthodes Mathématiques de la Morphogénèse*, 1972.
16. A. N. Godwin, Three dimensional pictures for Thom's parabolic umbilic, *Inst. Hautes Etudes Sci. Publ. Math.* **40** (1971), 117–138.
17. A. E. R. Woodcock and T. Poston, A geometrical study of the elementary catastrophes, *Lecture Notes in Mathematics* No. 373, Springer, 1974.
18. E. Hopf, Abzweigung einer periodischen Lösung von einer stationären Lösung eines Differential systems, *Ber. Verh. Sächs. Akad. Wiss. Leipzig, Math.-Nat. Kl.* **95**, No. 1 (1943), 3–22.
19. V. I. Arnold and A. Avez, *Ergodic Problems of Classical Mechanics*, Benjamin, 1968, Appendix 29: Parametric resonances.
20. R. Abraham, *Hamiltonian Catastrophes*. Quatrième rencontre entre mathématiciens et physiciens, Université de Lyon, 1972.

CHAPTER 6

GENERAL MORPHOLOGY

Μεταβάλλον ἀναπαύεται.
Fire rests by changing.
HERACLITUS

6.1. THE MAIN TYPES OF FORMS AND THEIR EVOLUTION

A. Static and metabolic models

The distinction in Chapter 4 between static and metabolic forms is justified for the following reasons.

Definition. Given a static (metabolic) model of local dynamics over an open set W, a *static (metabolic) form* is the set of points $x \in W$ controlled by an attractor c of the field. In the case of a static model (i.e., a gradient dynamic), c is a point attractor in the manifold M of internal states.

A static form has the following properties. The boundary of its support is generally not complicated, being locally polyhedral, and it is comparatively rigid and insensitive to perturbations. If a static form undergoes an interaction with an external system, the form remains, to begin with, isomorphic to itself by virtue of its structural stability. Increasing the intensity of the perturbation will result in the collision of the attractor with an unstable regime and its capture by an attractor of lower potential, at least locally in the region of maximum intensity of the perturbation or in a neighborhood of the boundary where the attractor is less stable. This process corresponds to an ordinary catastrophe or to a finite sequence of ordinary catastrophes, and the topology and evolution of the adjacent attractors are given by the topology of the fiber dynamic, as we saw in the examples of Chapter 5. The final situation may be topologically very complicated, but generally it behaves like a static form.

A metabolic form has the following contrasting features. The boundary of its support can have a very complicated topology and is generally fluctuating and very sensitive to perturbations. When the form is subjected to a sufficiently small perturbation, it will persist by virtue of its structural stability, although the topology of its support may be affected; but as the perturbation increases to the point of hindering the underlying metabolism and destroying the recurrence in the fiber dynamic, a new and abrupt phenomenon will occur and the form will disintegrate almost instantaneously into a continuum of elementary forms with simpler internal structures, static or metabolic forms of attractors of smaller dimension than the initial attractor c (a *catabolic catastrophe*). We shall discuss later the different topological types of this kind of catastrophe, which we call a *generalized catastrophe*. It should be added that a metabolic form always has properties of its internal kinetic connected with the dynamic of its attractor, for example, characteristic frequencies, and therefore it can resonate with metabolic forms of the same type. We shall interpret this in Chapter 7 by saying that a metabolic form is a *significance carrier*.

The typical example of a static form is a solid body—a stone, for instance; of metabolic forms, there are a jet of water, a wreath of smoke (these forms are defined purely by their kinematics), a flame, and (disregarding their complicated internal morphologies) living beings. However, the distinction between static and metabolic forms is an idealization that cannot be pushed to extremes: most static forms are only *pseudostatic* in the sense that their point attractor c can undergo changes that make the form metabolic; probably, according to the πάντα ῥεῖ of Heraclitus, every form is metabolic when the underlying phenomena ensuring its stability are examined in sufficiently fine detail. Perhaps even the conversion of matter into energy can be considered as a catastrophe of this type.

B. Competition of attractors of a Hamiltonian dynamic

Given a field of local Hamiltonian dynamics with fiber (M, X) over an open set W, suppose that the hypersurface at level $H = E$ has a finite number of vague attractors $c_1, c_2, \ldots, c_k$ with Liouville measures $m_1, m_2, \ldots, m_k$ as functions of E; then form the relative entropies $S_j = \log m_j$ and the corresponding temperatures $T_j^{-1} = \partial S/\partial E$. If there is also a universal thermodynamical coupling between the attractors c_j, there can be equilibrium only when all the phases X_j (controlled by attractors c_j) are at the same temperature. Suppose that the functions $T_j(E)$ are invertible; then $E_j(T)$ will be the corresponding energies of the c_j. Let $V_1, V_2, \ldots, V_k$ be the volumes of phases $X_1, X_2, \ldots, X_k$. Then

$$V_1 + V_2 + \cdots + V_k = \text{the total volume of } W,$$

$$E_1 V_1 + E_2 V_2 + \cdots + E_k V_k = \text{the total energy of the system}.$$

These two equations are insufficient to determine the V_j and the spatial division of the phase X_j; very probably this division is determined by s ›me minimality condition on the contact surfaces. Let A_{ij} denote the area of the shock wave separating X_i and X_j; then some expression of the form

$$P = \Sigma k_{ij} A_{ij} \quad (k_{ij} > 0)$$

must be minimized.

In such a case the topology of the fiber dynamic has little effect. The patterns encountered in many biological structures (butterfly wings, shells, etc.) probably depend on such a mechanism, for they have too much individual variety to be as strictly determined as the usual morphogenetic fields. But even supposing the existence of such secondary optima as this far from determines the topological and geometrical structure of the final configuration, and usually it happens that the *equilibrium configuration between phases in a state of thermodynamical equilibrium has an indeterminate character*. Obviously, if the dynamic has a pseudogroup of equivalences, the same pseudogroup will operate on the space of equilibrium configurations and will give rise to a continuum of solutions.

We now deal with a special case of the preceding problem, the appearance of a new phase.

C. Creation of a new phase; generalized catastrophes

Suppose that, for $t < 0$, one attractor c of a dynamic fiber governs exclusively on a domain D, and that at $t = 0$, as a result of internal or external variations in the parameters, it stops being the exclusive attractor. The trajectories freed from the attraction of c will then go toward other attractors, for example, c_1. Areas of regime c_1 will appear in D, generally of a fine and irregular character; these areas will join up and simplify their topological structure until a new regime of global equilibrium is established. This type of phenomenon, which we call *generalized catastrophe*, occurs frequently in nature. The topological appearance can be very varied, and we give here an overall qualitative classification, in which the important factor is the codimension of the nuclei of the new phase c_1. Formally a generalized catastrophe is characterized by the destruction of a symmetry or homogeneity; when a domain whose local dynamic was invariant under the action of a pseudogroup G ceases to exist, there will be a generalized catastrophe.

1. Lump catastrophes. To describe a generalized catastrophe we can, as a first approximation, consider a static model. Suppose that, for $t<0$, the potential V is constant on D; at $t=0$, V stops being constant and has, at first, an infinite number of small and random oscillations. (In biological dynamics, the beginning of a catastrophe might be determined by a pre-existing germ set.) Suppose that the minima of V, the attractors of field grad V, correspond to specks of the new phase c_1. Domain D, initially homogeneous, then appears to the observer at $t = 0$ to be disturbed through the precipitation of these specks. As t increases, the function V will, in general, simplify topologically, the number of oscillations of V will decrease, the corresponding specks will coagulate, and at the end of the catastrophe there will be only a finite number of lumps localized at the minima of V. This is the general model of coagulation or *lump catastrophes*, of which the condensation of water vapor into rain is a typical example.

2. Bubble catastrophes. When the original phase is dominated by the new phase c_1, there can occur a situation, topologically dual to the preceding one, in which the new regime is initially three dimensional and relegates the original regime onto surfaces (two dimensional), giving a *bubble catastrophe*. Domain D will become a kind of foam, each attractor of $-V$ defining a three-dimensional basin, with these basins, bounded by surfaces of the original phase c, showing singularities of the strata of conflict predicted by Gibbs' phase rule; as t increases and V simplifies, the number of bubbles decreases and they enlarge by destroying the intermediary walls (according to the pattern of the conflict strata of codimension three; see Section 5.3.D).

3. Laminar and filament catastrophes. It can also happen that the codimension of the new phase is one or two. The first gives a *laminar catastrophe*: at the beginning of the catastrophe domain D appears to be striated by a large number of thin bands of the new phase c_1, folded together, which may simplify and thicken by amalgamation. When the initial codimension of the new phase is two, there is a *filament catastrophe*: at $t = 0$ a large number of small, circular filaments appear in D which may again coagulate; in the most plausible dynamical model of a catastrophe of this type this capture occurs by an interchange between two filaments, leading to the formal process of crossing over in genetics.

Curiously, almost all of the changes of phase of inert matter seem to be of the first two types, but in biology laminar and filament catastrophes exist. (See Photographs 14, 15, 16, and 17.)

4. Catastrophes with a spatial parameter. Time has been given a privileged role in the preceding models of generalized catastrophes, but we can do the same for a space coordinate. In the idealized situation the plane $x = 0$ is controlled by the attractor c, but specks of phase c_1 appear for $x > 0$ and simplify by coagulation as x increases. This gives a topological configuration of one tree or several trees with trunks at large values of x and twigs to the right of $x = 0$. This type of phenomenon appears, in general, where the new phase c_1 undergoes transport; this phase is ordinarily governed by a gradient that does not control the initial phase c. A typical example of such a situation is given by an inclined plane of sand which is watered gently and homogeneously along the horizontal $x = 0$; many small rivulets form, which flow together as x increases. Another example is an electrical spark which ramifies from a pointed electrode toward a plane electrode (Photographs 18 and 19). This type of catastrophe—the *catastrophe with a spatial parameter*—plays an essential role in biological morphogenesis and can be seen in the form of dendrites in vegetation, and the circulation of blood in animals, to mention only two obvious examples.

D. Superposition of catastrophes

It can happen that two or more catastrophes, whose conflicting attractors are independent or very weakly coupled, exist over the same domain W. This can give rise to a complicated morphology; for example, the superposition of two periodic laminar catastrophes leads, by a *moiré* effect, to the formation of lump catastrophes occurring along the alignments and giving the impression of a new laminar catastrophe more widely spaced than the original ones. Similarly the superposition of a laminar and an appropriate lump catastrophe will give rise to a striated structure with the orientation of the strata parallel in all the lumps; such a structure resembles those found in the ergastoplasm of some cells (Golgi apparatus). There is almost no limit to the variety of configurations that can arise in this way (see Photograph 20).

E. Models for a generalized catastrophe; change of phase

We have seen the reason for associating with an attractor c, in a metabolic model with Hamiltonian dynamic, a pseudogroup $G(c)$ of topological equivalences of the dynamic which gives rise to the partial differential equation describing the evolution of the field toward equilibrium. However, as suggested by the example of crystalline phases, the attractors c divide into continuous families obtained by the action of the total group of displacements on the pseudogroup $G(c)$. Now observe that a phase of attractor c_1 cannot appear in the middle of a phase c unless the pseudogroup $G(c_1)$ is a sub-pseudogroup of $G(c)$. This appears to contradict the well-known fact that changes of phase are usually reversible; but although it is certainly possible that a phase c can give rise to a phase c_2 with $G(c_2)$ not contained in $G(c)$, this requires, as preliminary catastrophe, the dislocation of phase c. Thus, in the transformation solid→liquid, the liquid appears first on the boundary in a crystalline solid, which is an inevitable dislocation. Let S be the surface bounding a nucleus of the new phase c_1, where $G(c_1)$ is contained in $G(c)$, and let $\{g\}$ be the local isotropy subgroup at a point x, the set of $g \in G(c)$ such that $g \circ G(c_1)_x$

$= G(c_1)_x$ where $G(c_1)_x$ is the germ of $G(c_1)$ at x; then the set of elements of $G(c)$ conserving S is precisely $\{g\}$.

This simple rule allows us to specify, in many cases, the codimension of the new phase; for example, if the two phases c and c_1 are homogeneous [$G(c)$ and $G(c_1)$ are the group of displacements], S is a sphere. When $\{g\}$ is a discrete subgroup of the rotation group (the case of crystals), S is the polyhedron given by Wulff's rule, and the irregular nature of this form is most probably associated with the dendritic growth of crystals. If the new phase c_1 is governed by a gradient (but otherwise homogeneous), $\{g\}$ is the subgroup of rotations with the gradient as axis, and then the new phase appears in codimension two and S is a cylinder.

F. The formalization of a generalized catastrophe

The determination of a model for generalized catastrophes certainly represents one of the most difficult tasks for the theoretician, for a situation of generalized catastrophe is characterized essentially by a failure, phenomenologically, of Curie's principle: Any symmetry in the causes will lead to a symmetry of the effects. Thus a *generalized catastrophe is not a formalizable process* (in the sense of Chapter 1), for logical deduction satisfies the formal analogue of Curie's principle: Any automorphism of a system of premises of a formal system extends to an automorphism of the set of conclusions. Does this mean that no model is possible ? Here are two possible directions in which, it seems to me, we might attempt to construct models.

First, we might suppose that the catastrophe is well determined at its origin by the specification of its *germ set*, that is, the set of points where the new phase appears. Mathematically, considering the model of the lump catastrophe defined by a potential function $V(x,t)$, we can associate to $V(x,0)$ a highly pathological, measurable, discontinuous function $g(x)$ and then regularize this function by convolution with a kernel $m(x,t)$ whose support increases with t. Then the functions $V_1(x,t) = m(x,t) * g(x)$ start from a topologically complex situation and simplify. If x is a cyclic coordinate, V_1 will tend toward the harmonic of smallest frequency; this may explain the appearance of periodicity in the final state of the catastrophe.[1]

A second, more subtle model arises from considering under what conditions a metabolic field can become a static field through catastrophe, or, equivalently, how a dynamic whose associated attractor has recurrence can degenerate into a dynamic without recurrence, like a gradient field. The topological question is little understood; but, even in the simplest case of a vector field on a surface, the destruction of a closed stable trajectory by collision with an unstable trajectory is followed by an infinite number of ordinary catastrophes of static character (see Sotomayor [1]). This means that the subset, in the function space F of dynamics on M, consisting of the fields with recurrence is in general bounded by a bifurcation hypersurface H which is the frontier, on the side of fields without recurrence, of the set K of ordinary catastrophes. (In the case of a vector field on a surface, these catastrophes are links between saddle points of codimension one.) The transversal section of H and K by the section defined by the field shows that $s^{-1}(H)$, which is the hypersurface at the beginning of the catastrophe, is the limit of ordinary catastrophes $s^{-1}(K)$, which ramify more and more finely toward $s^{-1}(H)$ (see Figure 6.1). Although the set $s^{-1}(K)$ is not structurally stable in a model of this type, at least in a neighborhood of the hypersurface at the beginning of the catastrophe $s^{-1}(H)$, we might expect a certain structural stability of the set $s^{-1}(K)$ at the end; only this final situation can be predicted in practice and so be represented by a structurally stable morphogenetic field (in general, of a static model).

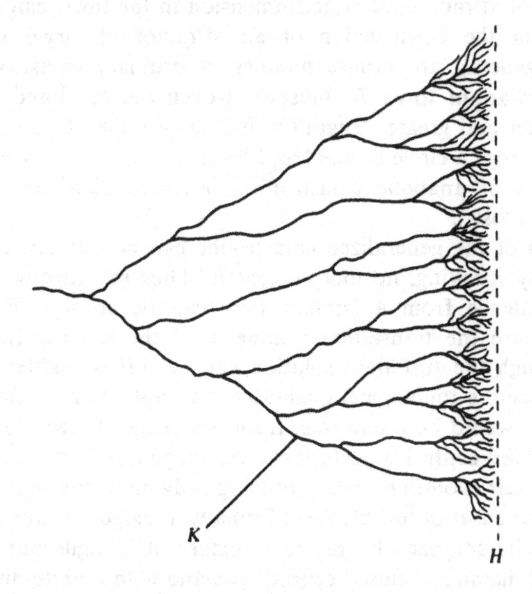

FIGURE 6.1. The ramified structure associated with the bifurcation set of a generalized catastrophe

The final state of a generalized catastrophe is often determined by the presence of *polarizing gradients*; for example, when one phase is denser than another, it will sink to the bottom at the end of the catastrophe. Even when such gradients are absent, the final state will often have a distinctive symmetry, usually a periodic structure of unknown origin; there are many such examples both in living beings (somites, phyllotaxis, feather buds) and in inert matter (Liesegang rings, the cusps in a beach at low tide, waves in a straight channel, Bénard's phenomenon) (see Photographs 21 and 22). The periodic elliptic breaking, instead of hyperbolic breaking, in the diadem of a splash (Photograph 12) is another example; the explanation of this in terms of a germ set of the catastrophe given above is not very convincing, and it seems rather that here is some deep phenomenon whose mathematical structure needs serious investigation. After all, if symmetries exist in nature, this shows that, despite their apparent instability, the process that generates them is structurally stable (see Note 2 and [2] and [3]).

An analogous catastrophe occurs when a high-dimensional attractor c disintegrates into attractors of smaller dimension (although not necessarily points); this gives a *catabolic catastrophe*. *Anabolic catastrophes*, leading to the formation of attractors of larger dimension in the fiber, can also exist, and here again the construction of an attractor of larger dimension demands, in general, an infinite number of ordinary catastrophes; for example, an invariant torus T^2 must be woven out of closed attracting curves of greater and greater length (excluding here the immediate formation of a torus from a circle by the Hopf bifurcation). The only plausible a priori examples of anabolic catastrophes occur in biological dynamics (telophase of mitosis).

The concept of the generalized catastrophe can be relevant even when there is, strictly speaking, no morphogenesis. Thus in hydrodynamics the onset of turbulence from a laminar flow can be considered as a catastrophe because the translation symmetry of the laminar flow is destroyed, although the turbulent solution can be differentiable; however, even in this case the topological qualitative description of the field is very complicated. It would be interesting, from this point of view, to study the variation of closed central trajectories of the tangent solenoid field at each moment, and their evolution toward topologically more complicated states. In this sense the onset of turbulence is formally analogous to the process of melting; it is characterized by the replacement of a vague attractor of a Hamiltonian dynamic, of closed central type and with a finite-dimensional pseudogroup of symmetries, by a large ergodic set whose associated pseudogroup is transitive, characteristic of a gaseous phase.

6.2. THE GEOMETRY OF A COUPLING

The theory of bifurcation of a dymanical system may be considered as the theory of dynamical systems with rigorous first integrals, for if a dynamical system is parametrized by points p of a manifold P, we can consider it as a global differential system (M, X) for which the coordinate functions of P are first integrals, or which is tangent along the fibers of a differentiable fibration $P : M \longrightarrow P$. In addition to this theory, we are led, by a natural weakening of the hypotheses, to consider dynamical systems (M, X) with approximate first integrals defined by the fibration $p : M \longrightarrow P$, and this leads to the idea of a mean field.

A. Mean fields

Consider first a dynamical system (M,X) which admits a system of global approximate first integrals defined by a fibration $p : M \longrightarrow P$; that is, the field X in $T(M)$ is of the form $X = X_0 + Y$, where X_0 is a vertical component tangent along the fibers of p, and Y a horizontal component that is small with respect to X_0. Associate to each point p of P the attractors of the corresponding dynamic (M, X_0), thus defining a space $\hat{P}$ stacked (*étalé*) over P; each leaf of this space is bounded by bifurcation strata along which the corresponding attractor is destroyed by catastrophe. To simplify matters, suppose that a stable regime of an attractor c persists as long as the representative point p of P does not encounter a stratum S of the boundary of the leaf corresponding to c in $\hat{P}$ (this is equivalent to saying that there is *perfect delay* in the transformation of the attractors); further suppose that, when an attractor $c(X)$ disappears at a point p of S, its basin of attraction is, in general, captured by another attractor $c_1(X)$, so that it jumps along S from the leaf c to the leaf c_1. [In the most complicated situation there might be competition between several attractors for the succession of $c(X)$; we leave this case until later.] Taking the average of the horizontal component Y at each point $\hat{p}$ of the leaf of $c(X)$ for all points of $c(X)$ [supposing now that the field is ergodic on $c(X)$] gives a vector field Y' on the leaf c that is characteristic of this leaf. This gives a vector field Y' on $\hat{P}$, called the *mean field* of the dynamic $M(X)$. It can happen that the mean field Y' itself has structurally stable attractors $c(Y)$; and, when there are other approximate invariant parameters, this construction can be iterated

Remarks. We shall not consider here the very difficult estimation of the error in replacing the actual evolution of a field by its mean evolution (the work of Arnold [4] shows the difficulties of this problem in the case of Hamiltonian dynamics) but will make only the following observation. In a dynamical system consisting of two weakly nonlinear, weakly coupled oscillators of the same frequency, the energy passes periodically between the two oscillators–the phenomenon of nonlinear beating. Here the mean field would lead to a constant energy distribution between the oscillators,[3] and the magnitude of the error, which is even qualitatively apparent, is due to the fact that the mean field is not structurally stable. Generally the smaller the horizontal component Y is with respect to X_0, the smaller the error will be.

B. Examples of mean fields associated with an elementary catastrophe

1. The fold. Suppose that S is a fold hypersurface bounding the domain of existence of an attractor c, and that, on S, c is destroyed by collision with an unstable regime and is captured by another stable attractor c_1; let Y and Y_1 be the fields associated with c and c_1, respectively. If S is defined by $z=0$, and if $z<0$ is the zone of existence of c, an interesting situation arises when the z component of Y is positive, and that of Y_1 is negative: the field undergoes a discontinuity along S, making it a reflecting boundary (Figure 6.2). When Y is oriented toward decreasing z, there is no interesting effect since S can never be reached (Figure 6.3).

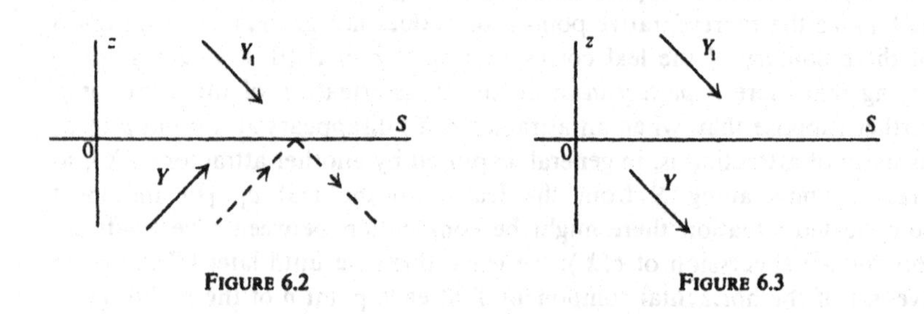

FIGURE 6.2 FIGURE 6.3

2. The cusp. Here we consider the Riemann-Hugoniot catastrophe and the semicubical parabola $4u^3+27v^2=0$; inside, $4u^3+27v^2<0$, there are two stable regimes in competition, and let Y_1 and Y_2 be the corresponding fields. At the origin these fields coincide; suppose that there Y points along the Ou-axis, in a symmetrical configuration. Then the interesting case occurs when the fields Y_1 and Y_2 have opposed v components, so as to make the branches of the parabola reflecting (as in the preceding paragraph). According to the sign of the u-component, there will be a global configuration, either funneling and ensuring the convergence of all points within the cusp toward O (see Figure 6.4), or dispersing directed away from O (when the u-components of Y_1 and Y_2 are negative; see Figure 6.5).

The great interest of this mean field, is to show that converging funneling mechanisms, with successive discontinuous corrections, a priori of a finalist nature, can occur naturally and in a structurally stable fashion. When Y has a small v-component at the origin, the only effect of this perturbation will be to open wider the mouth of the funnel. We have here an example of Waddington's *cusped canalization* [5].

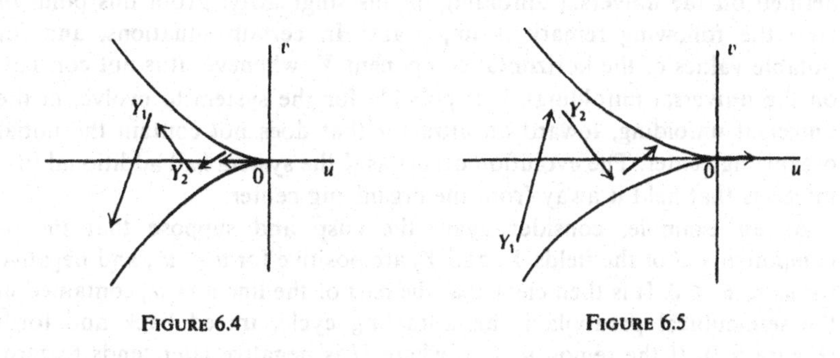

FIGURE 6.4 FIGURE 6.5

These are only the simplest examples of fields associated with elementary catastrophes. Let me give one more two-dimensional example, in which there are three stable regimes c_1, c_2, and c_3 in competition in the neighborhood of O. When there are three lines of folds S_{12}, S_{23}, and S_{31} along which the attractors are captured, $c_1 \rightarrow c_2$, $c_2 \rightarrow c_3$, and $c_3 \rightarrow c_1$, respectively, and if the mean fields are approximately constant with the directions given in Figure 6.6, this will give rise to an invariant attracting cycle; such a cycle is structurally stable.

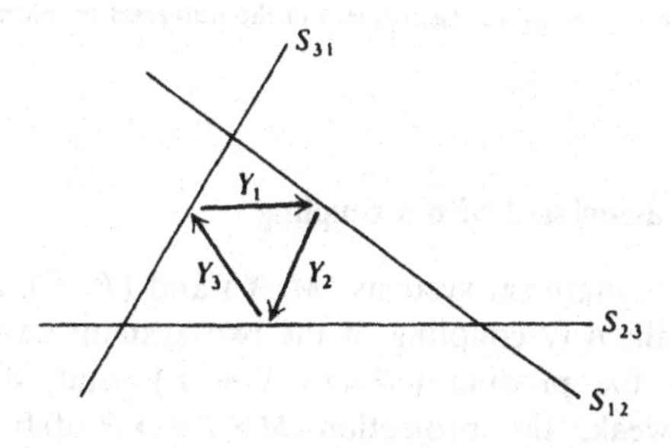

FIGURE 6.6. Attracting cycle defined by a field with discontinuities.

A complete topological classification of mean fields Y is obviously still out of reach (although fields with discontinuities have been studied under the name of *systèmes déferlants*), but we might assume, a priori, that the only fields of this kind that could occur naturally would be those associated with a structurally stable singularity in the fiber dynamic, a singularity that might be called the organizing center of the mean field defined on the universal unfolding of this singularity. From this point of view the following remark is important: In certain situations, and for suitable values of the horizontal component Y (whenever it is not constant on the universal unfolding), it is possible for the system to evolve, in the universal unfolding, toward an attractor that does not contain the initial organizing center. The evolution occurs as if the system had additional first integrals that held it away from the organizing center.

As an example, consider again the cusp and suppose that the u-components U of the fields Y_1 and Y_2 are positive for $u < u_0$, and negative for $u_0 < u < 0$. It is then clear that the part of the line $u = u_0$ contained in the semicubical parabola is an attracting cycle, traced back and forth (Figure 6.7). If the region $u > u_0$ where U is negative later tends to zero, the attracting cycle tends again to the organizing center, which can reconstitute itself.

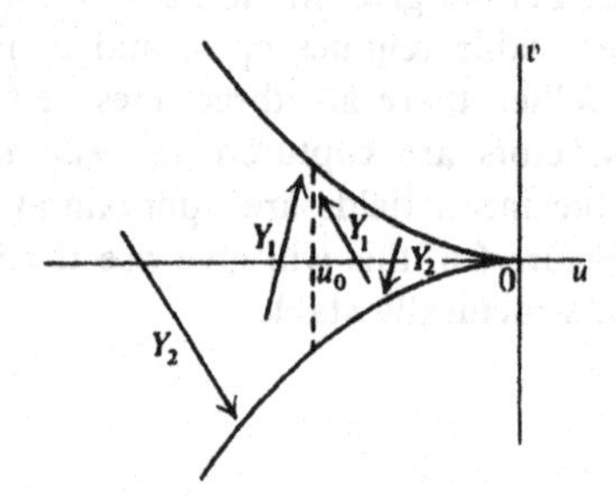

FIGURE 6.7. An attracting segment in the universal unfolding of a cusp.

C. Mean fields associated with a coupling

Consider two dynamical systems (M, X) and (P, Y), and suppose that $Y = 0$ or is small. Any coupling of the two systems can be defined as a perturbation of the product $(M \times P,\ X + Y)$; and, if $Y = 0$ and the interaction is weak, the projection $M \times P \rightarrow P$ defines a system of approximate first integrals of the perturbed product system that will, in general, give on P the mean fields Z induced by the coupling. The structurally stable attractors of these mean fields constitute stable configurations of the coupling on P, and they will remain stable under small deformations of the interaction.

Insofar as these mean fields arise from a structurally stable singularity of the fiber dynamic, these configurations are completely independent of the global topology of the dynamic (M, X). The form of these local configurations can thus be largely independent both of (M, X) and of the interaction.

In abstract terms, we can form the function space B of all dynamics (M, X), and this space has a stratified structure defining the bifurcation set H. Any interaction defines a map h of P into B as well as a mean field in the image $h(P)$; this gives a universal model generating all possible configurations of the mean field induced by interaction.

D. Mean field, scale, and catastrophes

We have seen how the dynamic defined by a mean field in a space P of approximate first integrals can have ordinary (gradient-like) structurally stable attractors. When this is the case, it will be impossible to reconstruct from a macroscopic observation of the process (i.e., from the morphology defined by the mean field) the original dynamic (M, X), whose existence can hardly be guessed. But when the mean dynamic has unstable structural phenomena and, in particular, generalized catastrophes, the evolution of these catastrophes can give information about the primitive dynamic (M, X). This evolution will depend on a germ set which, we can suppose, is determined by the fiber dynamic (M, X) at the beginning of the catastrophe. Photography plays a well-known part in scientific observation, and is not photography a controlled chemical catastrophe the germ set of which is the set of points of impact of the photons whose existence is to be demonstrated? The same is true of the bubble chamber or scintillation counter in the detection of elementary particles. Also, in embryology, the many generalized catastrophes constituting the development of the embryo are controlled by the fine structure of the nucleic acid of the chromosomes. This whole discussion shows again the theoretical and practical interest in generalized catastrophes, which form practically our only means of investigation of the infinitesimal.

6.3. SEMANTIC MODELS

Let P be a natural process taking place in an open set W in space-time. Any analysis of the process starts with the aim of determining in which regions of W the morphology of the process is structurally stable. Suppose that it is structurally stable in an open set $U \subset W$, and recall that this means there is a universal model $K \subset U_1$ such that P, restricted to U, is isomorphic to the local process P' induced from K by an embedding $j : U \to U_1$. In this case we say that P is defined on U by a morphogenetic field or, in the terminology of Waddington [5], that P has a chreod on U.

A. Definition of a chreod

A *chreod* c in space-time $\mathbf{R}^3 \times \mathbf{R}$ is specified as follows:

1. By an open set U in the hyperplane $t = 0$, called the *initiation set of* c.

2. As in all natural evolving processes, a point x of space-time can affect only the events in a cone $C(x)$ with vertex x, the successor cone of x (like the light cone in relativity), and the union of these cones $C(x)$ over all x in the initiation set of c is an open set W, called the *zone of influence* of the chreod.

3. By an open set V contained in W and containing U in its boundary, the *support* of the chreod, written as $|c|$, and a (static or metabolic) morphogenetic field defined, up to isomorphism, on V. The set W-V is called the bifurcation zone or *umbilical zone* of the chreod (see Figure 6.8).

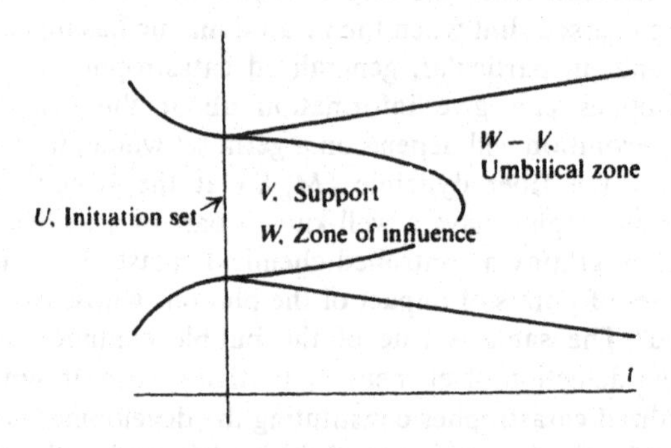

FIGURE 6.8. Plan of a chreod.

It can happen that the support of the chreod extends to $t = \infty$, and that the section of the morphogenetic field at $t =$ constant tends to a fixed limit as t tends to infinity; in this case the chreod is called *asymptotic* (see Figure 6.9).

Remarks. The idea of a chreod differs from the more general idea of the morphogenetic field only in the privileged role allotted to time and its orientation. Irreversibility of time is justified by the fact that for natural processes which depend on diffusion and are, at least partially, controlled by parabolic equations the possibility of qualitative "retrodiction" (reconstructing the past from the present situation) is much more restricted than that of prediction.

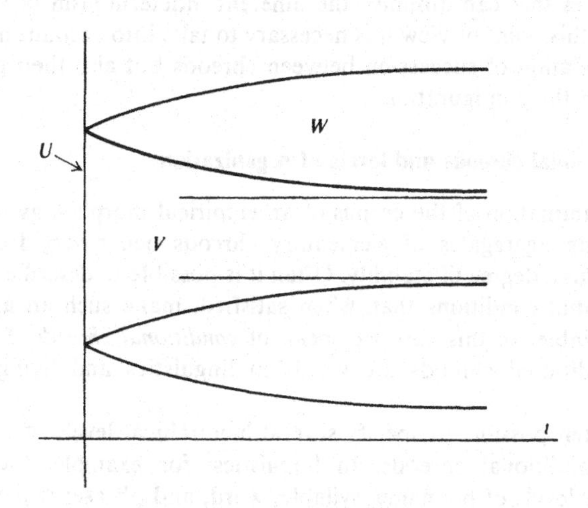

FIGURE 6.9. An asymptotic chreod.

B. A subchreod of a chreod

A chreod J is a *subchreod* of a chreod K if the initiation set and thus the support of J are contained in the support of K. The morphogenetic field on the support of J is given by the injection $|J| \rightarrow |K|$ of supports.

There is an infinite number of subchreods of K whose support contains a given point x in the support of K; in practice, every sufficiently small open neighborhood of x defines such a subchreod. When x is an ordinary catastrophe point of the morphogenetic field associated with K, there is a fundamental system of isomorphic subchreods containing x. Recall that such a system does not exist if x is an essential catastrophe point of the field. Conversely, any chreod is contained in a maximal chreod. If the process studied is contained in a unique chreod, it is deterministic and structurally stable.

C. The family tree of chreods

Say that a chreod g is the *successor* of chreods $k_1, k_2, \ldots, k_r$ if the initiation set of g lies in the intersection of the umbilical zones of $k_1, k_2, \ldots, k_r$. Generally the same chreod g can have several systems of parent chreods, and, conversely, the same parents can give birth to several configurations of chroeds. The complete description of a semantic model requires the specifying of all possible rules of succession and of all the known rules that can diminish the inherent indeterminism of the scheme, and from this point of view it is necessary to take into account not only the abstract relation of succession between chreods but also their geometrical position in the configuration.

D. Conditional chreods and levels of organization

The examination of the corpus of an empirical morphology often shows that certain aggregates of elementary chreods occur very frequently or exhibit a high degree of stability. Often it is possible to describe constraints on the initial conditions that, when satisfied, make such an aggregate of chreods stable. In this case we speak of *conditional chreods*. Examples of such conditional chreods are words in linguistics and living beings in biology.

It is often possible to specify several hierarchical levels of organization among conditional chreods. In linguistics, for example, there are the successive levels of phoneme, syllable, word, and phrase; while in biology there are supramolecular organization, cytoplasm, cell, organ, organism, and population.

The decomposition of a chreod at level $j + 1$ in terms of chreods at the lower level j is called its *structure*. A pair of levels $j, j + 1$ will be called formalizable if there is an abstract morphology M (generally finite) and a homomorphism of the chreods at level $j + 1$ into the aggregates of M such that the images of the component chreods at level j are the atoms of M (the homomorphism being compatible with the contiguity relations). This notion accounts for the ideas of function in biology and grammatical category in syntax. In generative grammar, for example, the structure of a sentence is represented by a tree such as the following:

```
                      Sentence
            SP                          VP
           /  \                        /  \
          A    N                      V    OP
          |    |                      |   /  \
          |    |                      |  A    N
          |    |                      |  |    |
         The  cat                 chases the mouse.
```

Here the abstract morphology M consists of symbols such as A (article), N (noun), and V (verb). Syntatical analysis of a sentence is then a homomorphism of the collection of words in the bottom line of the tree agreeing with the abstract morphology M. Thus the symbols A, N, and V represent grammatical functions (or, more precisely, categories).

Before discussing the general theory of semantic models, it is worth while to show the great generality of the scheme by means of some examples.

E. Examples of semantic models

1. The diffusion of a drop of ink in water. Recall Helmholtz experiment, quoted by D'Arcy Thompson [2]: After falling through a few centimeters, a drop of ink hits the surface of the water and, stopped by the liquid, gives birth to a vortex ring (corresponding to the inverse mushroom chreod). Then the vortex ring disintegrates into three or four droplets, each of which, falling, gives birth to' smaller rings; these are other inverse mushroom chreods succeeding the first after an intermediary umbilical zone which is evidence of the great instability of the process, for the number of droplets into which it disintegrates can vary greatly according to the initial conditions. In fact, this succession is typical of a generalized catastrophe where the metabolic form of the vortex ring decays into the static form of the droplets.

2. Feynman graphs in the theory of elementary particles. Here the chreods are the elementary interactions symbolized by the vertices of the graph, and all the unknown internal structure of the interaction is ignored. The successor relations between vertices are defined by the edges, and they are completely characterized by the qualitative nature of the particle (electron, photon, etc.) and its momentum. All configurations compatible with the combinatorial constraints imposed on elementary interactions are supposed to be possible in this theory, and an auxiliary aim, at least, of the theory is to evaluate the probability of each configuration.

3. In biology, almost all biological morphogenesis can be considered as a semantic model (see Chapters 9 and 10).

4. Animal and human behavior can be decomposed into functional fields acting as chreods. In particular, human language is a system described by a one-dimensional (time) semantic model whose chreods are the phonemes and, on a higher scale, the words.

This last example is highly significant; if, as Condillac says, it is true that "all science is a well-made language," it is no less true that all natural phenomena constitute a badly understood language. Also consider Heraclitus: "The lord whose oracle is at Delphi neither speaks nor conceals, but gives signs."

F. The analysis of a semantic model

Let us start the analysis with the particular case of language. A first method is the formal attack: neglecting completely the internal structure of each chreod (here, the meaning of each word), we describe the formal relationships among them. To this end the basic step is to collect a stock of examples sufficiently large to allow valid conclusions to be drawn about the formal relationships of sucession; this stock is the *corpus* of the linguists, and from it, in principle, one can deduce the *grammar* which governs the association of chreods in the configurations found in the corpus.

The foundation of the corpus is the primordial essential task of the experimenter, and many sciences, particularly biology, have scarcely passed this stage. In linguistics, thanks to our direct intuition of the meaning of words, we have been able to produce a formal classification of words associated with their grammatical functions and, in this way, to define formal conditions for a phrase to be grammatically correct, but not for it to be meaningful. The linguist's task is very difficult, for there is almost no connection between the written or spoken structure of a word and its meaning. The choice of the word corresponding to a given meaning is the result of a long historical process, a quasi-permanent generalized catastrophe. This is due to the fact that human communication has imposed rigid constraints on the structure of expression (see Chapter 13).

In the natural processes that do not aim at communication and that can be assimilated in language only by metaphor (but by a useful metaphor), we must expect the internal structure of each chreod to be relatively transparent; in such a language all words will be onomatopoeic. In fact it is a good idea to suppose, a priori, that a chreod contains nothing more than can be deduced by observation, that is, the catastrophe set, and to proceed to the dynamical analysis of the chreod which is the most conservative (*économique*, in French). From this point of view, the *significance* of a chreod is nothing more than the topological structure of the catastrophes it contains and its possible dynamical interpretation. This leads to the definition of the organizing center of the chreod.

G. The dynamical analysis of the chreods of a static model

We now restrict attention to chreods associated with gradient dynamics because these are the only ones that are susceptible to mathematical treatment; metabolic models lead, via catabolic catastrophes, to generalized catastrophes about which we know nothing at the moment. All the chreods associated with such a process on space-time have been described; what about their associations? Some of them occur more frequently than others; one important class of these configurations is that with organizing centers of codimension greater than the dimension of space-time. Suppose that z is a jet that is determinate but, for example, of codimension five; the universal unfolding of the singularity defined by this jet is five dimensional, and in it Maxwell's convention defines a universal catastrophe K. The intersection near the origin of K and a transversal $\mathbf{R}^4$ will then have a configuration of chreods that are structurally stable under small deformations of the global evolution of the process. This permits us to talk of the jet z as the organizing center of the configuration. The local stable configurations of chreods, the grammar of the process, is completely specified by the topology of the bifurcation set in the function space of potentials. A perturbation of the embedding of the $\mathbf{R}^4$ in the universal unfolding $\mathbf{R}^5$, sweeping across the origin, for example, will result in a modification of the stable configuration of chreods, and this allows us to speak of configurations being linked or dual with respect to the organizing center z.

It may seem difficult to accept the idea that a sequence of stable transformations of our space-time could be directed or programmed by an organizing center consisting of an algebraic structure outside space-time itself. The important point here, as always, is to regard it as a language designed to aid the intuition of the global coordination of all the partial systems controlling these transformations. We shall be applying these ideas to biology in Chapters 9 and 10.

APPENDIX

Spiral nebulae

The morphology of spiral nebulae. Let O denote the center of gravity of a mass of gas concentrated under gravitational forces, and suppose that a significant fraction of the gas is already concentrated in the neighborhood of O forming what we will call the core. We are going to study the velocity field of the gas in the neighborhood of the core, regarding it, to first approximation, as a point.

Let J be the total angular momentum of the gas. The velocity field $V(m)$ is zero at O; suppose that this zero is a generic zero of the vector field, so that the matrix M of linear parts of the components of V has distinct eigenvalues whose real parts are nonzero. Since the global process is a contraction, at least one of these eigenvalues has a negative real part. On the other hand, the only stationary states of the system are those for which the gravitational potential is that of a revolution around the Oz-axis parallel to J (see [6], theorem 3.5). Under these conditions, the only matrix compatible with a stationary state is that in which Oz is an eigenvector, and the two other eigenvalues c and $\bar{c}$ define a rotation in the orthogonal plane Oxy. Suppose that the given matrix M is, at least to begin with, close to this matrix associated with the stationary state.

Under these conditions, we can suppose that M is initially contracting both on Oz and in Oxy. Now the contraction on Oz, which is not opposed by the centrifugal force due to the rotation, will increase as the eigenvalue corresponding to the eigenvector along Oz decreases, and the field in Oxy will dilate (weakly) owing to the influx of material entering the core. The result will be that the singularity of V at O will undergo the Hopf bifurcation, and the field around O in the Oxy-plane will form a closed invariant circle, an attracting trajectory of V, corresponding to the formation of a ring in the nebular disc. The severing of the ring from the equator of the nuclear disc will have the effect, through reaction, of making the field V again contracting at O, and this will correspond to a bifurcation forming a closed repulsive trajectory; then the reappearance of a dilating situation in Oxy will lead to the formation of a new concentric ring within the first, and so on.

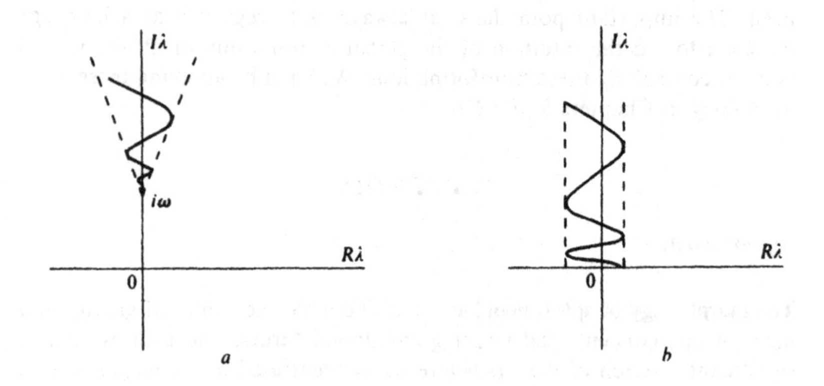

FIGURE 6.10

Geometrically, the characteristic value c, with imaginary part $i\omega$, ω being the angular velocity, oscillates from one side to the other of the imaginary axis. At the same time the rings that are formed move outward from the core and keep their angular velocity, consequently carrying angular momentum away from the core; therefore the imaginary part $i\omega$ of c will decrease as the process proceeds, and the curve traced out in the complex plane by c will look like Figure 6.10*a* or 6.10*b*. In case *a* there will be a limit cycle defined by a pure rotation ($c_0 = i\omega$ on the imaginary axis) corresponding to the case of regular spiral nebulae where the core has a relatively slow rotation at its boundary (this model makes no attempt to deal with the interior of the core, where more rapid rotations could probably be found). In case *b* the eigenvalue c finishes on the real axis at O or close to O, and the two coincident eigenvalues c bifurcate on the real axis: one, γ_1, becoming negative, gives a highly contracting eigenvalue, and the other, γ_2, positive, an eigenvector Y in Oy that is weakly dilating. The material of the nuclear disc thus contracts onto the Oy-axis, forming a kind of arm extended across a diameter yy' of the nuclear disc. Gravitational attraction, which acts on the peripheral ring, still rotating slowly, will capture this thread of material and give birth to the θ-configuration often observed in barred spirals. The arm $\alpha\alpha'$, deprived of its supply of material, will die and break up into cloudlets of stars (Figure 6.11), and then the active arm captured by the differential rotation will realign and finally adopt a rectilinear configuration imposed by the vector field of gradient type controlling the rotation field.

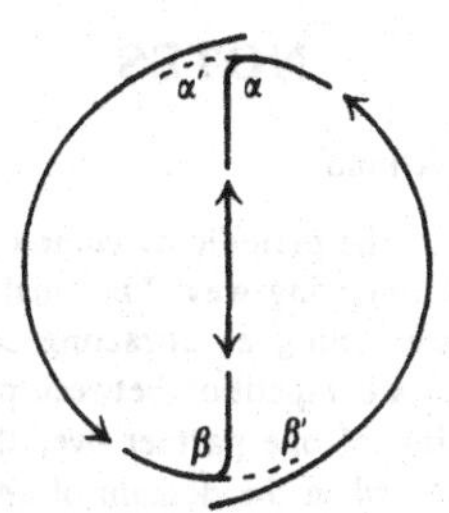

FIGURE 6.11

The spiral structure. The model described above leads to the case of normal nebulae with a concentric annular structure, and not to spiral arms. A finer model will probably have to take into account the form of the core and the structure of the equitorial breaking-up. Comparing the ball shape of the newest elliptical nebulae (state E_7, e.g., NGC 4530, Photograph 23) with an older state (see Photograph 24) like NGC 4594, one can hardly avoid thinking that a hyperbolic breaker is produced along the equitorial axis; the cusped section thus produced is unstable and breaks into closed or spiral rings. How can such a phenomenon occur? Suppose that two hyperbolic umbilic points m and m' appear first on the equator. Each of them is, a priori, free to move on the equator, and shortly afterward the first peripheral ring from a periodic parabolic breaking can perhaps appear. If the point m catches up with m', it can lead to the formation of a closed ring, but then a stable configuration of ejection points can form on the equator, for example, two antipodal points, each giving birth to a free arm. This situation could well be intrinsically unstable and susceptible to many topological variations, with the arms becoming captured and giving birth to many different branched configurations.

The model given above for barred spirals has a serious defect since, in principle, it requires the diametrical arm yy' to be fixed. It is experimentally plausible, however, that this arm might have a rotation about the center O; this could be the effect of a magnetic field, an axial magnetic field having the same effect on a charged mass as a Coriolis force due to a rotation. Also, the movement of the ejection points around the equator of the nuclear disc is not necessarily connected with the rotation of the core, just as similarly the displacement of the breaking point (the hyperbolic umbilic) on the crest of a breaking wave is not connected with the concomitant movement of the liquid.

NOTES

1. I owe this remark to V. I. Arnold.

2. *The stability of symmetries.* The periodic outcome of certain generalized catastrophes can perhaps be explained in the following way. The final state is the result of *competition* between lumps of the new phase, acting as attracting centers for the new phase. Now, anthropomorphically, the result of a competition between partners of equal force or skill will probably be a series of local gains, of one partner over the others, that are unstable and subject to reversals. Geometrically, when the domain of an attractor extends unduly from a point, it will form a protuberance at this point with a relatively weak gradient and a relatively long frontier, and this zone will be fragile and comparatively more exposed to enemy attack. It would obviously be worth while to construct a geometrical model for this type of process. See Weyl [3] for a more complete examination of this question.

3. This assertion is correct only when the interaction Hamiltonian has no quadratic term. Otherwise (and this is the usual case) the mean field leads effectively to a sinusoidal distribution of energy between the two oscillators.

REFERENCES

1. J. Sotomayor, *Estabilidad estrutural de primeira orden e variedades de Banach*, Thèse I.M.P.A., Rio de Janeiro, 1964.
2. D'A. W. Thompson, *On Growth and Form*, Cambridge University Press, 1945 (Liesegang rings, p. 660; Bénard's phenomenon, p. 505; Helmholtz' experiment, p. 396).
3. H. Weyl, *Symmetry*, Princeton University Press, 1952.
4. V. I. Arnold and A. Avez, *Ergodic Problems in Classical Mechanics*, Benjamin, 1968, paragraph 22: Perturbation theory.
5. C. H. Waddington, *The Strategy of the Genes*, Allen and Unwin, 1957.
6. S. Chandrasekhar, *Principles of Stellar Dynamics*, Dover, 1942.

CHAPTER 7

THE DYNAMIC OF FORMS

There is much here to excite admiration and perplexity.

LORD RAYLEIGH, *Collected Papers*, p. 661
(on Liesegang rings)

Most of the ideas put forward in this chapter are of a speculative kind, for they lack both a mathematical formulation and a rigorous mathematical justification. However, some of my readers may agree with me that the degree of interest or importance of a mathematical or physical idea rarely corresponds to its formal rigor; the rest should perhaps wait for a few years . . . .

7.1. MODELS OF MECHANICS

A. Limitations of classical and quantum models

Only two types of models of mechanics exist at present: classical and quantum. Recall that the classical model consists of a configuration space M, a differential manifold whose contagent space $T^*(M)$ is the phase space, and a Hamiltonian $H: T^*(M) \longrightarrow \mathbf{R}$; then the field X of the dynamic is the symplectic gradient of the Hamiltonian, that is, is defined by $i(X) \cdot a = dH$, where a is the canonical 2-form $\Sigma_i dp_i \wedge dq_i$ on $T^*(M)$. In the quantum model the states of the system are parameterized by vectors of a Hilbert space, and the evolution of the state vector ψ is defined by a self-adjoint operator H (the Hamiltonian) such that

$$\frac{d\psi}{dt} = iH(\psi).$$

These models have a characteristic in common: they are of interest only when the Hamiltonian H (classical or quantum) is sufficiently simple to allow explicit calculation or, at least, global qualitative evaluation of the dynamic. In fact each model has had a typical successful example: the classical for Newtonian celestial mechanics (with the gravitational inverse-square law) and the quantum for the hydrogen atom (with the Hamiltonian $\Delta - 1/r$), and these two applications have assured the success of the corresponding models; whenever any more complicated case has been tackled, however, both models have rapidly become less effective. The reason is simple: either the Hamiltonian is given a priori, when the integration of the evolution may become difficult or impossible (the three-body problem has resisted all attempts, classical or quantum), or an experimental situation is given and one tries to construct an appropriate (classical or quantum) model to fit it. This generally results in the problem having too many solutions, whereas the Hamiltonian has to be reconstructed ad hod, artificially, and without intrinsic properties connected with the configuration space. From this point of view it would be a good thing to enumerate all the natural classical dynamics, such as geodesic flows on special manifolds, Lie groups, and homogeneous spaces with canonical metrics admitting large isometry groups. It seems probable that some conservative natural phenomena give geometrically imposed realizations of these privileged dynamics; this is a possible reason for the presence of internal symmetry groups in the theory of elementary particles. But there are few spaces with a natural metric having a large isometry group. Also the field of application of natural classical models is very restricted, and the situation is not much different for quantum models.

B. Determinism

The classical controversy of determinism is grafted onto these known defects of mechanical models. It is generally supposed that classical mechanics deals with phenomena of the macroscopic world, which are therefore rigidly deterministic, whereas phenomena at the quantum level are fundamentally indeterministic. This pedantic view is, it seems to me, basically faulty. I will not consider here the question of indeterminism in quantum mechanics, except to say that the argument on which it is based —the uncertainty principle—depends only on the crude and inadequate model of a particle as a point; it would be surprising if incompatibilities and obvious paradoxes did not follow as a result of forcing an object into an inadequate conceptual framework. But in the macroscopic world, and even from a Laplacian viewpoint, eminently metaphysical or even theological, of a universal differential system parameterizing all the states of the universe, any such model can be subject to experimental control only under two conditions:

1. That the model has a localization procedure allowing the construction of local models.

2. That these local models are structurally stable.

Now it is an everyday experience that many common phenomena are unstable (see Chapter 1), and clearly there are no experimental criteria to distinguish between a structurally unstable process and a fundamentally indeterministic process if such an object exists. This is why the question of determinism, freed from its philosophical background, reduces, in phenomenological terms, to the following almost incontestable assertion: *there are phenomena that are more or less deterministic*. The degree to which a process is deterministic is essentially an expression of the degree of smoothness (differentiability) of the evolution of the process in terms of its initial conditions. Let me state immediately the following:

Hypothesis (LS): The degree to which a process is determinate is determined by the local state of the process.

Then any process satisfying (LS) divides into two type of region: the region in which the process is well determined and structurally stable, and the region of instability or indeterminism. This is how we introduced the semantic models in Chapter 6, with the regions of the first type, veritable islands of determinism, being Waddington's chreods, and the instabilities being the umbilical zones, often the sites of generalized, unformalizable catastrophes.

These models, which make determinism or instability a local property of the process under consideration, seem likely to apply before long to biological processes; may they one day be adapted also to quantum processes? One can at least express the hope.

7.2. INFORMATION AND TOPOLOGICAL COMPLEXITY

A. The present use of the idea of information

This chapter is directly inspired by the present state of molecular biology, whose specialists find themselves constantly writing the word "information," talking of "the information contained in the nucleic acid of chromosomes," "the flux of information sent by DNA in the cytoplasm via the intermediary of the RNA messenger," and so forth. Such biologists calculate the number of bits of information contained in the *Escherichia coli* chromosome and compare it with the corresponding value for a human chromosome, implicitly postulating that all combinations of nucleotides in the DNA chain are equiprobable, a hypothesis that would be difficult to justify either practically or theoretically. However, even when the present use of "information" is incorrect and unjustifiable, the word does express a useful and legitimate concept. Here we set ourselves the problem of giving this word a scientific content and releasing it from the stochastic prison in which it is now held.

B. The relative nature of complexity

The central thesis of this chapter is that usually, when we speak of "information," we should use the word "form." The scalar measures of information (e.g., energy and entropy in thermodynamics) should be geometrically interpreted as the topological complexity of a form. Unfortunately it is difficult, with the present state of topology, to give a precise definition of the complexity of a form. One difficulty arises in the following way: for a dynamical system parameterized by a form F, the complexity of F is rarely defined in a way intrinsic to F itself, for it is usually necessary to embed F in a continuous family G, and then define only the topological complexity of F relative to G. Thus energy appears as the complexity relative to the largest system in which the given system can be embedded and is that complexity which retains its meaning in every interaction with the external world; it is the *passe partout* parameter and so contains the least information about complexity. Taking G to be a smaller subfamily which captures more of the particular characteristics of F gives a narrower and more specific complexity that can also be considered as information. Another example can be cited from biology: plants take in through their chloroplasts the grossest complexity of light, namely, energy, whereas animals extract, through their retinas, the correlations of forms, or the information that they need to obtain their food, and thus their energy.

Let us now deal with the technical difficulties of defining the complexity of a form.

C. Topological complexity of a form

Consider the typical case of a differentiable map between manifolds X and Y. We have seen how there is a locally closed hypersurface with singularities H, the bifurcation set, in the function space $L(X, Y)$ such that each connected component of $L(X, Y) - H$ defines a structurally stable form of maps. When f and g are two forms such that f can be deformed continuously to g, there is a differentiable path joining f and g, and almost all of such paths cross the hypersurface H transversally; we write $d(f, g)$ for the minimum number of intersections of such a path with the strata of H of codimension one. This defines an integer-valued distance on the set of forms, because $d(f,g) = d(g,f)$ (symmetry): if $d(f,g) = 0$, then f and g have the same form (reflexivity); and if f, g, k are three structurally stable maps, then $d(f,k) \leqslant d(f,g) + d(g,k)$ (triangle inequality). We next designate a ground form f_0 of the family, which is considered as the topologically simplest form, and define the complexity of any structurally stable form f to be the integer $q(f)=d(f_0,f)$. For a map g which is not structurally stable, we can generalize the definition of complexity to the upper bound of $d(f_0,g')$, where g' is a structurally stable form arbitrarily close to g. In this way the determination of the complexity of a form depends, above all, on the choice of the simplest ground form, and in this choice lies a problem almost untouched by present-day topology.

D. The choice of a ground form

Let $m=\dim X$, $n=\dim Y$. In one case, at least, the choice of the ground form needs no discussion, for when $n \geqslant 2m+1$ there is a simplest form in the set of all homotopy classes of maps from X to Y, namely, the embedding of X in Y; any two such embeddings are isotopic, and so the topological type of an embedding is unique. When $n<2m+1$ we can say nothing, and there is no reason to believe that there is a unique simplest form; the most that we can hope for is that there will be at most a finite number of candidates. When $n<m$ it is natural to consider the maps without singularities as simplest. Now (at least when X and Y are compact) these maps are the differentiable fibrations of X over Y; but as it is not known whether the number of these fibrations (up to homeomorphism) is or is not finite (nor can we expect that it is, as the case of coverings of the circle $S^1 \rightarrow S^1$ seems to indicate that it is not); perhaps homology or homotopy criteria will specify the simplest form or forms—I merely raise the question. Another natural way of defining the complexity of a map is to take account of its singularities; supposing that the complexity of a closed set K can be measured, to a first approximation, by the sum of its Betti numbers $\Sigma_j b_j(K)$, we can define the complexity of a map f as the sum of the Betti numbers of the set of singularities of f. For a differentiable function $f: V \rightarrow \mathbf{R}$ we can define the complexity of f to be the total number of critical points of f, supposed quadratically nondegenerate. The known results on the topology of the graphs of characteristic functions of the equation of a vibrating membrane (or the Schrödinger equation) justify this point of view. Unfortunately, in general, there is an infinite number of different topological types of functions on a compact manifold V having the minimum number of critical points required by Morse theory. Some of these have gradient cells in more complicated positions and are, so to speak, more knotted than others that could be thought of as simpler.[1] Nevertheless, in situations of classical dynamics, such as are described by the equation of a vibrating membrane, the complexity seems to be describable directly as a function of the singularities.

Finally, a remark on the complexity of nonstructurally stable forms is in order: maps having singularities of infinite codimension have necessarily an infinite complexity, for they can be approximated locally by maps of arbitrarily large complexity. Thus the real-valued function $f: \mathbf{R}^n \to \mathbf{R}$, which is constant on an open set U in $\mathbf{R}^n$, has a singularity of infinite codimension at each point of U and so is of infinite complexity. As every function f can be continuously deformed to the zero function, we can join any function f to any function g by a path having for each value of the parameter one of the three forms f, 0, and g. Thus the zero function is a frontier point of all the strata of structurally stable forms in the function space $L(\mathbf{R}^n, \mathbf{R})$.

E. Complexity in a product space

Let $F = L(X, Y)$, $G = L(A, B)$ be two function spaces, and write $H = L(X \times A, Y \times B)$, and let $i: F \times G \to H$ be the canonical injection defined by

$$i(f \times g)(x, a) = [f(x), g(a)].$$

If f and g are two structurally stable maps in F and G, respectively, the image map $i(f,g)$ is also structurally stable in H. Thus there is a class of open sets in H formed by the products of open sets of structurally stable maps of the factors; these forms, called *decomposable*, are characterized by the property that, if one coordinate (a or x) is fixed, the corresponding section (in Y or B, respectively) is structurally stable. If the spaces F and G are configuration spaces of static models, and if the systems are coupled by interactions defined by a map $h: X \times A \to Y \times B$, we shall say that there is a *weak coupling* if the graph of h is decomposable and a *strong coupling* if it is not decomposable.

In H, as in all function spaces of differentiable maps, there is a bifurcation set K which contains the image under i of the product of the bifurcation sets of the factors F and G. If the ground form in H is defined to be the image of the product of ground forms of F and G, it is easy to see that the complexity in H of a decomposable form (f,g) is less than the sum of the complexities in F and G of the factor forms f and g, because, writing (f_0, g_0) for the base forms in F and G, we can join any decomposable form h of H of type (f,g) to the base form (f_0, g_0) by a differentiable path formed of the following:

1. A path transforming the given map h into a map of the form $i(f,g)$, which does not involve any change of form; this is possible as h is a decomposable form.
2. A path of the form (f_t, g_t) joining f and g to the base forms (f_0, g_0), intersecting the bifurcation sets of F and G transversally at a finite number of points, and realizing the sum of the complexities $q_E(f) + q_F(g)$.

Since there may exist other paths in H that join h to the ground form, we have

$$q_H(f, g) \leqslant q_F(f) + q_G(g).$$

Thus the topological complexity in a product space is subadditive, reminiscent of the analogous classical property of negentropy or the information associated with two random processes. It is clear that we could conceive of more general definitions of complexity than that of the topological type of differentiable maps, given here, which would have the same property.

7.3. INFORMATION, SIGNIFICANCE, AND STRUCTURAL STABILITY

A. Free interaction

The idea of stored information or memory leads, from the point of view of general dynamics, to a curious paradox. If M is such a stock of information, the structurally stable states of M must somehow define the information contained in M because, if this information is to remain as time elapses, M must be in a structurally stable state; but a memory is only useful when it is consulted, and so M must be in interaction with a receptor system A and must be able to stimulate large variations in A without itself undergoing a perturbation taking it outside its original structurally stable class. What are the constraints imposed on the structure of M and A by such asymmetrical behavior?

It is clear, a priori, that, whatever the systems M and A are, we can always devise interaction potentials that can give rise qualitatively to the required behavior; therefore, if we want to minimize the role of the interaction, the coupling must be as weak as possible, and for this reason we introduce the idea of the *free interaction* of two dynamical systems, a weak perturbation in a random direction from the given position in the topological product of the two systems (this is the case of null interaction). Under these conditions, if M and A are structurally stable static forms (due to gradient dynamics), so will the topological product be, and the free interaction of these two systems will have no qualitative effect. Therefore *no exchange of information is possible in the free interaction of two static forms*. On the other hand, a free interaction can have considerable qualitative effects between metabolic forms, because generally the *topological product of two structurally stable dynamical systems is not structurally stable*. For example, two constant fields on the two circles S_1 and S_2 define a parallel field on the torus $S_1 \times S_2$ which is not structurally stable; when the rotation number is rational, the field degenerates into a structurally stable field with closed attracting trajectories (the phenomenon of *resonance*), and when the rotation number is irrational, a small perturbation can make it rational. Thus there are always resonances between two systems having recurrence, and the stability and the global qualitative effect on the evolution of the product system is the more pronounced, the smaller the denominator of the ratio of resonance (the rotation number). Consequently two metabolic forms in free interaction always exchange significance. The case of the product of a static and a metabolic form has not been studied mathematically, but it seems probable that here there will not be any active interaction because of the absence of resonances. Let us now apply these ideas to some examples.

Examples. To make two given systems interact freely it is sufficient, in principle, to bring them close together or, better, to put them into contact. The typical example of a static form is a solid body (a crystal), though we shall immediately see what restrictions are imposed by this statement, and applying the theorem above, that there is no free interaction between static forms, we observe that it is difficult to establish a chemical reaction between solid bodies. As in the case of gunpowder, an intermediary metabolic form, like a flame, is necessary to start the reaction.

In another example dealing with solids, the key-lock system dear to biochemists, there will be an interaction only if the key is mobile (in all the positions allowed in the complement, the lock). A key carries significance for a lock only when it can be inserted and turned. This is a typical example of a situation where, when the key and the lock are considered to be static, almost all of their interaction structure lies in the motor agent, the individual who turns the key—that is, in the interaction potential; ıı one wishes to adopt a view that eliminates the interaction, it is necessary to make both systems mobile.

Consider now the system consisting of a salt crystal (NaCl), on the one hand, and water (H_2O), on the other; when these systems are put into contact, the salt will dissolve in the water. Even considering the liquid as a metabolic form, by virtue of the thermal agitation of its molecules, do we not have here a long-scale interaction between a crystal, presumed to be a static form, and a liquid? The origin of this phenomenon is well known: the NaCl and H_2O molecules ionize into Na^+ Cl^- and H^+ OH^- and make up a kind of *ménage à quatre*, undergoing the reversible reactions $H^+ + Cl^- \rightleftharpoons HCl$ and $Na^+ + OH^- \rightleftharpoons NaOH$. Here we have a typical example of a system which, though static in appearance, is actually undergoing infinitesimal oscillations of which one characteristic frequency resonates with a characteristic frequency of another system. It is even conceivable that the first system, becoming metabolic, could remove energy from the second and thus make considerable modifications to it; in this way it might be said that the salt is a significance carrier for the water, but this is not a memory system, as it is completely destroyed by the interaction, and this interaction is only partially reversible, as it is necessary to cool the system to recover the salt crystal. A form is called *pseudostatic* if it appears static but can be excited with relative ease into infinitesimal variations; this example shows that a pseudo-static form can be a significance carrier for a metabolic form.

Another example is the system of book + reader. The book can transmit information only when it is illuminated; hence the intervention of a metabolic field, the light field, is necessary, and we really have to consider two interactions, book⇌light, and light⇌reader. The latter interaction is between two metabolic forms and involves no difficulties of principle, even though it does involve the use of a very specialized receptor, the eye. The former uses, in an essential way, the electromagnetic interaction of photons and electrons that rebound elastically on the white parts of the paper but are absorbed by the black. Whatever the precise nature of this action, it can be interpreted as before: the system consisting of the book undergoes very small deformations, absorbing the energy from the other system in, qualitatively, a very specific way (spatially precise, in this case), but these deformations degrade the energy, then absorbed as heat without affecting the global form of the letters, which remain invariant because of their statistical nature, unaffected by the large number of small local interactions. This appears to be the most perfect kind of memory⇌receptor system interaction where the memory is almost completely unaffected by the interaction, all the energy of the interaction being provided by the receptor, which is, at the same time, metabolic (recurrent) and dissipative (the energy passes from the source to infinity). It would be interesting to know how far these characteristics of the interaction hold in general; it seems, in any case, indispensable that the receptor be an open system permeated by a qualitatively diverse flood of energy. It is probable that no receptor could read a printed page placed in a box with perfectly reflecting walls (a black body), and the existence of a continually active luminous source seems to be necessary.

The basic biological system of DNA⇌(the metabolism of small molecules) seems to work on analogous principles. Under certain conditions, necessitating regulation, the flood of energy from the metabolism follows a particular cycle that excites, by resonance, a geometricochemical deformation of a part of the chromosome. The cycle thus started extracts energy to sustain its own action, and the end products exercise the corresponding regulating action. This, further, gives a regulating action of a finalist character, which is undoubtedly connected with the fact that here is a natural memory perhaps created and certainly maintained by the ambient dynamic.

B. The entropy of a form

There is, mathmatically speaking, only one way of specifying the idea of a *free interaction* between two dynamical systems: this is to define a measure on the function space B of the dynamics of the product system and then define the idea of a *weak random perturbation*. Despite the work of many mathematicians (e.g., Wiener and Segal) and the integration techniques of some physicists (e.g., Feynman), the theory of integration on function spaces is far from clear, and in any case it depends drastically on the type of function space B (Hilbert, Banach, etc.) under consideration. I myself incline to the following procedure to define the measure. The function space B of a dynamical system frequently occurs as the inductive limit of nested Riemannian manifolds $X_1 \subset X_2 \subset \cdots \subset X_k \subset \cdots$ forming what is frequently called the *spectrum* of the function space. Given a structurally stable form F, take the i-dimensional measure $M_i(F)$ of the open set in X_i of points having topologically the form of F; then a sum like $m(F) = \Sigma m_i(F)/i!$ (where the convergence factor $i!$ is probably more or less irrelevant) could represent the total measure of elements of form F in B, which could be called the entropy of the structurally stable form F (not to be confused with the microcanonical entropy defined formally in Chapter 3).

There is probably a relation between this entropy and the topological complexity of a form (the larger the topological complexity of a form, the larger is its entropy) because it is intuitively clear that a highly topologically complex form can resist a perturbation of given amplitude less well than a less complex form. Thus, returning to the typical case of a constant field on the torus T^2, resonances with "weak denominators," whose attracting cycle is embedded in T^2 in a topologically simple way (for a given homology base), are much more stable than those with "large denominators," whose attracting cycles are embedded in a more complicated way. In this case the privileged role of the base circles of the torus appears in the position of the spectrum in the space B of vector fields on T^2, and this fact gives a reason for identifying the scalar measure of information contained in a form with its topological complexity, since this complexity and the negentropy vary in the same sense.

When the index i of the manifold X_i in the spectrum is proportional to the energy E, the measure $m_i(F)$ can be expressed in terms of a differentiable function $g_i(E)$ and then

$$m(F) = \int g(E)\, dE.$$

In this way the function $g_F(E)$ will completely characterize the thermodynamic of dynamics of the form F. Loosely, one might say that the thermodynamic of a system is determined by the position of the spectrum X_k with respect to the bifurcation set H.

With a theory of this kind, the integral cannot be expected to have all the properties that hold in finite dimensions, but this is not very important because, in practice, only special kinds of open sets need be considered.

C. Competition of resonances

When two dynamical systems with recurrence A and B are put into free interaction, the product system $A \times B$ is structurally unstable; generally, if C and C' are the attractors of A and B at the instant of coupling, the product attractors $C \times C'$ will undergo a catabolic catastrophe and degenerate into an attractor K of smaller dimension, contained in the product $C \times C'$. This is the most general form of the phenomenon known as *resonance* in dynamics. Usually the system must choose from several possible resonances, and this *competition of resonances* has never been studied mathematically, even though it seems to be of the greatest importance; it seems possible, a priori, to define an entropy for each resonance K in such a way that it is large when K is embedded in a topologically simple way in $C \times C'$. In the case of fields of local dynamics with metabolic fields A and B this can result in a generalized catastrophe. Let me add further that the final result can depend crucially on the kind of perturbations present at the beginning of the coupling;[2] in particular, some quasi-static forms can, through their infinitesimal vibrations, play a controlling part in the choice (e.g., the loop described in Section 5.4.D). This, I think, is the type of process operating in the control exercised by chromosomal DNA during development (see Section 10.5).

D. Information and probability

Classically, when an event with probability p occurs our information is increased by an amount $k \log p$, where k is a negative constant. I shall now show how to give the formula a simple geometrical interpretation when the event in question can be described by an elementary catastrophe (on an external space of any dimension whatever), provided that the information is identified with a certain topological complexity of the generic morphology resulting from the catastrophe.

As we shall see in Chapter 13, in connection with the spatiotemporal interpretation of syntactical structures, each event is essentially the occurrence of a catastrophe. Now a catastrophe, in the everyday sense of the word, is basically an improbable, nongeneric event. Consider the collision between two automobiles at a crossroad; if the two roads, meeting at right angles at O are taken as axes Oxy, and if (a, a') denotes the position and the velocity of the first car on Ox and (b, b') the second car on Oy, supposing the two cars to be represented by points, there will be a collision if and only if $a'/a = b'/b$. This means that the catastrophe set in the phase space $\mathbf{R}^4$ of initial conditions is a quadratic variety Q of codimension one. Since the cars are not, in fact, points but have a certain size $\frac{1}{2}\epsilon$, we are led to replacing the variety Q by a tubular neighborhood of radius ϵ; hence the probability of collision will be proportional to ϵ (in the unit cube of $\mathbf{R}^4$). More generally, for a dynamical system to enter into catastrophe during the natural course of its evolution, its initial conditions must lie in a certain subvariety Q of codimension q; however, allowing for a certain size of material points, we consider a tubular neighborhood of radius ϵ of this variety whose volume will be proportional to ϵ^q. The more improbable a catastrophe, the greater will be the codimension of the organizing center of the catastrophe and the more complex the resulting generic morphology once the stable generic situation is again achieved.

If we suppose that the catastrophe is an elementary catastrophe arising from an isolated singularity of a potential $V(x)$, it is known that the codimension q of this singularity is given by the *Milnor number* M of the singular point of V (the rank of $\mathbf{R}[[x]]/\text{Ideal } V_x$) minus one. Now the generic perturbed function in the unfolding of this singularity will have $M = q + 1$ critical points (in the complex case), and very probably (although this has not yet been proved, to my knowledge) this is also the maximum possible number of real critical points of a generic perturbation of V. Hence it is reasonable to define the topological complexity of a generically unfolded germ, and hence of the catastrophe, to be this number M; and M is proportional to the logarithm of the probability, with the factor of proportionality $\log \epsilon$ negative since ϵ is small.

To say that two catastrophic situations are locally independent is to say, first, that the two organizing-center varieties cut transversally, and, second, that during the catastrophe each germ unfolds independently of the other, on its own parameter space. Thus the codimensions of the two center varieties add, just as does information. If the two catastrophes are allowed to interact freely, the induced morphology will have $(q_1+1)(q_2+1)$ critical points, and the complexity will behave multiplicatively; I do not know whether there is any concrete realization of this situation.

7.4 ENERGY AND SPATIAL COMPLEXITY

A. The spectrum

Consider a classical dynamical system defined on a domain D of Euclidean space, for example, a vibrating string or membrane. The configuration space is a function space L, for example $L(I, \mathbf{R})$ in the case of a vibrating string. As is well known, the vibration can be decomposed into a sum of harmonics, and so we can consider the set X_p of positions obtained by restricting the movement to the sum of the first p harmonics. This will give a nested family of finite-dimensional manifolds $X_1 \subset X_2 \subset \dots \subset X_p \subset \dots$ in L forming the spectrum of the movement. The existence of such a spectrum is well known for linear self-adjoint systems and probably does not depend on the linear nature of the differential operator and the configuration space, as is often believed, for there is a spectrum in any problem involving a variational principle of extremal action or Lagrangian satisfying suitable local conditions. There is a functional A on this function space L (or on a hypersurface of L defined by a normalization condition like $\int|\psi|^2=1$) which, in the absence of an external potential, reduces to the total curvature of the graph of the map. Next determine the critical points of the functional A and when they are quadratic with a finite number of negative squares in the second variation (the usual case of functionals satisfying Legendre's condition), it is possible to form the gradient trajectories of A ending at these critical points. This defines a packet of nested cells, as in classical Morse theory, defining the spectrum of the dynamical system.

In a situation of this type a critical point of A is never a point of the bifurcation set K of L, for in any neighborhood of a form $f \in K$ there are forms g of different topological types arbitrarily close to f, among which there will, in general, be a most complex and a least complex with different total curvatures. For example, in the graph $y = g(x)$, the point of inflection $y=x^3+1$ can be approximated by $y = x^3 - a^2x + 1$ (with two bumps) and by $x^3 + a^2x + 1$ (without bumps), and manifestly the total curvature of the first approximation is greater than that of the second (there can even be a nonzero first variation A with a normalization condition like $\int|\psi|^2 = 1$). The *stationary elements*, or critical points of A, then correspond to structurally stable forms whose topological complexity increases with A.

B. Sturm-Liouville theory in several dimensions

Little progress has been made with the mathematical problem that consists, in the most classical case, in characterizing the increase in the topological complexity of the eigenvectors of the equation $\Delta u = Eu$ as a function of the eigenvalues E, except in the case of dimension one, for vibrating strings. We know in this case, from Sturm-Liouville theory, that the eigenvalues E are characterized by the number of nodes of their corresponding eigenfunctions and thus by the number of critical points of the eigenfunctions. We canot expect so simple a result for dimension $n > 2$; for example, solving the equation $\Delta u = Eu$ on the square $|x|$

$\leqslant \pi$, $|y| \leqslant \pi$ gives, for eigenvalue $E = -50$, the two eigenvectors $u_1 = \cos 5x \cos 5y$ and $u_2 = \cos 7x \cos y$ with different numbers of critical points (25 for the first and 7 for the second in the first quadrant). We may hope that this type of coincidence is an accidental superficial effect whose importance will disappear asymptotically for large volumes.

Let us suppose, then, the existence of a positive increasing function $q = h(E)$, determined by the geometric characteristics of the system and relating the complexity of the eigenfuction to the value E of the energy, and leave for later its mathematical justification. Suppose now that we are dealing with a free quantum system on a domain U whose wave function ψ satisfies Schrödinger's equation,

$$\frac{d\psi}{dt} = i\,\Delta\psi;$$

then the energy of the system in stationary state ψ_α, $\Delta\psi_\alpha = E_\alpha\psi_\alpha$, can be evaluated as the function h^{-1} of the complexity of the graph of ψ_α. In this sense the energy is defined by the position of ψ_α in $L(U, \mathbf{C})$ with respect to the bifurcation set K; what then is the evolution of the system in general? In the case of a stationary energy state E_α, ψ_α admits the frequency $\nu_\alpha = E_\alpha/\hbar$, and considering Re ψ_α, the real part of ψ_α, we see, that it changes topological type ν_α times each unit of time. This suggests giving as the average value of the energy during time T the quotient of T by the number of times the graph of Re ψ changes topological type during T. Of course these changes of form are not physically real phenomena; they have only a virtual existence and can be detected only by projection, that is, by interaction with another system. Such an interpretation will at any rate be compatible with two facts.

1. The impossibility of measuring the energy of a system at a given instant, as a consequence of the uncertainty principle. To obtain an average value of the density of changes of form, it is necessary to observe the system for a longer time, which, for stationary states, must be not less than the period $T_\alpha = 1/\nu_\alpha = \hbar/E_\alpha$, where $\hbar$ is Planck's constant; the longer the time of observation and the closer the system to a stationary state, the better the approximation will be.

2. The average time complexity (the average time density of change of form) of a product of two weakly coupled quantum systems S_1 and S_2 is the sum of the average complexities of the two factors; this again gives the result that the energy of a coupled system is the sum of the energies of the constituent systems. The spatial complexity of the product graph will generally decrease if the coupling becomes strong, leading to a corresponding decrease in the average time complexity; the visible energy will decrease, and this can be interpreted as the storing of linking energy

associated with the setting up of a *bound state* between S_1 and S_2. If this coupled system, at the end of the collision, evolves toward a decomposable system, this linking energy will be released.

C. Aging of a dynamical system and the evolution of a field toward equilibrium

Consider a field of local dynamics, on a domain W, whose fiber dynamic has an attractor K with Hamiltonian properties (like an Anosov system). This attractor can be assigned a Liouville measure and thus an entropy S, a function of the energy E; this will associate two reduced fields with the given field, one defining the local energy $E(x)$ and the other the local entropy $S(x)$, $x \in W$. The local energy $E(x)$ will tend, by diffusion, to a constant over the domain; the same will be true of the entropy if the field is thermodynamically self-coupled.

Equilibrium cannot be established, however, unless $S(E)$ is such that $\int S(E_0)\, dx$, the final value of the total entropy, is greater than the total initial value of S. When this condition is not satisfied, the field will change its attractor locally, but this either will be scarcely discernible, occurring by almost continuous steps, or will jump suddenly to a disjoint attractor; then a new phase will appear through a generalized catastrophe. This is apparently a paradox: that the evolution of the system toward spatial homogeneity can cause the appearance of what is, in fact, an inhomogeneous situation, namely, a mixture of two phases.

In the case of the almost continuous sliding of an attractor, which we shall describe in detail in Section 12.2, it might happen, for an open system, that $S(E)$ has a maximum for the average value E_0; if this condition is satisfied, comparatively large transfers of local energy can occur without seriously affecting the equilibrium. It is conceivable that this condition plays a part in the continuous flux of energy across a living tissue. The evolution of $S(E)$ toward a curve with a sharp maximum at E_0 is characteristic of *aging* and the loss of competence in embryology.

7.5. FORMAL DYNAMICS

When the configuration space of a dynamical system is a function space, for example, the space $L(X, Y)$ of maps of a manifold X into Y, it is possible to define dynamics associated naturally with this space, because in this case L contains a canonical, closed subspace H, the bifurcation set. Suppose that a dynamic is defined by a potential $V : L \rightarrow \mathbf{R}$ which is constant on L-H and has discontinuities (or at least large variations) on the strata of codimension one of H. Such a dynamic, whose potential V is defined by the topological form of the graph of the map $f : X \longrightarrow Y$, will be called a *formal dynamic*. In most cases V is a monotonically increasing function of the topological complexity of the form of the graph, and thus the simplest topological forms are the attractors. (This is especially the case in linear dynamical systems, such as vibrating strings and membranes.) The appearance of formal dynamics seems specifically characteristic of biological phenomena, but there exists at least a suggestion of formal dynamics in certain phenomena of inert matter; the models that we describe allow us, perhaps, to imagine how formal dynamics (charged with significance) can appear from undifferentiated classical dynamics, and thus partially to fill the irritating gap between energy and information which, thermodynamically, separates the inert from the living world.

A. The origin of formal dynamics

Let us return to a classical oscillating system, such as a vibrating membrane or the free Schrödinger equation on a domain W; the eigenvectors, or eigenfunctions, are found by considering the extrema of the function $I = \int |\text{grad } \psi|^2$ on the hypersurface $\int |\psi|^2 = 1$. Now, as in Chapter 3, it is easy to see geometrically that an ordinary point of the bifurcation set H cannot be an extremum of I, for one can pass from one side to another of any point of a stratum of codimension one of H in such a sense that the topological complexity either increases or decreases, giving nonzero first variations for I. The eigenforms that constitute the spectrum of the dynamical system are therefore structurally stable. Even though the dynamic is not formal in the sense defined above, the system evolves in a way that avoids the bifurcation set H, showing the elementary origin of the formalization.

The most probable origin of this formalization can perhaps be found in certain types of coupling, of which the typical model is that of a process of *percolation*. The original scheme is a mesh of points in $\mathbf{R}^3$ with integer coordinates lying between two planes $z = \pm m$, and one seeks the expected value N_0 such that, for $N > N_0$, there is a connection along the edges between the two planes after adding at random N edges of length one to the graph. If there is a potential difference between the two planes, and the edges are conductors, then, as soon as a connection is established, a current will flow between the two planes, causing a flow of energy in the system. This is the prototype of a situation in which the energy state depends essentially on a topological property of the configuration (namely, a connection between the electrodes). Analogous situations could very probably exist in which the factor releasing the flow of energy would be not a connection, but a nonzero relative homology group of dimension greater than one, or some other topological property.

In any case, we can formalize this kind of situation as follows: there are two coupled dynamics, over an open set W: the first is a gradient dynamic (the *motor* dynamic), defined by giving on the boundary $S = \partial W$ of W a locally constant potential function V, which gives rise to a current in the interior of W defined by a vector field X—the total flux of this current defines the contribution F of energy in W due to this dynamic; the second dynamic plays the part of a *filter* with respect to the first and is defined by a field of Hamiltonian dynamics on W characterized by the competition between attractors c_i, where, if the attractor c_i controls a region A_i of W, c_i has an associated conduction coefficient r_i. Then the field X of the motor dynamic is defined to be the gradient of U, where U is the solution of a Dirichlet problem on W defined by the extremum

$$\delta \sum_i \int r_i \operatorname{grad} U = 0 \qquad \text{and} \qquad U = V \text{ on } S.$$

It is reasonable to suppose further that the motor dynamic acts by interaction on the filter dynamic, and in the absence of more precise hypotheses we can suppose that the interaction favors the states of the filter dynamic for which the entropy formation is minimum, following Le Chatelier's principle. Moreover, it is not unreasonable to interpret the entropy formation as proportional to the total curvature of the field X; the privileged states would then be those which lead to a more direct, less winding circulation of the current of the motor dynamic. This inverse action of the motor dynamic on the filter can also lead to a periodic evolution of the filter; a simple model is given by a helix placed in a fluid flow, which turn with an angular velocity proportional to the velocity of the flow. For a biological example, it is possible that living matter consists essentially of two types of molecules, large molecules (nucleic acids, protein) and small and average-sized molecules (water, amino acids, nucleotides), and that, to a first approximation, the local configurations of the large molecules play the part of the filter dynamic with respect to the biochemical kinetic, which affects the small molecules that play the part of the motor dynamic. This is reasonable because the proteins have an enzymatic effect which can considerably alter (by a factor of the order of ten) the speed of a reaction among small molecules. In this way, the duplication of a living cell can be considered as a periodic phenomenon affecting the large molecules and induced by the reaction of the motor dynamic, that is, the metabolism of the small molecules.

Finally the reaction on the filter dynamic can lead to a genuine spatial circulation of the areas affected by each of the attractors c_i; in anthropomorphical terms, we say that the energy circulation of the motor dynamic coincides with a *circulation of information*, which governs it. Nothing prevents us from iterating this construction, with this circulation controlled by a new dynamic filtering it, giving rise to second-degree information, and so on. We shall discuss these mechanisms in more detail with respect to biological finality and the relationships between genetic and thermodynamic regulation. Let me show now how certain phenomena are qualitatively possible.

B. Phenomena of memory and facilitation

Suppose that the Hamiltonian filter dynamic has the following topological form on the energy interval $a \leqslant E \leqslant b$ (see Figure 7.1): for $E = a$, two

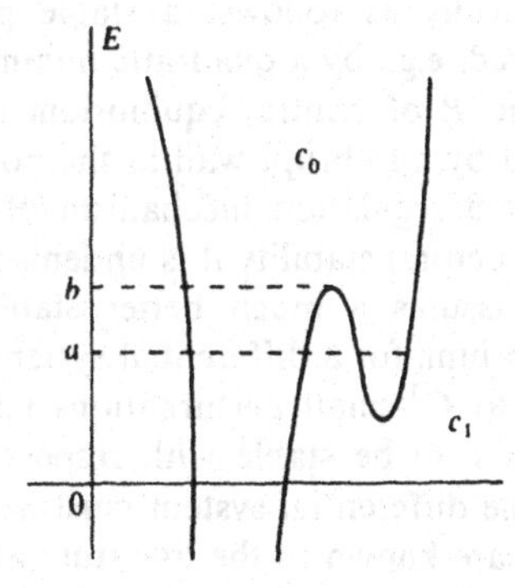

FIGURE 7.1.

disjoint attractors, one, c_0, representing a large disordered phase (gas) and the other, c_1, of small Liouville measure, representing an ordered state (crystal); as E tends to b, the attractor c_1 is absorbed by c_0, corresponding to melting of the crystalline state. Next suppose that the motor dynamic has a singularity s in a subregion U of the space, characterized by a large local variation of the potential V, and that the presence of phase c_1 in U has the enzymatic effect of decreasing the production of entropy in the neighborhood of s. Then it is likely that phase c_1 will appear in U, for it will be, by reaction of the motor dynamic, an attractor in the strict sense. A small area of phase c_1 will then appear in U, first excited to an energy E close to b. If, later, the external conditions of the ambient metabolism vary and the singularity s of the motor dynamic disappear, c_1 will cease to be attractive and the local energy E will decrease; it is then possible that the local area of c_1 will remain, isolated in the large phase c_0, and this fragment of phase c_1 will become metabolically inert, appearing as a relic of the metabolic singularity s. If, much later, conditions in the metabolism (the motor dynamic) appear close to s, the fragment of phase c_1 can play the part of a germ and can facilitate, by local catalytic action, a return of s in U. If this return is effective, c_1, becoming attractive again, can grow. Thus we can conceive that certain metabolic regimes can have a kind of malignity; once realized, the difficulty of their reappearance decreases as they manifest themselves more often. The phase c_1 will then appear as a system of inducing particles of this malign transformation.

C. Canalization of equilibrium

Another phenomenon occurs frequently in living dynamics and can be described geometrically as follows: a stable point of equilibrium of a system (characterized, e.g., by a quadratic minimum of potential) is finally replaced by a zone P of neutral equilibrium (where the potential V is constant), bounded by an abrupt wall in the potential, this reflecting wall being the result of a regulation mechanism of genetic origin. From the point of view of structural stability it is undeniable that an arrangement of this second type assures a much better stability than the usual thermodynamic equilibrium, for a differential system can be structurally stable only with respect to C^1-small perturbations (small, and with small first derivatives), whereas, to be stable with respect to perturbations that are only continuous, the differential system must necessarily have discontinuities (these systems are known in the literature as *déferlants*).

The formation, during evolution, of areas of neutral equilibrium bounded by potential walls seems to be an effect arising from the coupling between a filter dynamic of genetic character and a motor dynamic (the usual metabolism): suppose that the metabolism has an attractive cycle z whose stability is assured by the mechanisms of biochemical kinetics. Under the effect of inevitable random perturbations of relatively small amplitude, the whole of a tubular neighborhood T of z will, in fact, be described by the system. The filter dynamic itself will be deformed and will give a facilitation effect throughout T; but, since this effect will not occur outside T, it will lead to the formation of the reflecting wall around T. This can be seen by putting a heavy ball on a sandy basin and shaking the system randomly; thereupon, a relatively flat area will surround the ball, bounded by a steep wall (Figure 7.2). Very probably an analogous mechanism, not of physiological regulation but localized spatially, can account for organogenesis in the course of evolution. This mechanism shows that in many cases the appearance of catastrophes (discontinuities) is the price that must be paid to ensure the best possible stability of a process.

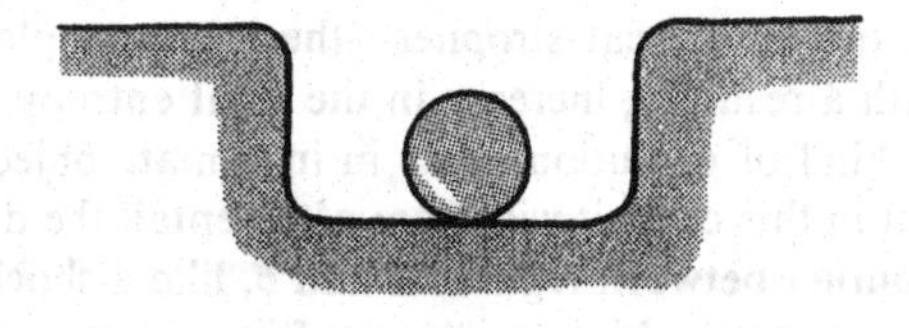

FIGURE 7.2.

D. Threshold stabilization

The phenomenon of *threshold stabilization* is one of the typical features of biological morphogenesis; singularities that should never have appeared, because their codimensions are too large, appear in a stable way. Here is a possible formal mechanism to account for this behavior. A dynamic (M, X) undergoes bifurcation, in the sense of Hopf; this has the effect of increasing the dimension of the attractor (what we call a *competence gain*, a silent catastrophe in the sense of Chapter 11). Let (P, Y) be the new fiber dynamic thus created above the attractor c of $(M\ X)$, where, for example, P is a k-dimensional torus. Then singularities of codimension $\leqslant 4 + k$ can occur structurally stably on the product $M \times P$; if such a singularity projects to $p \in P$, the system must almost always be sent into a neighborhood of p. If the coordinates of p have biological significance, this will mean that there are homeostatic mechanisms directing the system toward p. It is conceivable that this mechanism occurs because of a mean coupling field between the slow dynamic (P, Y) and the rapid dynamic (M, X), and this will then control not only the formation of new reaction cycles (P, Y) but also their resonances with the already existing dynamic (M, X).

E. Threshold stabilization and the theory of games

This phenomenon of threshold stabilization is somewhat analogous to a classical theorem of the theory of games, von Neumann's minimax theorem: when two stable regimes a and b, separated by a threshold s, are in competition on a domain U, the threshold stabilization will be characterized by the evolution of almost all of U toward s, with only a fragment of the domain remaining free to oscillate between a and b. The process will behave as if two players a and b compete against each other at each point of U, each adopting the common strategy minimizing their losses. Anthropomorphically, it is as if *all conflict evolves so as to minimize the damage that results*, for after all we can interpret the damage, in dynamical terms, as the total density of local catastrophes of the domain. Such an evolution will be dominated by an overall evolution of the forms to those having the fewest catastrophes—the least complex and thus the most stable—with a resulting increase in the local entropy.

Probably this kind of evolution exists in inanimate objects as well as in living beings, but in this case[3] it will more often entail the disappearance of its generating conflict between regimes *a* and *b*, like a shock wave damped by friction. Conversely, in biology, the conflict is imposed by the global regulation of the living organism and the periodic character of its reproduction; this conflict cannot disappear, and the organism will evolve genetically, through heredity, toward the most stable situation compatible with maintaining the conflict, that is, the regulating function.

F. Other formal aspects of a coupling; coding

Given two Hamiltonian systems whose configuration spaces are function spaces L_1 and L_2 (e.g.,, vibrating strings), suppose that they are coupled by an interaction potential V_{12}; they will be formally coupled if this potential has discontinuities, or at least very large variations, on certain strata of the product of the bifurcation hypersurfaces H_1 and H_2 of L_1 and L_2, respectively. If, for example, V_{12} has a minimum on certain connected components of $L_1 \times L_2 - H$ of the form (g_1, g_2), where g_1 and g_2 are eigenforms, as soon as system L_1 is in state g_1 system L_2 will have a large probability of being in state g_2; the product system will behave as if there is a *code* associating stratum g_2 of the second system with g_1 of the first. Of course it is not necessary that the interaction potential be strictly formal for such a coding effect to occur; there will generally be privileged associations, often in competition, and the coding will happen only with a certain noise and a greater or less probability of error. Consider a simple case: let L_1 and L_2 be two identical oscillators with an attractive interaction potential of the form $V = k(x_1 - s_2)^2$ (or any analogous even function); then V is minimum on the diagonal of $L_1 \times L_2$—that is to say, if L_1 is in eigenstate g_1, L_2 will find itself attracted toward the same eigenstate, and the form g_1 will resonate and extend spatially.

Resonance also explains why changes of phase in a medium are accompanied by discontinuities in the properties of the medium. The correlations and periodicities of a crystalline condensed phase generate effects by resonance in every coupled dynamical field on this space, and these effects disappear on melting during the change to the disordered gaseous state.

7.6. FORM AND INFORMATION

It is sometimes said that all information is a message, that is to say, a finite sequence of letters taken from an alphabet, but this is only one of the possible aspects of information; any geometric form whatsoever can be the carrier of information, and in the set of geometric forms carrying information of the same type the topological complexity of the form is the quantitative scalar measure of the information. If we are used to imparting information as a sequence of letters, the reason is, I believe, above all a need to accomplish technically the transmission of this information. The easiest method (and perhaps, in the last analysis, the only method) of spatially duplicating a form is to use the mechanical phenomenon of *resonance*: if the source S and receptor R are two identical, symmetrically coupled oscillators, and S is excited in eigenstate a, then R will be also excited in the same eigenstate; after waiting for the effects of a to die away in R, eigenstate b can be transmitted, and so on. Thus the one-dimensional oriented character of a message is a reflection of the one-dimensionality and irreversibility of time.

Another possible type of propagation would be by contact or diffusion, by putting a polarized local field of dynamics into contact with an undifferentiated dynamic and producing a spatial diffusion of the eigenform associated with the first dynamic. This process seems to allow very limited spatial extension, however, for as soon as a dynamical regime extends spatially, it reacts on the eigenform of the regime, whose topological characteristics change (at least this seems to be the case for dynamics founded on chemical kinetics, as in living things); the spatial diffusion perturbs the geometric clarity of the eigenform and spoils the transmission of information. If, in biology, the nucleic acids with their linear structure seem to be the almost exclusive carriers of the genetic endowment, perhaps the reason is that this static form is indispensable for carrying, for (relatively) large distances, the information normally present in metabolic structures (attractors of biochemical kinetics), whose topological complexity permits only a very limited spatial diffusion. As in the swinging of a pendulum, the energy oscillates between kinematic (at the bottom) and potential (at the top), so in the periodic process of the division of a cell the complexity oscillates between a metabolic form (in interphase) and a static form (in metaphase), although here the transition is discontinuous, by catastrophes.

To conclude it must not be thought that a linear structure is necessary for storing or transmitting information (or, more precisely, significance); it is possible that a language, a semantic model, consisting of topological forms could have considerable advantages, from the point of view of deduction, over the linear language that we use, although this idea is unfamiliar to us. Topological forms lend themselves to a much richer range of combinations, using topological products, composition, and so on, than the mere juxtaposition of two linear sequences. We shall return to this topic in Chapter 13.

APPENDIX 1

Conservation of energy and the first law of thermodynamics. Let E be an isolated dynamical system (i.e., a system having no interaction with the outside world), whose dynamic is invariant under the transformations $t \longrightarrow t_0 + t$, $t \longrightarrow t_0 - t$; the first law of thermodynamics asserts that the time-average complexity of the representative graph, and the spatial complexity, do not change with time. Thus stated, the law would have little interest if there were no means of comparison of the complexities of disjoint systems. Observe that the associated complexity of two disjoint or weakly coupled systems can be taken to be the sum of the complexities of the two components. Now there exist, as a matter of experience, very important dynamical systems for which the only relevant energy is the kinetic energy (e.g., solid bodies), and for such systems complexity can be considered as a given function $q = h(E_k)$ of the kinetic energy E_k. Let S be a dynamical system and suppose that, by interaction of S with a solid body B, the system can be taken from state a to state b. At time t_0 the system $S+B$ is weakly coupled with total complexity $q(a)+h[E_k(B);t_0]$, and after interaction (which can include strong coupling), at time t_1, the total system will be weakly coupled with complexity $q(b)+h[E_k(B);\ t_1]$. Equating these two values gives

$$q(b) - q(a) = h[E_k(B);\ t_0] - h[E_k(B);\ t_1].$$

If h is linear and independent of B, we see that $E(B)(t_0) - E(B)(t_1)$ depends only on states a and b of S; it is this real number that is used to define the difference of energy between the states, and thus the kinetic energy is used as a universal standard of the variations of internal energy of a system. Hence, for the first law to be of any use, it is necessary that almost all coupling between a dynamical system and a material body evolves toward a decomposed or weakly coupled state; in the classical system of gas + piston, so well known in thermodynamics, it is necessary to suppose that the system never stops being decomposable (as would happen, e.g., if the gas attacked the piston chemically). This extraordinary stability of material bodies must be accepted as a fact of experience; without it the first law would be almost void of significance, and it would be impossible to give a scalar measure of energy.

It seems, a priori, to be most improbable that two coupled systems should return again to a situation of weak coupling after a strong interaction; the decomposed maps form an infinite-codimensional subspace in a function space of type $L(X \times A,\ Y \times B)$, and the probability that a system should evolve toward such a state must, a priori, be zero. Nevertheless, in quantum mechanics it seems accepted that many collisions between particles (corresponding to an open set of initial conditions) evolve toward a decomposed state. It does not seem possible to me to explain this most mysterious fact by a Hamiltonian reversible in time, unless such a Hamiltonian is far more rigid and insensitive to external perturbations than is commonly supposed; perhaps the representation of a composed state, like the tensor product of the constituent particles, is too redundant, containing unnecessary parameters. In other words, the pseudogroup left invariant by the Hamiltonian will be much larger than the classical permutation group imposed by the principle of indiscernibility of the particles.

APPENDIX 2

Topological complexity of a dynamic. We have seen the great difficulties encountered in the definition of the topological complexity of a function or a map; these difficulties are even greater when defining the topological complexity of a dynamical system on a configuration space M. There are only two cases in which a manageable definition may be possible.

1. When the dynamic X on M is a gradient field, $X = \text{grad } V$, where V is a real potential $V: M \to \mathbf{R}$. In such a case it is natural to define the complexity of the dynamic X as that of the real function V.

2. When M is a symplectic manifold and X a Hamiltonian dynamic on M; $X = i \cdot \text{grad } H$, H a real Hamiltonian $H: M \to \mathbf{R}$. In this case we can associate a discrete set $J(E)$, whose points correspond to ergodic sets of the dynamic X on the hypersurface $H = E$, with each value E of the first integral H. This gives a space $\hat{E}$ over $\mathbf{R}$ which can be assigned a topology (in general, non-Hausdorff) indicating which sets take the place of an ergodic set when it disappears (more precisely, this topology is that inherited from the quotient space M/X identifying trajectories with the same sets and limits); the complexity of this space $\hat{E}$ (a kind of non-Hausdorff graph) can measure the complexity of the Hamiltonian dynamic H. In many cases the hypersurface $H = E$ has a large component of ergodicity containing almost all the manifold (and corresponding to the gaseous phase) and a small number of invariant sets of small (Liouville) measure, satellites of closed trajectories of central type, or, more generally, invariant manifolds of central type (corresponding, in general, to the condensed solid or crystal phases). The graph $\hat{E} \to E$ and the functions $S_i(E)$ (the entropies associated to each of the points of $\hat{E}$ above the given value of E) characterize entirely the thermodynamic properties of the dynamic under consideration, together with those of the associated pseudogroups G_i of invariance.

APPENDIX 3

Infinite complexity of geometrical forms. A strange ambiguity is connected with the idea of complexity. From a naive point of view, the simplest of all real-valued functions on U is the constant function, whereas here it has an infinite complexity because it can be approximated arbitrarily closely by functions whose graphs have an arbitrarily large topological complexity. (This is reminiscent of the famous paradox in quantum field theory, where the *energy density of the vacuum is infinite.*) The reason for this phenomenon is as follows. In the set of all the forms of maps from X to Y the identifiable forms consist only of structurally stable or almost stable forms; they include also geometric forms, given by direct definition, for example, a map defined by the composition $X \rightarrow Z \rightarrow Y$, where $\dim Z < \dim Y$ and the two constituent maps are the base forms of $L(X,Z)$ and $L(Z,Y)$, respectively. Although these maps are well defined, they have infinite topological complexity and belong to strata of infinite codimension in the bifurcation set $H \subset L(X,Y)$, where they are limits of the nonform forms described in Chapter 1. If the representative graphs of the reduced fields (Chapter 3) tend in general toward geometric forms, and not to nonforms, in the course of thermodynamic evolution toward equilibrium, this must be considered as an expression of the homogeneity of the space; the dynamic generally admits a group or pseudogroup, usually transitive, of important symmetries, this generally being a residue of the relativistic or Galilean invariance (whichever is relevant) of elementary interactions. Now there is the following obvious result.

The symmetry principle. If a system S is maintained in interaction with an ambient constant thermostat system T, and if the initial conditions and all interactions are invariant under a pseudogroup G, the final equilibrium state, *if it is unique*, is also invariant under G.

As elementary interactions generally satisfy Galilean or relativistic invariance, defining a transitive invariant pseudogroup, the equilibrium state, if it is unique, is frequently constant or at least constant on the orbits of the group G of automorphisms of the system. This makes the reduced graph of the equilibrium state have a geometric form of infinite complexity.

Now the thermodynamical limit has no reason to be unique in such a situation. For example, when a spherical balloon filled with water vapor is cooled below saturation point and without any external gradient (of temperature or gravity), the liquid phase will start by condensing on the walls, but the final division between the liquid and gaseous phases will be indeterminate and will almost never have spherical symmetry. The symmetries of the initial conditions will be broken in a process of this type, a case of a generalized catastrophe described in Chapter 7. It may happen that the regime of local interactions of the pseudogroup G stops being stable and gives way to a new regime with pseudogroup G', although the system S tends to its equilibrium state defined by a G-invariant graph; a new phase with pseudogroup G' will appear out of a topologically very complicated situation. This will again be a generalized catastrophe. Such oscillations between possible limit regimes are found frequently in nonisolated systems (called "open" in thermodynamics) and particularly in living beings.

The space of forms of a function space. For any two structurally stable forms f and g in a function space $L(X, Y)$ we defined, in Section 7.2, the relative topological complexity $d(f,g)$, which gives a distance on the discrete set of structurally stable forms; it might be possible to embed this discrete space isometrically in a metric manifold W. We call such a manifold a *space of forms* of $L(X, Y)$, and it will take the part of the dual space in the theory of Fourier transforms. If a fluctuating dynamical process is defined in $L(X, Y)$, we can associate with each structurally stable form f the probability that it will appear in a unit of time, giving a function $g: W \rightarrow \mathbf{R}^+$ such that $\int_w g = 1$. The complexity of this function g gives a measure of the entropy of the system; neglecting the phenomenon of resonance, the system will develop as if g tends, through diffusion, to a constant on W as in a fluid. However, the oscillator with eigenform f tends perferentially to exchange energy through resonance with certain other forms, the so-called multiples of f in W, and these forms are generally far from f in W; therefore the behavior of g cannot be compared with the continuous flow of a fluid because there can be an instantaneous transport of the complexity g of a point f to a point f' far from f. We might try to modify the topology of W to take account of this phenomenon, but this would probably no longer give a Hausdorff space and, furthermore, the relation "f' a multiple of f" is not symmetric.

I would not have carried these very hypothetical considerations so far if they did not give a good representation of the behavior of nervous activity in the nerve centers, where an excitation (called a stimulus in physiology) remains relatively canalized until it results in a well-defined motor reflex; here the role of diffusion seems to be strictly controlled, if not absent. It is known (the *Tonusthal* theorem of Uexküll) that, when the associated first reflex of a stimulus is inhibited by artificially preventing the movement, there is a second reflex which, if inhibited, leads to a third reflex, and so forth. This seems to suggest that diffusion of the excitation is, in fact, present, but that, as soon as the excitation finds an exit in an effective reflex, all the excitation will be absorbed in the execution of this reflex. This gives a curious analogy with the mysterious phenomenon of the *reduction of a wave packet* in wave mechanics.

NOTES

1. This phenomenon already occurs in the case of functions with four critical points on the product $S_2 \times S_2$ of two spheres. I owe this remark to D. B. A. Epstein.

2. The models of reproduction in jointed pieces by L. S. Penrose [*Ann. Human Genetics*, 23 (1968), 59–72] illustrate this phenomenon; a kind of nesting of two linked pieces A and B in a complex (AB) is interpreted as a degeneration of the phase space $D(A) \times D(B) \longrightarrow D(AB)$, where D denotes the Euclidean group, and so as a catabolic catastrophe or resonance. Since there are two ways of linking the pieces, two resonances are in competition. The final state of the multiple system of isolated pieces A and B thus depends essentially on the initial conditions, and hence on the first nested complex put into the system.

3. The phenomenon of threshold stabilization is also found in geology, where, for example, the vault of a Jurassic anticline wears away under erosion and gives rise to a comb pattern eroded in the softer underlying material. Formally this is similar to the duplication in a shock wave through the formation of a stable transitional regime, described in Chapter 10. Similarly it often happens that a threshold, or a col, is eroded away locally by sheets of water, leading to a permanent formation; an example is Lake Chexbres, to the northeast of Lausanne, which flows both toward the Aar and the Rhine to the north, and to Lake Léman and the Rhône to the south. See Figure 7.3.

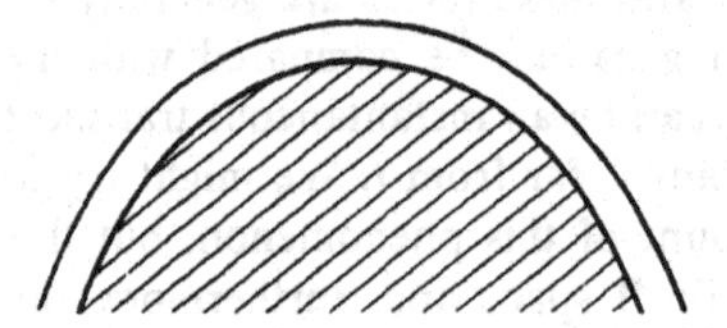
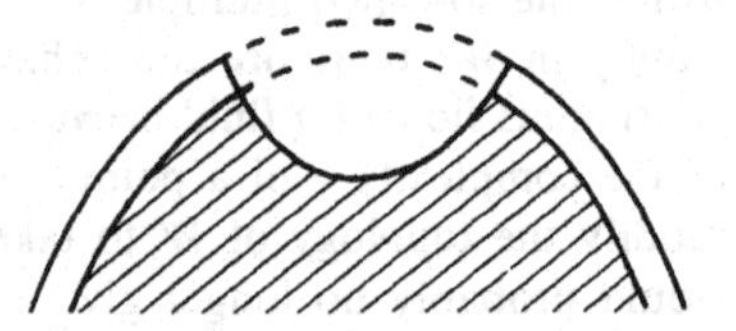

FIGURE 7.3.

CHAPTER 8

BIOLOGY AND TOPOLOGY

Prendre possession de l'espace est le geste premier des vivants, des hommes et des bêtes, des plantes et des nuages, manifestation fondamentale d'équilibre et de durée. La preuve première d'existence, c'est d'occuper l'espace.

LE CORBUSIER
(taken from F. Le Lionnais, *Les grands courants de la pensée mathématique*)

8.1. THE TOPOLOGICAL ASPECT OF BIOLOGICAL MORPHOGENESIS

A theory of morphogenesis must, of course, be applicable to biological morphogenesis; in fact, I was led to develop the qualitative dynamics presented here through reading works on embryology (particularly the books of C. H. Waddington), and my aim was to give mathematical sense to the concepts of the *morphogenetic field* of embryologists and the *chreod* of Waddington. I know very well how these concepts are at present denigrated by biologists, who criticize them for giving no chemical explanation of epigenesis. I think, however, that from an epistemological point of view an exclusively geometrical attack on the problem of morphogenesis is not only defensible but perhaps even necessary. To declare that a living being is a global structure is merely to state an obvious fact and is not to adopt a vitalist philosophy; what is inadmissible and redolent of vitalist metaphysics is to explain local phenomena by the global structure. Therefore the biologist must, from the beginning, postulate the existence of a local determinism to account for all partial microphenomena within the living being, and then attempt to integrate all the local determinisms into a coherent, stable global structure. *From this point of view the fundamental problem of biology is a topological one, for topology is precisely the mathematical discipline dealing with the passage from the local to the global.*

Pushing this thesis to its extreme, we might look upon all living phenomena as manifestations of a geometric object, the *life field* (*champ vital*), similar to the gravitational or electromagnetic field; living beings would then be particles or structurally stable singularities of this field, and the phenomena of symbiosis, of predation, of parasitism, of sexuality, and so forth would be the interactions and couplings between these particles. The first task is, then, the geometrical description of this field, the determination of its formal properties and its laws of evolution, while the question of the ultimate nature of the field—whether it can be explained in terms of known fields of inert matter—is really a metaphysical one. Physics has made no progress in the understanding of the ultimate nature of the gravitational field since Newton; why demand, a priori, that the biologist should be more fortunate than his colleagues, the physicist and the chemist, and arrive at an ultimate explanation of living phenomena when an equivalent ambition in the study of inert matter has been abandoned for centuries?

Of course all the following models that I propose in biological morphogenesis do not require the introduction of this hypothetical life field, but use only static or metabolic fields, and so can be formally interpreted in terms of biochemical kinematics. I shall not attempt these interpretations, because my knowledge of biochemistry is very rudimentary, and, furthermore, the extreme complexity of the reactions involved in the metabolism of living matter probably would make this task extremely difficult. The essential point of view advanced here is this: the stability of every living being, as of every structurally stable form, rests, in the last analysis, on a formal structure—in fact, a geometrical object—whose biochemical realization is the living being. Here we are interested only in explaining this geometrical structure, and we only offer models of it. The realization of these models as configurations of biochemical kinematics (if this is possible) belongs to the future.

8.2. FORM IN BIOLOGY; THE IDEA OF A PHENOTYPE

A. The spatial form

The fundamental idea of topology is that of homeomorphism: two spaces A and B have the same form, or are *homeomorphic*, if there is a bijective, bicontinuous map of A onto B. Stated thus, the idea is not without interest in biology, for consider two animals of the same sex and the same species; there will then, roughly, exist a homemorphism of one animal onto the other that is compatible with the histological specializations of cells, and will map eye on eye, tooth on tooth. Observe that every animal is topologically a three-dimensional ball, the only possible exceptions arising with primitive animals living in colonies; is there an a priori reason for this topological simplicity in living beings? One obvious consideration is the necessity of spatial duplication, which is much easier to realize with a ball than with a more complicated figure, such as a solid torus.

The situation is altogether different for plants, where we can speak of homeomorphism only between organs taken in isolation, such as leaves, stems, or roots, whereas, in principle, no global isomorphism exists between organisms. Even in the animal kingdom the idea of homeomorphism is not what biologists really want, for it is both too fine and too coarse: too fine because a homeomorphism between two animals of the same species cannot take into account the finest details of their anatomy (e.g., two cats will not necessarily have the same number of hairs), and too coarse because two animals of neighboring species can be homeomorphic to as good an approximation as two animals of the same species (recall, in this connection, the classical diagrams of D'Arcy Thompson identifying fishes of different species). It is necessary to replace the notion of homeomorphism by that of G-equivalence in order to define the spatial aspect of the relationship of belonging to the same species, where G is a group of equivalence associated, in practice, with each biological form or phenotype. A formal definition of this group G is not at all easy; G must approximately preserve the metric (between animals of the same age), the vertical, and perhaps also a finite set of horizontal planes. But probably each phenotype has its own equivalence group, and it is likely that these groups have little in common.

We have so far used only the spatial structure of an organism to define the idea of its phenotype, but the word is often used in a wider sense; for example, we could consider that certain ecological or physiological characteristics of an animal form part of the phenotype of the species. In bacteria, enzymatic behavior and resistance to antibiotics are important components of the phenotype (perhaps more from considerations of experimental convenience than for intrinsic reasons), and it is important to realize that the definition of phenotype as a class of forms modulo an equivalence group G retains its validity even in this extended sense. For example, a given bacterium placed in a specific environment containing streptomycin at concentration c will give birth to a colony whose development will be expressed by a function $N = F(c, t)$, and the dependence on (or resistance to) streptomycin is given by the topological type of the function $F(c, t)$. Abstractly, the phenotype can be defined as the set of responses of the organism to all possible perturbations of its environment and the qualitative classification of these responses, and hence it is practically impossible to give a formal definition of phenotype. The subject called formal genetics rests wholly on the concept of the normal or wild phenotype, which is almost impossible to formalize. But such a situation occurs frequently, even in more exact disciplines, and a mathematician cannot take exception to it.

B. The global form

We must note, however, that the spatial structure of an individual is only one aspect of the global biological structure of the species considered. We can obtain a much more essential form by considering the geometric figure, in space-time, obtained by starting with an individual X at age a and finishing after a time T with the formation of descendants X_j, isomorphic to X and aged a; the spatial form of an individual is then only a section F_t of this global figure F by a hyperplane t = constant. Even then, F would far from constitute the whole of the geometric structure associated with the species, for there should be associated with it all the stabilizing mechanisms assuring the multiplication of the species under approximately normal conditions. In one of the global models that I propose, the set of living local states is parameterized by a space U (a finite or infinite-dimensional function space); the figure F is embedded in U, but not canonically, encapsulated in some sense in a tubular neighborhood with reflecting walls R_j, and it is the collection of reflecting walls (which describe the mechanisms of epigenesis, regulation, homeostasis, and even of morphogenetic regeneration) that forms, strictly speaking, the total geometric structure associated with the species. In the end the spatial form F is only a kind of necessary realization, a compulsory pathway of a growth wave which develops in a space U of many more dimensions than space-time $\mathbf{R}^4$. We return in more detail to this model in Chapter 10.

8.3. MOLECULAR BIOLOGY AND MORPHOGENESIS

A. The inadequacy of biochemistry

We should not let the recent spectacular progress in molecular biology conceal the fact that the problem of the stability of the spatial structure of living beings (which can, with reason, be considered as the essential problem of biology) has not yet been dealt with. Of course it is important to know how a certain protein forms—from what constituents, and following which cycle of reactions; but it is even more important to know *where* and *when* this protein will form and why it forms at one part of a cell while not forming in another part of the same cell. This whole geometrical and spatial aspect of biochemical reactions eludes the power of biochemical explanation; usually the realization of an enzymatic reaction *in vitro* is considered a great success if it has so far been thought to be specific to living matter, when it should rather be deplored because, when all is said and done, an animal will never be a test tube.

It is almost certain that the spatial organization already appears in the macromolecular state, as the relatively rigid organization of certain organelles (chromosomes, centrosomes, mitochondria, etc.) testifies. In this respect biology may perhaps labor under the same delusion as physics: the belief that the interaction of a small number of elementary particles embraces and explains all macroscopic phenomena, when, in fact, the finer the investigation is, the more complicated the events are, leading eventually to a new world to be explained in which one cannot discern among the enormous set of new phenomena the relevant factors for macroscopic order. It seems difficult to deny that statistical simplifications must appear in the epigenesis of a large animal, whereas a virion is a crystalline form, bound to a precise, discrete geometrical configuration; we can say that a virion is simpler than a mammal, but all mathematicians know that simple (or irreducible) structures (e.g., "simple" groups) can actually be very complicated, and the determination and classification of all possible simple structures constitutes, in general, a problem of immense difficulty. Moreover, even if the geometric structure of a virus (in the virion state) can be considered to be simple, its biological significance (i.e., its interaction with its host cell) certainly is not, as can be seen by considering the reproductive cycle of a virus (in the nonlysogenic state). No formal proof of the morphological continuity between a viral particle infecting a cell and its descendant found in the lysate cell exists, to my knowledge; even conceding the possibility of this continuity, it cannot be denied, from considerations of genetical recombination, that it implies an extremely complicated molecular intermingling. We are far from the topological simplicity of the reproduction by eggs of Metazoa.

Of course I am not saying that present researches in molecular biology are useless; on the contrary they are indispensable, for they alone allow us to reconstruct the dynamic underlying the biochemical morphology that they describe. Our ignorance of the mechanisms controlling the macromolecular associations is too great for us to despise experiments and their results; who could have foretold the constitution and behavior of chromosomes, to mention only the leading puzzle of contemporary biological dynamics? But I want to underline the following point of view: since the discovery of Mendel's laws and the progress in the analysis of macromolecules, there has been a tendency to underestimate the dynamical and continuous nature of living phenomena and to overestimate the importance of chromosomes as director-elements of all living metabolism;

might we not consider chromosomes as organelles that are similar in all respects to other organelles and whose ambient metabolism ensures, at the same time, stability, duplication, and eventual variation?

There is a well-known difficulty in the general program proposed here (to interpret dynamically the form of cellular ultrastructures), in that we are completely ignorant of the nature of the forces assuring the operation of cellular organelles and causing mitosis, meiosis, crossing over, and so forth. This should not surprise us; considering the difficulties of a mathematical study of a condensed structure as simple as that of the Ising model in the theory of solidification, it is not surprising that we do not know how to estimate the effect of the presence of a macromolecular structure on the ambient medium of living cytoplasm and the effect, by reaction, of this medium on the structure. This is why we propose to use the opposite approach: in place of explaining the overall morphogenesis by modifications of the cellular ultrastructure, we try to explain the cellular ultrastructure by dynamical schemes similar to those of global morphogenesis, but on the cellular level. Certainly this program cannot be attempted without great arbitrariness in the dynamical interpretation of cellular organelles—this is a danger inevitable in any new attempt.

B. Morphology and biochemistry

At this point we might ask whether, instead of explaining morphogenesis by biochemistry, we should not be taking the opposite viewpoint; the problem of finding the tertiary structure of a protein as a function of its primary structure is a typical problem in the dynamic of forms, in the sense of Chapter 7, and is concerned with finding the minima of the function U, the free energy, on the function space L of all configurations of the molecule. No one would deny that the bifurcation set S of L, representing the set of positions where the molecule crosses itself, acts as a singularity for U, and that the relative topology of the basins of the different minima of U have an essential part in the enzymatic, or motor, activity of the molecule. Recent studies of several enzymes (e.g., lysozyme) have shown the obviously morphological aspect of their enzymatic reactions; we can see the molecules fingering, pinching, twisting, and tearing themselves apart, almost like living beings. We should not be surprised. Insofar as a biochemical reaction reflects a local incident in the spatial competition between two different regimes, the overall topological constraint of structural stability requires these local events to simulate the global catastrophes of morphogenesis on space-time $\mathbf{R}^4$. In many of these cases, these molecular mechanisms arose because they provided the best local simulations of the global biochemical catastrophes that gave birth to them. The fact that some of these mechanisms persist *in vitro* should not hide their biological origin and significance; they have a powerful internal stability with respect to their environment, but they lose the polarized, oriented character that they possess in a living environment when they occur in the test tube.

In this connection, it is striking to note that the genetic reactions affecting the duplication of DNA (which some want to make the very key to life) are relatively coarse and resistant, whereas the fundamental reactions of metabolism, taking place in the plastids and mitochondria, depend on a morphology that so far cannot be reproduced in the laboratory.

8.4. INFORMATION IN BIOLOGY

I discussed in Chapter 7 the abuse of the word "information" in biology; it is to be feared that biologists have attributed an unjustified importance to the mathematical theory of information. This theory treats only an essentially technical problem: how to transmit a given message in an optimal manner from a source to a receiver by a channel of given characteristics. The biological problem is much more difficult: how to understand the message coded in a chain of four letters of DNA (the usual description of which, it seems to me, needs some improvement). Now, all information is first a form, and the meaning of a message is a topological relation between the form of the message and the eigenforms of the receptor (the forms that can provoke an excitation of the receptor); to reduce the information to its scalar measure (evaluated in bits) is to reduce the form to its topological complexity (in the sense of Chapter 7), and to throw away almost all of its significance. For example, it is too simple to say that oogenesis is an operation of transmitting information from a parent to its descendants, for the transmitting channel (the egg) and the receptor (the embryo emerging from the egg) are identical.

It has often been pointed out that theoretical biology has never rid itself of anthropomorphism, its most deep-seated flaw; ever since Descartes' theory of animal-machines, living beings have been compared with man-made machines. The present version of this deplorable tendency is to compare a living being with an electronic computer provided with means of control and autoregulation. This point of view is not completely false (we return to it in Section 12.1, Finality in Biology). The Cartesian imagery is perhaps more true and more enigmatic than the cybernetic comparisons of today; it is beyond question, for example, that the heart works as a pump and the lungs as bellows, and at the level of the ultrastructure the shell of the bacteriophage functions as a syringe. There has never been a satisfactory explanation of this extraordinary convergence, observed by Bergson, between products of technology and biological organs (are they both optimal solutions of the same technical problem [1]?). The present language of biochemistry is peppered with anthropomorphism: "information," "copy of information," "message," and so forth. These concepts have been (and probably will remain) very useful for developing an intuitive idea of the way in which living systems work, but we must sometimes attempt a dynamical justification of life, even at the expense of some abstraction; then information will be no more than the geometric parameterization of a stationary regime of local dynamics, the copy of this information a spatial extension of this regime, and the message a preferential subordination between two local regimes. What sense should we give to the message carried by the DNA of chromosomes? Here I must reply by analogy: in the geometry of differential systems we know that the form of a structurally stable dynamical system (like a gradient field) is almost completely determined by the singular points of the field (the equilibrium points of the system, where the field is zero). Now let us consider a cell in the process of division as a dynamical system with structurally stable duplication. There are *spectral elements* or singularities in this duplication, consisting of the set of points where the spatial duplication starts; these correspond to the chromosomes and the molecules of nucleic acid. Any sufficiently strong modification of the dynamic of self-reproduction can be interpreted as a change of structure (geometrical or chemical) of these spectral elements. This is the sense—and, I believe, the only sense—in which DNA can be called the support of genetic information.

Phenomena of genetic transformation, observed in certain bacteria by injection of external DNA or by transfer by a lysogenic virus, lead to the belief that it is possible to modify freely the chromosomal stock of a living being; I think that this is a delusion. The replacement of a segment of chromosome by an exogenic homologous segment can occur only if the effect of the substitution is thermodynamically favorable to the ambient metabolism. Evidently we are far from knowing in what circumstances this overall effect is favorable: there is no reason to believe that a locally favorable substitution will have a globally favorable effect on the metabolism; otherwise a viral infection would never be successful.

APPENDIX

Vitalism and reductionism. Contrary to what is generally believed about the two traditionally opposed theories of biology, vitalism and reductionism, it is the attitude of the reductionist that is metaphysical. He postulates a reduction of living processes to pure physiochemistry, but such a reduction has never been experimentally established. Vitalism, on the other hand, deals with the striking collection of facts about regulation and finality which cover almost all aspects of living activities; but it is discredited by its hollow terminology (e.g., Driesch's organizing principle and entelechies), a fault accepted and exaggerated by subsequent teleological philosophers (Bergson, Teilhard de Chardin). We must not judge these thinkers too severely, however; their work contains many daring ideas that those who are hidebound by mechanistic taboos can never glimpse. Even the terminology of Driesch is evidence of the mind's need to understand a situation that has no analogy in the inanimate world. In any event, the dispute is really pointless; many of the physiochemical properties of matter are still unknown, and realization of the ancient dream of the atomist— to reconstruct the universe and all its properties in one theory of combinations of elementary particles and their interactions—has scarcely been started (e.g., there is no satisfactory theory of the liquid state of matter). If the biologist is to progress and to understand living processes, he cannot wait until physics and chemistry can give him a complete theory of all local phenomena found in living matter; instead, he should try only to construct a model that is locally compatible with known properties of the environment and to separate off the geometricoalgebraic structure ensuring the stability of the system, without attempting a complete description of living matter. This methodology goes against the present dominant philosophy that the first step in revealing nature must be the analysis of the system and its ultimate constituents. We must reject this primitive and almost cannibalistic delusion about knowledge, that an understanding of something requires first that we dismantle it, like a child who pulls a watch to pieces and spreads out the wheels in order to understand the mechanism.

Of course, in some cases, the catastrophic destruction of a system can reveal some interesting facts about its structure, but for this the destruction must not be so brutal as to reduce the significant factors into insignificant pieces, as almost always happens in the chemical analysis of living matter. In this way, in experimental embryology, the study of interactions between living tissues (after dissociation and recombination of the constituent layers) has led to a body of knowledge undoubtedly much more interesting than the results of all previous biochemical analyses.

Our method of attributing a formal geometrical structure to a living being, to explain its stability, may be thought of as a kind of *geometrical vitalism*; it provides a global structure controlling the local details like Driesch's entelechy. But this structure can, in principle, be explained solely by local determinisms, theoretically reducible to mechanisms of a physicochemical type. I do not know whether such a reduction can be carried out in detail; nevertheless, I believe that an understanding of this formal structure will be useful even when its physicochemical justification is incomplete or unsatisfactory.

NOTES

1. We might say, within the framework of this theory, that if an organ and a tool are isomorphic, the reason is that they both have homologous positions in the universal unfolding of the same archetypal chreod.

CHAPTER 9

LOCAL MODELS IN EMBRYOLOGY

Καὶ ὁ λόγος σὰρξ ἐγένετο.
And the word was made flesh.
The Gospel according to Saint John

9.1. THE VARIETY OF LOCAL MECHANISMS OF MORPHOGENESIS IN BIOLOGY

In the few cases in which biologists have been able to shed some light on morphogenesis, they have undertaken to establish carefully the immediate reason for a given morphogenetical process. Thus the formation and curvature of the stalk of a plant can be attributed to the presence, at certain points in the tissue, of a substance, auxin, stimulating the local growth; and the rigidity of the notochord of a vertebrate embryo can be explained by the hydrostatic pressure of a liquid contained in the vacuoles of the cordal cells. But this information, however precise it may be, only sidesteps the theoretical problem of organization; why does the auxin only synthesize in precisely the place where it is necessary, and why are the vacuoles of the cordal cells in a state of hypertension? One conclusion to be drawn from this type of information is that, although the problems that have been solved by biological organogenesis are relatively uniform, they have been solved by organisms in an enormous variety of different ways. For instance, the various solutions of the problem of the spatial transport of a tissue include the following (a list that is very far from complete):

1. The displacement of individual cells by pseudopodia and flagella (e.g., the dissociation of neural crests in amphibia).
2. The use of auxiliary cells emitting tractive pseudopodia (gastrulation in the sea urchin).

3. The secretion of inert exudates acting as a support (in some algae and fungi, a spider's web, a man's bicycle).
4. The global movements of certain tissues (gastrulation in amphibia).
5. Differences in mitotic or growth rates.
6. Adhesion to an animal of another species (bees carrying the pollen of plants, a man riding on a horse).

What is most striking is that, despite the almost infinite variety of means, the functional interpretation of the morphology of living beings rarely presents any difficulties, and relatively simple finalistic considerations can usually justify the observed phenomena. The situation is quite the reverse in the case of inert matter, where the local accidents of morphogenesis are generally fairly uniform and few in number, but the global structure that arises is often indeterminate, chaotic, and rarely open to interpretation (e.g., change of phase, growth of crystals). We might reasonably expect of a theory of morphogenesis that it should describe explicitly at each point the local cause of the morphogenetical process, but the models that we offer cannot, in principle, satisfy this demand. To consider such a problem would require a *fine theory* of the metabolic states of cells, and the factors governing the choice of local mechanisms of morphogenesis are, first, the *scale* of the process considered and, second, the *age* of the cells and their *genetic constitution*. Moreover, often a very slight variation in the external conditions is sufficient to change completely the course of evolution of a catastrophe, although the initial and final states may remain unchanged.

This is a serious deficiency in our model, because it makes experimental control very difficult. Hence, for the moment, we shall offer a conceptual construction essentially of only speculative interest.

9.2. THE PRESENTATION OF THE MODEL

We are now going to present the model, taking the cell as the basic unit, since this will perhaps allow for a more concrete interpretation. This is not a necessary feature, however, and later, particularly in Chapter 11, we shall take the cellular cytoplasm directly as our base space.

We suppose that the instantaneous biochemical state of a cell can be parameterized by a point g of a (locally) Euclidean space G, and that the average biochemical state (over a sufficiently long period, of the order of mitosis, e.g.) can be parameterized by a point u of another (locally) Euclidean space U. At each instant there is an average state, and hence a canonical map $h: G \rightarrow U$, which we suppose to be locally a fibration whose fiber, a differential manifold M, parameterizes the instantaneous states of the cell compatible with the average state $u \in U$. At each given moment t, the set of biochemical states of the cells of a living being A define sets of points in G and U; and, since A is a three-dimensional ball B^3, we can suppose that this image set is defined by a map $F_t: B^3 \to G$ and by $h \circ F_t$ in U. These maps define what we call the *growth wave* of A, and we suppose that they are at least piecewise continuous (and even differentiable). We can consider G locally as a product $U \times M$, so that a cell starting from point $g = (u, m)$ gives birth, through mitosis, to two successor cells defined by (u_1, m_1) and (u_2, m_2), and we suppose (this is obviously the only justification for introducing the spaces G and U) that the representative points of these successor cells are topologically close to the representative point of the parent cell.

Now, let us fix the average state u of a cell and vary the second coordinate m in the fiber M. Considering a mitosis from (u, m) to (u_1, m_1) and (u_2, m_2) and neglecting the variation in u gives the map $d(u)$: $M \to M \times M$, defined by $d(u): m \to (m_1, m_2)$. Writing p_i for the projection of $M \times M$ onto the ith factor ($i = 1, 2$), we suppose that the transformations $p_1 \circ d(u)$ and $p_2 \circ d(u): M \to M$, associating to each cell the successor cells at the corresponding moment in the mitotic cycle, are actually neighboring diffeomorphisms of the manifold M. If the diffeomorphism $p_1 \circ d(u)$ has an attracting fixed point k, we say that k defines a *local stationary regime* compatible with the average biochemical state u, and since k is attracting, such a regime will be stable; only the stable regimes need be considered, since an unstable regime will be destroyed by the slightest perturbation. Now we associate with each point $u \in U$ the set of stable regimes compatible with the average state u, thus defining a space $\hat{U}$ mapped onto U by a local homeomorphism (a space stacked over U). At points where an attractor k stops being stable, there will be catastrophes of the corresponding regime, and the set H of points in U where at least one of the attractors k enters into catastrophe is called the catastrophe set in U. Under optimal conditions the growth wave, the image $F(B^3, t)$, will meet H in a set $H' = F^{-1}(H)$, defined up to an ϵ-homeomorphism, and this will give rise to a system of chreods on $B^3 \times T$, a semantic model describing the development. In Chapter 10 we shall describe the specification of the global structure of this model; here we are mainly interested in the first singularities occurring in embryonic development, particularly in the process of gastrulation. Later we give a presentation of the development of a limb and finally, in an appendix, a description of neurulation and the formation of the spinal column in vertebrates.

We can suppose, to first appr ximation, that the catastrophe set H is defined by a static field in U. Then the strata of H that will intervene first will be those of codimension one, strata of fold points, and those of codimension two, strata of cusp points. As we saw in Chapter 6, the strata of fold points have, in principle, a morphogenetic effect only when they separate a stable regime from the empty regime—these are the boundary strata, and they correspond here to the boundary of a cavity or to the external epidermis. It is known that, in general, cavities play an important but rather mysterious part in embryology, for what can be their role when they have only a passing existence and possess no functional significance in the adult? Thus the neurocoele clearly belongs in an essential way to the neural chreod; during an artificial induction in a developed ectoderm, the neural tubes with their neurocoele will form without ever opening to the outside. The explanation might be that, if the mechanical properties of a tissue are completely determined by its average biochemical state (the image of the growth wave F in U), and the elasticity, growth, speed, and direction of mitosis are determined by this average local state, the integration of all these local situations into a three-dimensional mass of tissue has every chance of leading to an overdetermined problem. If V denotes the velocity vectors of the tissue at a point m, and ρ the density, V will satisfy a partial differential equation of Liouville type:

$$\frac{d\rho}{dt} + \rho \, \Delta V = 0,$$

such that the data of V and ρ lead to overdetermination. The simplest way of avoiding this overdetermination is to concentrate the determination of V on a two-dimensional surface F, with the rest of the tissue, off F, following this development more or less passively. In fact, there is an almost general rule that the support of a morphogenetic field in embryology is always a surface whose thickness does not exceed two or three layers of cells, and even when the final organ is solid and three-dimensional, like the eye lens, it is built up by growth on the surface, starting from the equator inside the lenticular vesicle.

When this hypothesis is added to the model, the support of a morphogenetic field (the inverse image of a stratum of codimension one of the bifurcation set H) can be of only two possible types: either a fold stratum, separating a stable regime from the empty regime, or a conflict stratum, a shock wave separating two regions of two regimes that are in competition. Now it is difficult to achieve a tight control of shock waves in primitive embryology, because such control would depend on the comparative thermodynamical states of the two competing regimes, and hence it would be necessary to modulate the metabolic cycles of the characteristic reactions of each regime by catalyzing enzymes; this would require both a highly differentiated structure of cytoplasmic organelles (ribosomes, mitochondria, etc.) and a strict polarization of the underlying tissue. Thus conflict strata appear after boundary strata requiring only the control of a single regime, and such a boundary field requires a cavity for it to bound.

Having said this, we now observe that histological specialization occurs in three different geometrical ways.

1. By an inequality in growth rates, with some tissues dividing more rapidly than others.
2. By relative movements of tissues. In embryology, some tissues move faster than others and give rise to shock waves on the surfaces of velocity discontinuity (the velocities being then tangential to these surfaces).
3. By histological modification of an already existing tissue, like the ectoderm transformation of neural tissue under the effect of an "inducing agent."

Here again the transformation will have geometrical consequences, either in the grouping together of cells (with the formation of palisade tissues) or by the later movements of new tissues. I will leave aside the question of determining to what extent the biological specialization of already existing cells precedes its geometrical manifestation, although this question is obviously important in embryology; it seems a priori clear that all spatial morphogenesis presupposes a phase of biochemical preparation, even if this preparation retains a relatively labile and reversible character until a very late stage.

In conclusion, we shall be assuming that embryonic development is directed by *director layers*, which we suppose to be wave fronts, and so these layers have the stable singularities of wave fronts that are the origin of catastrophes triggering later differentiations. Surface shock waves, whose qualitative appearance is given by a rule of the same type as Maxwell's convention, occur only exceptionally, in particularly compact and thick tissues such as yolk.

In this way we are forced into identifying the singularities of embryological morphogenesis with the elementary catastrophes of Chapter 5. Of course this identification may appear arbitrary, but I think that it can be justified in proportion to the size of the animal, and that in a large animal the statistical approximations, defined by differentiable maps, represent fairly well the underlying mechanisms of epigenesis. However, as with smaller animals, like the sea urchin, we are dealing with morphogenetical processes that are not very finely controlled in their detail (e.g., the formation of primary mesenchyme by migration of micromeres from the vegetative pole of the blastocoele), and it would be unreasonable to submit such a migration of individual cells in all its details to the control of a structurally stable plan. Instead, we should consider such a migration as a generalized catastrophe in which only the final result is relatively well defined, while the individual path taken by each cell is comparatively indeterminate and subject to the hazards of the local situation. It is only right to remark that even in mammals there are morphogenetical movements, some of them of great importance, which have this appearance of individual cellular migration, for example, the primitive layering of the hypoblast and epiblast, the dissociation of neural crests, and the migration of germinal cells in the gonads. It is not unreasonable to think that some transformations which initially had a precisely controlled evolution could later have evolved in this blurred way (see the models for gastrulation in vertebrates, later in this Chapter) particularly with the increase in the regulative capacities of the eggs.

A new complication, apart from the appearance of generalized catastrophes, soon occurs if the development is followed beyond gastrulation. In our model, this is the appearance of nongeneric structurally stable singularities—a general phenomenon called threshold stabilization, the appearance of new constraints resulting from the successive triggering of new homeostatic mechanisms. It can be described mathematically as follows. After a first differentiation with universal unfolding space U, the final situation is characterized by a stabilized growth wave $F: B^3 \longrightarrow U$; a second differentiation which follows is defined by a potential V_1, which is specified by composition from a real function $G: U \longrightarrow \mathbf{R}$, so that $V_1 = G \circ F$. The external variables of the first chreod act as internal variables of the second. Now such a function V_1 can have a nongeneric singularity at a point x, and we might only allow those deformations V' of V_1 that can be factorized in the form $V' = G' \circ F'$, where G' and F' are deformations of G and F, respectively. We can obtain a reasonable model of neurulation (and the formation of the vertebral column) only if we impose a bilateral symmetry of this kind. In this way, during development, there will be a progressive confusion between the internal variables of the differentiation catastrophe and the external variables, the coordinates of the unfolding which, representing epigenetic gradients, have spatial significance. This evolution achieves its most extreme form in the metric-like chreods, such as those of the eye and the limbs, where it seems that the internal space is practically isomorphic to the Galilean space of the mechanic of the external world.

We shall be considering this gradual transformation of external variables into internal variables in the property of self-reproducing singularities postulated in Section 10.2.E. There also the change from one generation to the next will be formally described by the fact that a system of external variables becomes a system of internal variables.

Before starting on an explicit description of gastrulation, we shall discuss the classical schemes that have been proposed in biology for embryonic development: the mosaic scheme and the theory of gradients.

9.3. A DISCUSSION OF HISTORICAL THEORIES

The theoretical schemes are as follows.

A. Development of mosaic type

The conceptually simplest model of epigenesis is clearly that called *mosaic*, in which each part of the egg has its destiny fixed from the very beginning. In our geometrical model, the trajectory of the growth wave $F(x, t)$ in U will be defined by a vector field on U, which we can suppose is constant, and which projects positively on the time axis in $U \times T$. The differential equation controlling development will then be

$$\frac{dF}{dt} = a,$$

where the constant a may possibly depend on x if part of the germ is subjected to special treatment during the course of time. The wave F will meet the catastrophe or conflict strata in U in points whose inverse images by F are the surfaces on which the germ differentiates into different tissues.

There is almost no strictly mosaic situation in embryology, and phenomena of regulation seem always to occur to a greater or less degree. This model can be improved so as to account for certain experimental results.

It is known that, right from the beginning, the egg is polarized in at least one direction, the animal-vegetal axis. Let y denote a function taking values $-\infty$ at the vegetal pole and $+\infty$ at the animal pole, and suppose that y is a coordinate in our space U, playing here the part of the epigenetic landscape of Waddington. The addition to the environment of substances with an animalizing (or vegetalizing) effect can then be expressed by a correcting term in the differential equation of the form

$$\frac{dy}{dt} = c,$$

where c is positive for an animalizing substance and negative for a vegetalizing substance, and y denotes a component of the map F.

Now, still supposing that U is (locally) Euclidean, we can account for certain regulation phenomena by postulating an evolution in F of the type of the heat equation, with a term expressing the effect of diffusion:

$$\frac{dF}{dt} = a + m\,\Delta F,$$

where Δ is the Laplacian, and m is a scalar coefficient proportional to the extent that the tissue is labile. Then, if part of the egg is ablated, the initial data $F(x, 0)$ will be discontinuous on the remaining part. Because of the diffusion term $m\,\Delta F$, the solution will become continuous and even differentiable, and for this reason some organs whose presumptive tissues had been removed might still appear (see Figure 9.1 for a model in which U is one-dimensional).

I will not insist further on this far too simplistic model, beyond saying that it does contain a certain local validity in the neighborhood of the center of each morphogenetical chreod of static type.

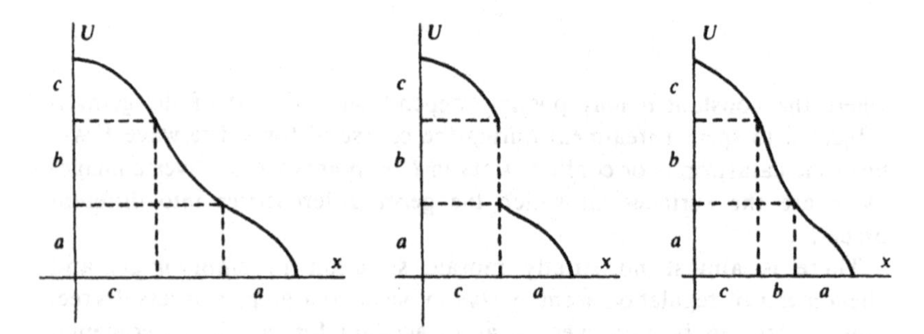

FIGURE 9.1. Regularization by diffusion.

B. Gradient theories

The so-called gradient theory (Boveri, Child, etc.) can be formulated in three ways.

1. *The most general formulation*: All morphogenesis is locally controlled by a system of local coordinates realized locally by biochemical magnitudes (in general, of an unknown nature).
2. *The less general formulation*: There are biochemical magnitudes in the embryo (and the adult animal?) enabling each point inside the organism to be localized.
3. *The restrictive formulation* (*Child*): Epigenesis is controlled by the real function defined by the global intensity of the metabolism, and the gradient of this function is nonzero on the cephalocaudal axis.

Insofar as formulation 1 is concerned, it is almost a methodological necessity to suppose its validity. As soon as we suppose that morphogenesis occurs according to a local model (and this is the idea of Waddington's chreod and of the general morphological field), we have to suppose that the local coordinates of this model have biochemical significance.

Formulation 2, which has a global characteristic, is much more delicate. Although all authors seem to suppose the existence of a cephalocaudal gradient in the embryo (and perhaps in the adult, in species with highly developed regenerative facilities), the existence of two other gradients (dorsoventral and bilateral) seems much more doubtful.

Child's metabolic formulation 3 evidently arose from a perfectly justified desire to describe an epigenetical gradient by an experimentally measurable parameter. To the extent that epigenetical gradients exist and are defined biochemically, the global intensity of the metabolism is a function of these values, and it can sometimes be taken as a local coordinate. But this is in itself such a limited measurement, containing so little information, that it is manifestly unreasonable to expect the whole of epigenesis to depend on it. (We must not forget that all local epigenesis requires three gradients, not one.)

Whatever the defects in the historical theory of gradients, it has at least one advantage over the more modern theories of the differential activation of the genes. It is a theory with a geometrical character and, mathematically speaking, shows evidence of the basically correct intuition of its originators.

9.4. MODELS IN PRIMITIVE EPIGENESIS OF AMPHIBIA

Let us first consider the case of amphibia; here textbooks of biology give a relatively complete description of the gastrulation process. Suppose that, in the model, the space U into which the growth wave is mapped, and which parameterizes the average biochemical states of each cell, is a four-dimensional space $\mathbf{R}^4$ identified with space-time. This is not so restrictive an assumption as it may seem at first sight, for, if U actually has many more dimensions and we suppose that the growth wave $F(B^3, t)$ describing the evolution of the embryo is an embedding, the only effective part of U (in normal epigenesis) will be a four-dimensional domain. At the blastula stage, the egg will be a thickened sphere $S^2 \times I$, and we may suppose that the external boundary of the blastula corresponds to a folded stratum of the set H, precisely that separating the regime a of the egg dynamic from the empty regime; in the interior, the blastocoele cavity is practically independent of any control during epigenesis and corresponds to no stratum of H. This situation persists up to a time T_1, the beginning of gastrulation; then, between times T_1 and T_2, a circle c of swallow's tails will appear on H, and at $T_1 + \epsilon$ the growth wave $F(B^3, t)$ will meet c at two points a_1 and a_2, the ends of a double line d on the sphere $\partial F(B^3, t)$. The H set will have a cusp line in $F(B^3, t)$, made up of two arcs g and g' joining a_1 to a_2 inside the blastula. As we saw in the theory of the swallow's tail (Chapter 5), one of these arcs, say g, will be a Riemann-Hugoniot set for the conflict between the two attractors, while the other will have a virtual role. There will be a shock wave F, bounded by g and d, separating the basins of these two attractors, and so the double line d (forming the bottom of the blastoporic fold) will be extended into the interior by the surface F and will separate endoderm from ecto-mesoderm (Figure 9.2); the double line d will then curve into a handle shape and form the lip singularity outside. After some time the two points a_1 and a_2 will again meet at the lowest point of the blastoporic circle, d will form the blastoporic circle (Figure 9.3), and the shock wave F, bounded on one side by the blastopore and the other by a Riemann-Hugoniot circle inside c, will be topologically a cylinder $S^1 \times I$; but, for $t > T_2$, the circle c will force its way rapidly into the interior and the surface F will extend, also toward the interior, and separate the vitelline endoderm from the mesodermal mantle.

According to this description, it seems that the mesodermal mantle separates from the vitelline endoderm (or, more precisely, from a so-far-undifferentiated tissue on the inside boundary of the blastula, almost in contact with the blastocoele) by delamination. This point has been closely studied by Vogt [1], who remarks that many sections show a state of topological continuity between the mesodermal layer and the endoderm (at the internal boundary of the mesodermal mantle). However, he rejects this interpretation, remarking that the two layers are always very precisely differentiated except in two regions in the plane of symmetry of the germ (the prechordal plate above, and the ventral lip below). Other experiments made with stained markers show that the material in the mesodermal mantle always comes from the lateral regions outside the blastoporic circle. Probably two, not at all incompatible, mechanisms are at work in the formation of the mesoderm: the very rapid increase of the mesoderm through invagination of surface ectoderm, and the differentiation from endoderm along the Riemann-Hugoniot circle c. Evidently, at the beginning, only the second mechansim is operating. Then the lateral arcs of the circle c will sink rapidly inward, while the median arcs will remain very close to the higher and lower points of the blastopore, and differentiation in the plane of symmetry will progress very little into the interior. The lateral arcs will move away until they rejoin underneath, after they have completed a circuit of the germ, and the circle c will then give rise to two circles c_1 and c_2, where c_1 (above) defines, at least partially, the undifferentiated zone where the prechordal plate will join the endoderm, and c_2 (below) defines the horse-shoe region of the ventral lip of the

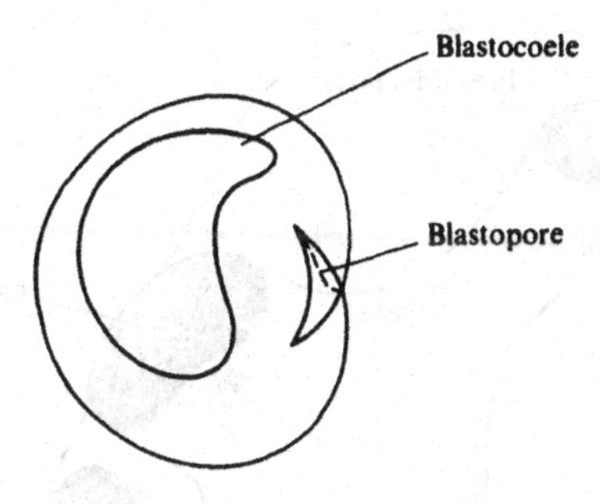

FIGURE 9.2. Gastrulation in amphibia.

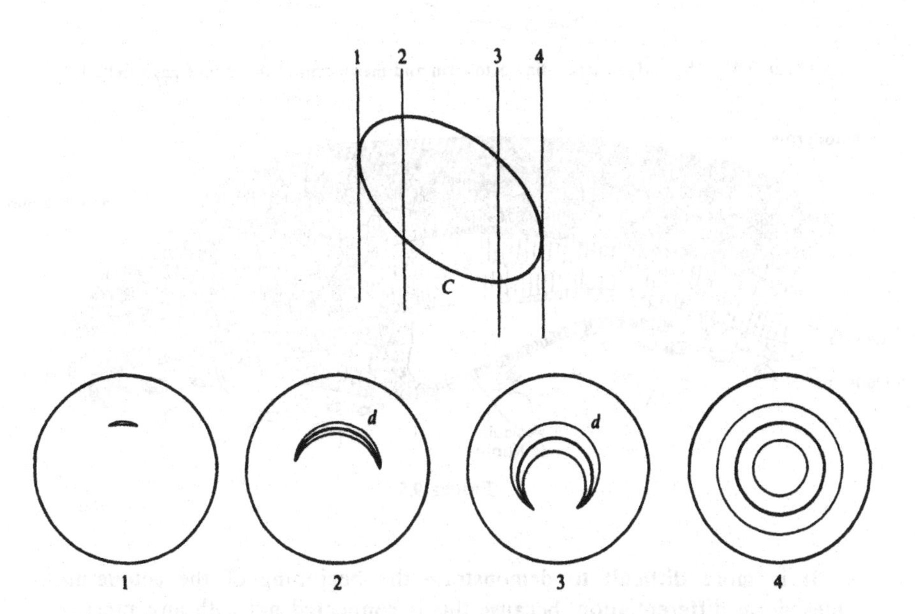

FIGURE 9.3. Theoretical plan of gastrulation defined by the swallow's tail.

blastopore where the mesoderm and endoderm still join. Some arcs of c_1 might not be covered by $F(B,t)$, and there the lateral mesoderm could have a free boundary (see Figure 9.4 for a picture of the shock wave F at the end of gastrulation).

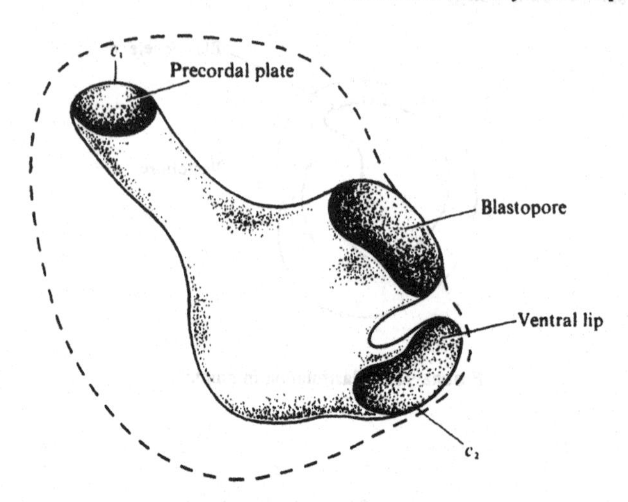

FIGURE 9.4. The surface separating ectoderm and mesoderm at the end of gastrulation.

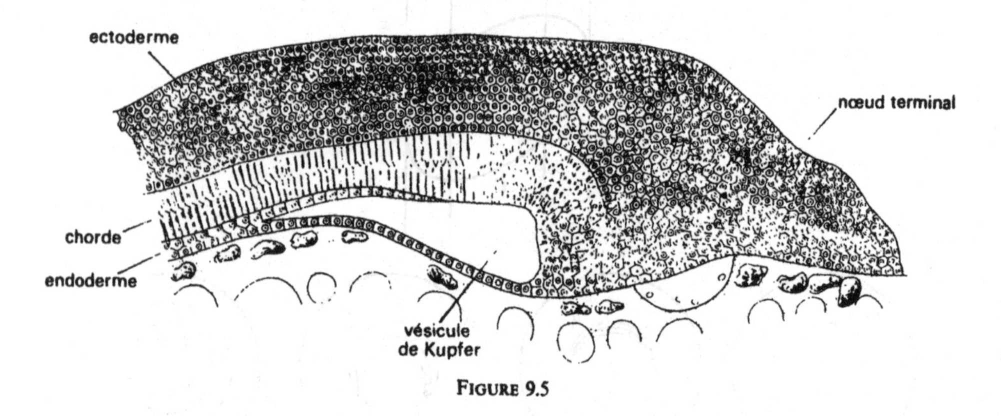

FIGURE 9.5

It is more difficult to demonstrate the beginning of the ectoderm-mesoderm differentiation, because this is connected not with any superficial phenomenon (like the appearance of the blastoporic groove), but rather with an internal phenomenon whose initial stage is very fleeting. The first differentiation probably appears as early as the beginning of gastrulation, as a surface S of Riemann-Hugoniot type inside the dorsal lip of the blastopore, and is connected with the flattening of the residual blastocoele separating the two layers. This surface S extends rapidly, first in the lateral zones and then inside the germ and underneath the ectoderm layer, and very quickly assumes the form of a surface of velocity discontinuities, with the ectoderm converging toward the blastopore, and the mesoderm moving in the opposite sense in the interior. The cusp line bounding the surface is effectively the locus of points where the direction of the movement reverses. Then the surface S completely separates the ectoderm layer from the mesoderm, except in the ventral region of the blastopore, where complete differentiation occurs only at the end of gastrulation.

Whereas the ectoderm-mesoderm differentiation in amphibia is somewhat confused by these kinetic phenomena, the situation in fish is much more precise. Here a block of undifferentiated cells, the terminal node, develops longitudinally through division, and a surface with a free boundary (Riemann-Hugoniot) appears in the mass, separating the ectoderm from the mesoderm (see Figure 9.5 taken from Gallien [2]). The mesoderm-endoderm differentiation seems also to be connected with a line of swallow's tails on the boundary of a cavity, the Kupfer vesicle.

Even primitive morphogenesis (gastrulation, neurulation) shows an enormous variety of epigenetical movements, according to the animal being considered, but much of this diversity might only be apparent, since seemingly very different movements might arise, one from the other, by very simple transformations. If, in each particular case, we were able to go back to the dynamical singularities of epigenesis, we might then be able to compare the different epigenetical maps (in our model, the spaces U and the bifurcation manifolds H) and perhaps reconstitute the successive foldings of the dynamic generating the morphogenesis. Then, ordering these maps by increasing topological complexity, we should be able to reconstruct the evolutionary sequence of development. Where is the topologically minded biologist who might attempt this program?

As a first study, we now compare gastrulation in birds and amphibia.

9.5. MODELS FOR THE PRIMITIVE STREAK

A. Gastrulation in birds

As is well known, the primitive epigenesis of higher vertebrates (birds and mammals) is noticeably different from that of amphibia. In birds, the embryo is of the blastodisc type, and, after cleavage, it appears as a flat disc on the subgerminal cavity. The first differentiation is that separating the endoderm (below) from the ectoderm (above); this separation, called delamination in textbooks, cannot be described with geometrical precision but rather seems to be due to individual migrations of the large vitelline cells toward the base, with the epiblastic cells staying in their places. No doubt the very blurred nature of the separation, which we might consider as a generalized catastrophe, is connected with the highly developed regenerative capacities of bird and mammal embryos. In order to allow for belated regulation, the different types of cells necessary for later development must remain intimately mixed, and so their eventual segregation has to take place as late as possible. Geometrically precise phenomena occur only with the formation of mesoderm issuing from the primitive streak. The kinetic of this rather strange process can be explained if it is looked at in the following way.

Suppose that each of the two discs, epiblast and endoderm, becomes polarized but in different ways: on the epiblast a tangential gradient Y is defined in polar coordinates (r, θ) by the orthoradial component $r\theta'$ (such a gradient is preserved under rotations of the disc), while on the endoderm the polarization has a privileged direction d, defined by the radius Op, where p denotes the point of the disc representing the future posterior extremity of the germ. Let q denote the point of Op (close to the lower extremity p) where the primitive streak will appear; $u(s)$, a bump function, equal to one in a neighborhood W of q, (Figure 9.6); and Oxy, a coordinate system such that $x = 0$, $0 > y > -1$, defines the radius Op. Also suppose that, because of diffusion or resonance between ectoderm and endoderm, each cell of the epiblast in the neighborhood W will be under the effect of a gradient of type $Y' = u(m)\cdot xY$, and also suppose that, once a cell comes under the effect of this gradient Y', it will remain so throughout its movement. Then each cell of the epiblast in the neighborhood W will converge toward the axis Ox in a neighborhood of q; and, because it continues to be affected by a gradient of the same sign, it will continue to move in the same sense in the lower layer, the mesoderm. As a result, the primitive streak will appear to be the meeting point of two opposing flows, and in the geometric model that we shall be considering the primitive streak will be a segment of self-intersection points of a disc immersed in $\mathbf{R}^3$ whose two extremities are cusp points. Such a dynamical scheme will account for phenomena of induction of the endoderm on the epiblastic disc, observed particularly if one disc is peeled off the other and then rotated and replaced.

By introducing a more complete and precise model, we obtain a description of the later development of the mesoderm. Consider the surface S in $\mathbf{R}^3$ (coordinates u, v, w), defined as follows. Take two circles c and c' with equations

$$u^2 + v^2 = 1, \quad w = 1 \qquad \text{and} \qquad u^2 + v^2 = 1, \quad w = -1,$$

respectively, and associate the point $m = (u, v, 1)$ of c with the point $m' = (u, -v, -1)$ of c'. Then S is the surface consisting of the segments *mm'*. This surface, whose equation is

$$w^2(1 - u^2) = v^2, \quad w^2 \leqslant 1,$$

is bounded by the circles c and c', and it has the segment $|u| \leqslant 1$, $v = w = 0$ as a double line whose two extremities $(-1, 0, 0)$ and $(1, 0, 0)$ are cuspidal points α and α'. The coordinates (u, v, w) represent biochemical gradients: u, the cephalocaudal gradient (increasing from $u = -1$, the tail, to $u = A < 1$, the head); v, a lateral coordinate, a left-right gradient; and w, a residue of the animal-vegetal gradient, with $w > 0$ representing first the ectoderm and then the notochord, $w < -1$ the endoderm, and the slice $-1 \leqslant w \leqslant 0$ the mesoderm with nonaxial vocation (Figure 9.7).

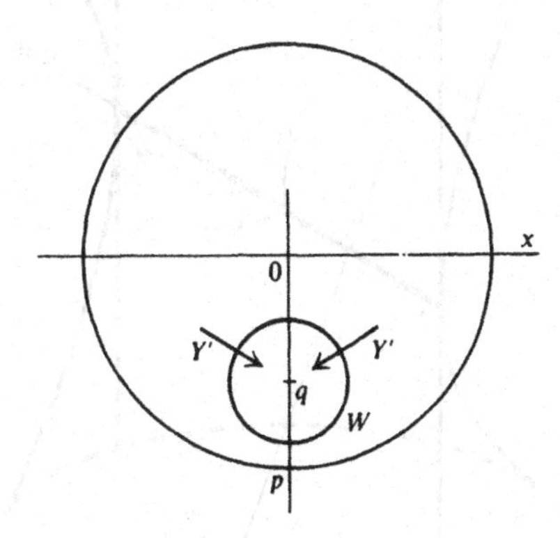

FIGURE 9.6. Plan of the cellular kinetic giving birth to the primitive streak.

The formation of mesoderm can then be given the following qualitative interpretation. During the evolution of the growth wave $F(\mathbf{R}^2, t)$ in the space, first $F(\mathbf{R}^2, 0)$ is embedded in $\mathbf{R}^3$ in the half-space $w > 0$, where it is almost horizontal, with a minimum of w on the axis of symmetry $w = 0$. As t increases, this surface undergoes a translation parallel to the Ow-axis and descends. At a time $t = 0$, for example, it touches the double segment of S at a point q, the beginning of the primitive streak; and, as the descending translation continues, a double line bounded by two cusp points, the intersection of two ectoderm layers and two mesoderm layers, bounded below by a circular curve c_t, appears on the counterimage of the surface S by F. This situation persists, but the right-hand cusp point moves up to $u = +1$ (and even beyond), while the left-hand one moves, more

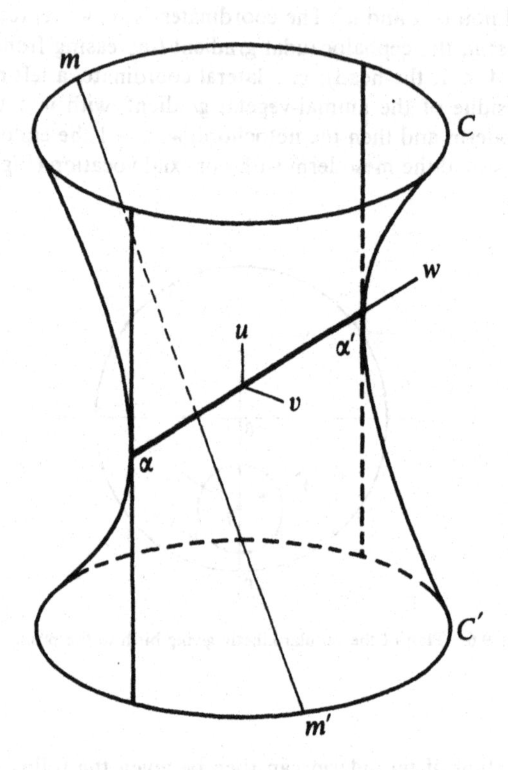

Figure 9.7. The surface S with two cuspidal points.

slowly, to $u=-1$. At a time $t=t_h$, the part of the surface in $1-a<u<1$ (a a small positive number) starts to move up toward positive w as if undergoing a new attraction toward the ectoderm; and, when the surface F again passes the point α' ($u=+1$), the intersection curve of S and F, bounding the mesoderm layer, has a cusp point called Hensen's node. Then, as F continues to rise, the intersection curve will develop a loop, and we might consider the interior of this loop, in $w>0$, as representing an axial tongue of mesoderm giving rise to a head process (Figure 9.8).

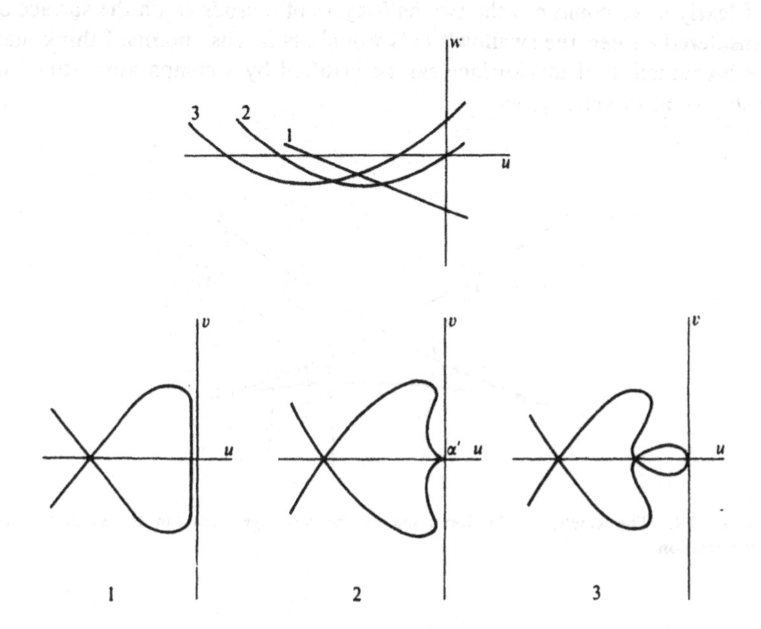

FIGURE 9.8. Diagram of the head process.

We can also interpret the primitive streak, as well as the blastopore of amphibia, as a double line of the epiblast layer bounded by two swallow's tails. More precisely, there will be a lip singularity in the epiblast plane bounded by two Riemann-Hugoniot cusps. Here, however, the Maxwell convention no longer applies; and, although the regime associated with the mesoderm is at a potential higher than the epiderm, we must suppose that the correcting jump does not occur because there is a situation of perfect delay (Figure 9.9). It will be sufficient for this to suppose that the jump to the lower layer stabilizes the regime of the mesoderm. The base of the swallow's tail corresponds to a saddle point with only a virtual existence. The vector fields Y_1 and Y_2, mean coupled fields, converge toward the axis of symmetry, defining the kinematic of the process. At the beginning of the catastrophe, the convergence of the cells being pulled along by the field Y produces à local thickening, but later the circulation organizes itself more economically with two currents eroding a path below and forming a shallow furrow. We shall show in the appendix how to interpret the Hensen node and the later formations.

Clearly if we could put the two half-layers of mesoderm on the surface S considered earlier, the swallow's tails would act as cusp points. I think that the introduction of this surface can be justified by a comparative study of gastrulation in vertebrates.

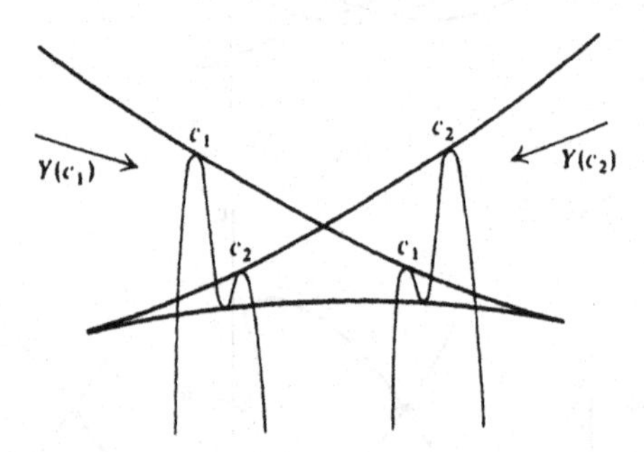

Figure 9.9. The kinetic of the formation of the primitive streak in the swallow's tail interpretation.

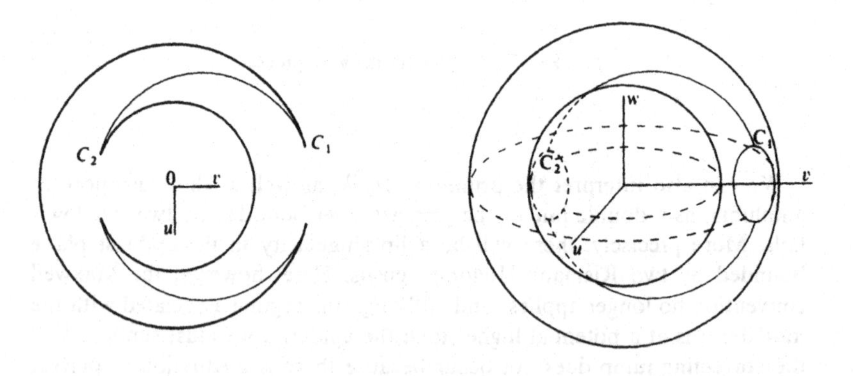

Figure 9.10

B. The comparative topology of gastrulation in vertebrates

Let us return to the case of amphibia at the blastula stage; there is a two-sphere S^2 whose surface is divided, according to the development, into an animal ectoderm cap, a mesoderm crown (bounded by the two parallel circles c_1 and c_2), and a vitelline endoderm cap. We might represent gastrulation by saying that the circle c_2 sinks inward (Figure 9.10), so that after gastrulation of the mesoderm there will be a cylinder joining two circles, c_1 on the ectoderm and c_2 on the endoderm. However, the topological continuity is very quickly broken on the circle c_2 (which is at the beginning of the surface of swallow's tails defining the blastopore), although it persists longer on c_1. Now consider an egg of blastodisc type, of a bird or mammal. According to many writers [3], the very irregular cavity separating the two layers after the separation of the endoderm from the epiblast must be thought of as the blastocoele, just as if all of the blastula had been squashed onto the vitellus, but our model will interpret this situation slightly differently.

Returning to the typical situation of amphibia, we can represent geometrically the state after gastrulation as follows: there are two concentric spheres of radii r and $R (0 < r < R)$ in $Ouvw$, and the outer sphere, supporting the ectoderm, is perforated along the circle c_1, situated in the plane $v = R - h$, h small, while the inner sphere, supporting the endoderm, is perforated along the circle c_2, situated in $v = -r + h$. The mesoderm is then represented by a cylinder C with boundaries c_2 and c_1, situated between the two spheres. The effect of evolution on gastrulation is represented by a rotation of the endoderm sphere around the Ow-axis, and the cylinder C will follow this rotation continuously, possibly penetrating to the interior of the inside sphere. A rotation of $\pi/2$ toward $u > 0$ gives the type of situation shown in Figure 9.11, which represents reptilian gastrulation (of the tortoise), with a chordal mesodermal canal directed toward negative u and flowing into the subgerminal cavity; here the u coordinate represents a cephalocaudal gradient increasing in the direction from the head to the tail. Now, continuing the rotation of the endodermic sphere about Ow, a rotation of π from the position representing amphibia will move the circle c_2 to a position c_2' in front of c_1 (Figure 9.12) with equation $w = r - h$. The circles c_2 and c_2' will have opposite orientations in this plane, however, and therefore the cylinder C with boundary $c_1 - c_2$ (or $c_1 + c_2'$) will have the form of the surface S with a double segment bounded by two cuspidal points of Figure 9.7 (the surface of minimum area satisfying these conditions).

At least one of the two gradients, head-tail and left-right, underlying the epi- and hypoblastic layers must be reversed. Normally the cephalocaudal gradient is considered to be much more stable than the left-right gradient (which, in fact, appears only after the formation of the heart and the consequent breaking of the bilateral symmetry); therefore the circles c_1 and c_2' are to be identified with preservation of the cephalocaudal gradient, and the double line of S, corresponding to the primitive streak, will be parallel to this gradient.

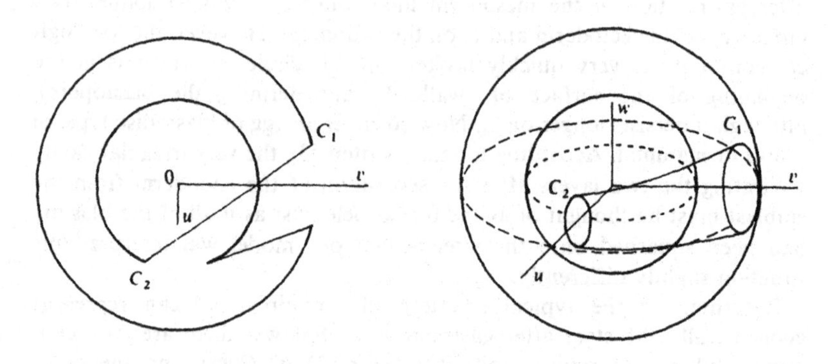

FIGURE 9.11

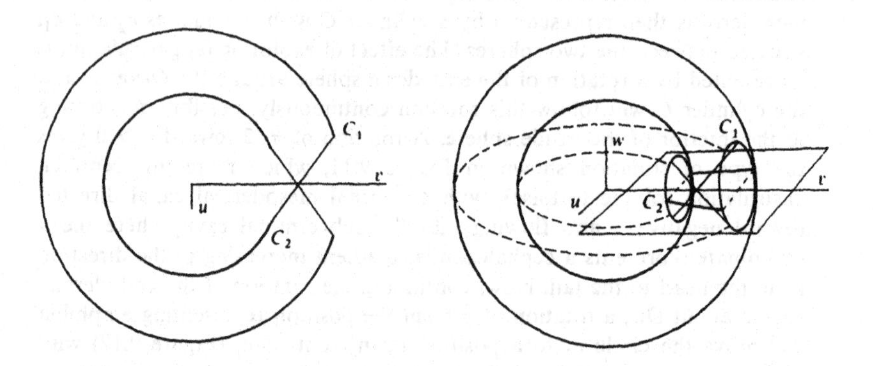

FIGURE 9.12

This scheme for the blastodisc of a bird's egg differs from that suggested in most textbooks in that we suppose that the embryonic knob is initially covered by an totipotent layer having the latent possibility of developing into three layers (the trophoblast of mammals), and that the ectoderm-endoderm delamination occurs directly only on the blastodisc (the *area pellucida*), as if the last stage of gastrulation of amphibia were being achieved directly.

In this way epigenesis in birds and mammals has replaced amphibian gastrulation, with its beautifully precise geometry, by a generalized catastrophe forming some kind of short cut between the undifferentiated state, with its high regulative capacity, and the postgastrula stage, and has jumped over the blastula-gastrula stage. This example shows how relative is the recapitulation law: ontogeny recapitulates phylogeny. Just as, when learning, it is not necessary to follow the whole historical process giving rise to a subject in order to learn that subject (only the essential points are needed), so the dynamic of epigenesis makes links and simplifications when certain mechanisms, inherited from the past, have lost their reasons for existence.

We might also ask why, in this scheme, the ectoderm and endoderm are connected with relatively rigid spheres, while the mesoderm is constrained to follow their variations. If the (u, v, w) coordinates have biochemical significance, this probably amounts to saying that at the pregastrula stage the ectoderm and endoderm are already relatively strongly polarized, whereas the mesoderm, which is only at the stage of the organizing center, preserves great flexibility. In any case, the scheme is defined only up to homeomorphism. We also note that the junction of the mesoderm with the endoderm (curve c_2) has only a fleeting existence below the primitive streak. It would be very interesting to continue the construction of this epigenetic polyhedron (see Section 10.2.B) for later differentiations (notochord, somites, etc.), although probably the comparison of these differentiations in the different branches of taxonomy will not be such an irritating problem as that of gastrulation. The appendix gives a model for neurulation and the formation of the spinal column.

Now we describe some typical situations of mean epigenesis.

9.6. MODELS IN MEAN EPIGENESIS

A. Induction crossed by resonance: glandular scheme

We have seen how the first differentiations of morphogenesis appeal to the Riemann-Hugoniot and swallow's tail elementary catastrophes on space-time, because these first specializations do not involve any choice; they have only a single parental regime, a single parental chreod, and this leads to a decomposition into layers, expressing the effect of tearing shock waves. However, epigenesis requires many other kinds of geometrical arrangements besides this one; also, the static catastrophes considered so far do not, in principle, allow catabolic catastrophes with a decrease in the dimension of an attractor. This is why generalized catabolic catastrophes, induced by resonance, appear as early as neurulation. Almost every excreting organ of gland type arises from a reciprocal induction between an epithelium and a mesenchyme, and this has been extensively studied in recent years. Let us consider an interpretation.

Suppose that the mesenchyme M is represented by an attractor of a static gradient field $V: M_1 \rightarrow \mathbf{R}$, with coordinate u on M_1, and the epithelium e is represented by an attractor of the gradient field $W: M_2 \rightarrow \mathbf{R}$, with coordinate v on M_2. Then, in a noninducing region, neither of these tissues reacts on the other; this is expressed dynamically by the fact that there is no interaction potential $H(u, v)$ on either tissue, whereas in an inducing contact zone an interaction potential $H(u, v)$ appears (probably by resonance between attractors of metabolic type), where H is large with respect to V and W. This means that the mesenchyme M and the epithelium e will enter into a generalized catastrophe provoked by the appearance of a new phase m of the attractor grad $(V + H) = 0$ in the mesenchyme, and a phase e' with attractor grad $(W - H) = 0$ in the epithelium. First the appearance and then the extension of the interaction (resonance) region might, a posteriori, be controlled by a mechanism of the following type. Suppose that the new phase m in the mesenchyme is sensitive to a gradient forcing it toward the epithelium; then (Chapter 6) a generalized catastrophe with a spatial parameter will appear, and this new phase will flow toward the epithelium and form streams that will capture each other and finally flow into relatively small areas p of the epithelium (the future outlet of the future gland). However, the new phase e' of the epithelium will be affected by this gradient in the opposite sense (since H has a negative sign), and hence the epithelium will infiltrate the mesenchyme and flow back down the streams of the catastrophe from the resonance areas p. This inverse catastrophe will meet the real catastrophe of the mesenchyme, which will then condense out around the streams. The process will generally condense at the half-way stage, by a stable transitional regime between m and e', leading, after coupling with the vascular system of the mesenchyme, to the formation of organs (of variable physiological type: the alveole of the lungs, the glomerule of the kidney, etc.) where the excretion itself takes place.

If a microfilter, which will not prevent the establishment of induction by diffusion but will prevent actual contact, is interposed between the two tissues, each tissue will enter into catastrophe on its own account, but there will be no spatial correspondence between the organs that the two tissues construct.

We might ask, concerning this model, whether the catastrophe of the new mesenchyme actually exists. I think that it might only have a virtual existence, and that the actual movements of the mesenchyme might be restricted to a condensation into tubes around the threads of epithelium. This raises the interesting question of how far the topological arrangement of the branching of the gland is genetically determined. Probably the dendritic arrangement is only weakly determined, and traces of inductor mesenchyme might lead to very different dispositions of the final organs, which could be no less functional. Remorseless believers in biological finality often cite the vascular system or the arrangements of the bronchial tubes as examples of very complicated structures that are highly adapted to their purposes. These forms are generated from catastrophes with spatial parameters, however, and although they are of great topological complexity, they are generated by comparatively simple mechanisms and, by themselves, contain very little "information." Ulam [4] has given examples of dendritic forms (like snowflakes) that can be generated by relatively simple recurrence laws, and to characterise forms that appear through generalized catastrophes in a homogeneous medium is an open problem in mathematics.

In these examples, the induction has appeared in a symmetric form, in which it is not specified which is the inductor and which is the induced. Generally, the epithelium is the inductor; this amounts to saying that it is often much more strongly predetermined and less variable than the mesenchyme, which frequently keeps its highly competent abilities.

B. Morphogenesis of a vertebrate limb

1. The morphogenetic field of a bone. As opposed to structures of dendritic type with only a feeble local control of their spatial catastrophes, the morphogenesis of a bone and its attached muscles presents a problem that, even on the conceptual plane, is of far greater difficulty. The model now to be given, which I do not think can be much simplified, conveys some idea of this, and here I will even ignore the theoretical problem, raised by the model, of knowing how the chreod associated with the bone can have an underlying metrical structure. Although the existence of local metrical coordinates connected with the bone is apparent once the formation is finished, the model requires that these coordinates be present long before ossification, at the precartilage stage at least. The biochemical support of these coordinates, and polarization, are discussed in Section 10.5.3.

Having mentioned the problem of this rigid character of the morphogenetical chreods of the bone, we will consider another troublesome aspect of the epigenesis of the skeleton. This is the problem of explaining how the joint surfaces between successive bones can form before the joint has become functional, and when one bone has not yet moved with respect to the other. The model here proposed rests on the intuitive idea that the morphogenetical field of an organ O, qua geometrical entity describing the biochemical kinetic of the organ, contains all the information required not only for the geometrical realization of O, but also for the virtual expression of all the later physiological capacities of O. In this sense the morphogenetical field of the bone O will determine not only the spatial form of O, but also all the possible positions of O relative to the organism; therefore the field associated with a bone such as the humerus or the tibia *will contain not only the form of the bone considered, but also the set of all possible displacements of this bone with respect to the immediately proximal bone:* the scapula for the humerus, the femur for the tibia, and so on. Then the set of positions of the distal bone with respect to the proximal bone will be given by a domain of the group of rotations, $SO(3)$, completely characterizing the geometry of the joint between these bones.

At the beginning of the formation of the limb bud, a characteristic swelling called the apical cap appears on the ectoderm, and the complete or partial ablation of this cap during development will completely or partially arrest the distal bone development. The inducing function of this cap has been studied extensively. In our model it will assist the development only in some purely geometrical sense, by giving the underlying mesenchyme a "polarization," a purely formal assistance in the separation of the bone fields, all the information required for the formation of the skeleton and its joints being supported in the mesenchyme but being realizable only through interaction with the cap.

The internal space of the mesenchyme, parameterizing the local biochemical states, will be a manifold M, and that of the cap a manifold Q isomorphic to the group of Euclidean displacements. The bones will be ordered in the proximal-distal sense, and in this way we can assign a number to each; if two different bones have the same rank, they will be called homologous (e.g., the tibia and the fibula, or phalanges in the fingers and toes). Suppose that the attractor associated with a bone of rank k lies in a submanifold of $M \times Q$ of the form $M \times q_k$, where q_k in Q is associated with the integer k, and that two different homologous bones of rank k have different attractors in $M \times q_k$. Suppose also that the separation of these incipient homologous bones is the result of a coupling of the metabolic regime defined by the attractor in M with the cephalocaudal gradient in the entire organism. Then we may suppose that at first the attractor of the bone regime is of the form $c \times Q$, where in Q, the factor describing the effect of the cap, the representative point at first moves about ergodically. Then, after a lapse of time, the situation degenerates through a catabolic catastrophe, and the manifold Q transforms into a variable gradient field with attractors $q_1, q_2, \ldots, q_k$ according to the distal position along the limb (where this degeneration might be the effect of a hypothetical proximal-distal gradient). Then the regime with attractor

(c, q_k) will result in a condensation of mesenchyme in this region of incipient bone.

We shall be most interested in the transitions between bones with adjacent ranks $k, k + 1$, that is in the formation of the corresponding joint; let A be the proximal bone with rank k, B the next distal bone with rank $k + 1$. The joint surface may, in principle, be considered as a shock wave, the conflict surface between regimes (c, q_k) and (c, q_{k+1}), even though (for reasons that we give later) the conflict zone between two regimes in competition will very often be fluctuating. We can suppose further that the stabilization of the regimes in Q will take place in the proximal-distal sense, so that the regime (c, q_k) may have stabilized in A, although the attractor of B continues to move throughout a domain W in Q containing q_{k+1}. In fact, this domain W will determine all the possible positions of B relative to A, according to the following mechanism: bone A will be supposed to be determined by a rigid chreod with coordinates (x, y, z) (biochemical parameters of unknown nature), while B will be defined in this coordinate system by an equation $S(x, y, z; q) = 0$, where S denotes the excess of relative entropy of B over A, and q the variable attractor of the regime B. Then the basic hypothesis is that, as a result of an appropriate biochemical coupling, if the point q in Q is of the form $q = Dq_{k+1}$, where D is a displacement, then

$$S(x, y, z; Dq_{k+1}) = S(Dx, Dy, Dz; q_{k+1}).$$

The result will be that the shock wave limiting B, in the chreod belonging to A, will vary according to the position of q in the domain W, the regime A will be stable only in the part of the chreod that never comes under the influence of B, and the effect will be as if the form of bone A arose from polishing by moving B through all the positions of W. This set of positions will contain the subset of the rotation group consisting of the functional positions of B relative to A, although, at least at the beginning, W might be even larger and contain, in particular, nonfunctional positions as if B were dislocated; this would explain the formation of stops on the surface of A limiting the movement of B (as on the elbow). Later domain W will shrink to that part of the rotation group defined by the normal play of the joint.

This mechanism provides an explanation of the adaptive nature of the joint surfaces. The formation mechanism for the other bone surfaces—in particular, the sides of long bones— is probably much simpler and belongs to the theory of the stabilization of a conflict stratum; therefore it can occur by successive accretion and destruction.

2. The formation of muscles attached to a joint. While the joint of two bones A and B is forming, we can suppose that the transitional regime between regimes A and B, characterized by the domain W of the rotation group, spreads by diffusion through the neighboring mesenchyme, which eventually becomes muscle. The transformation of a mesenchyme cell into a muscular cell (myoblast) is characterized geometrically by the appearance of a definitive polarization in this cell, denoted by a vector Y. The myoblasts in a muscle are grouped into myofibrils, tangent at each cell to Y, and this vector Y can also describe the force exercised by the contracted cell on its extremities. The differentiation of embryonic cells of the mesenchyme into myoblast occurs in laboratory culture and so requires no specific inductor; however, it is not unreasonable to suppose that the presence of a joint regime in the mesenchyme cannot but be favorable to this differentiation.

At the instant that we are considering, when the A-B joint is in the process of being constructed, an important phenomenon occurs in the transitional regime (A, B): at the beginning, this regime is characterized by the ergodic variation of a representative point in a domain W of the rotation group $SO(3)$ representing the set of functional positions of B with respect to A; shortly afterward, this ergodic variation undergoes a catastrophe. This catastrophe can be described as follows: W can be covered by the trajectories of at most three one-parameter subgroups (since W is at most three-dimennsional), with corresponding vector fields X_1, X_2, and X_3; then the movement of the representative point m in W will be obtained by superposition of oscillatory (swinging) movements along the directions defined by X_1, X_2, and X_3, each movement with a well-defined amplitude and period but with the phases being arbitrary. Then we suppose that the polarization of a cell in myoblast with axis Y has an effect of biochemical coupling with the joint regime which can be described as follows.

Let O be the center of rotation of a joint, and suppose that in a rigid chreod with center O the representative point m of the biochemical state, when the myoblast at point p is traversed in the direction Y, under a variation in $M \times Q$ whose projection on Q is given by $dm = Y \wedge pO$, where dm can be considered as an infinitesimal rotation with axis dm and so as a tangent vector to $SO(3) \supset Q$; but (and I insist on this point) this linear relation between Y and dm must be considered as a *biochemical coupling simulating a mechanical relation*. Then the variation in m will be parallel to one of the vectors X, by virtue of the joint regime, and because of resonance the polarization Y will, in the first place, be such that $Y \wedge pO = kX_i$. This means that the first cells to differentiate into muscle will be those in the planes perpendicular to the vectors X_i; they will organize themselves into myofibrils approximately joining A to B, and each vector X_i will give rise to a pair of incipient muscles, acting in opposite senses (according to the sign of X), flexor or extensor. Then, supposing that the polarization Y is a transient phenomenon propagated along a fibril with a constant speed, there will be, among all the fibrils parallel to Y and joining

A to B, one of such a length that if its end point at A is in regime q_A, then its B end point is always in regime q_B (at least in the projection on Q). Such a fibril, with length proportional to the period T of the swinging movement with axis X, will have a polarization system synchronous with the ambient joint regime and so will resonate and attract the neighboring fibrils of the same length; this will concentrate the emerging muscle. In this way, at most three pairs of muscles will form for each simple joint.

Of course this model could be improved in many ways. In particular, it is possible that the formation of a muscle is much more genetically determined, and the attachment points of the tendons are probably more closely controlled than we have indicated. Furthermore, this model seems difficult to apply to the case of very long muscles joining noncontiguous bones. However, even if the actual mechanisms are controlled genetically (but how?), a model like this will retain its value for explaining the evolutive origin of the functional adaptation of muscles to bones.

Finally, we observe that the formation and orientation of the muscles of a joint A lead to the extinction of the transitional regime (A, B) based on the joint zone of the two bones. If the effective functioning of this joint is then prevented (e.g., by an innervation fault or mechanical blockage), the local joint regime will be captured by the bone regime, and its degeneration will cause the joint to disappear. This is anchylosis.

9.7. LATE EPIGENESIS: SOME ARCHETYPAL CHREODS ASSOCIATED WITH UMBILICS

A. Capture chreods

Setting aside embryology for the moment, we shall describe the dynamical origin of the chreods by which a living being absorbs external food; here we shall describe in a mathematical framework the phenomena of phagocytosis and pinocytosis in unicellular organisms and their organic equivalents in Metazoa.

We return to the universal unfolding of the parabolic umbilic (with the addition of fourth-order stabilizing terms), and to begin with we consider the situation arising from the generic deformation of the elliptic umbilic (Figure 9.13). There will be an area (7) in the form of a hypocycloid with three cusps, and this concave zone, triangular in section (analogous to the neural plate in vertebrate embryology), will be the hinge of the whole process. With the approach of the prey, this elliptic state, in tension, will evolve toward a hyperbolic situation, with the formation of a hyperbolic umbilic at h (Figure 9.14), and then, by a concomitant catastrophe, the capture of regime 7 by the simpler, more relaxed regime 3; the two external lips of 7 will move outside the organism and join together to capture a piece of the external environment. Thus the prey p will be imprisoned in a vacuole which will become a digestive vacuole; the situation will revert to the elliptic state, as in Figure 9.15, and the part of regime 7 will then be taken by the (initially living) prey p, the process of digestion and progressive annihilation of the prey being described by the transformation of the hypocycloid (7) into the point circle (elliptic umbilic) (Figure 9.16).

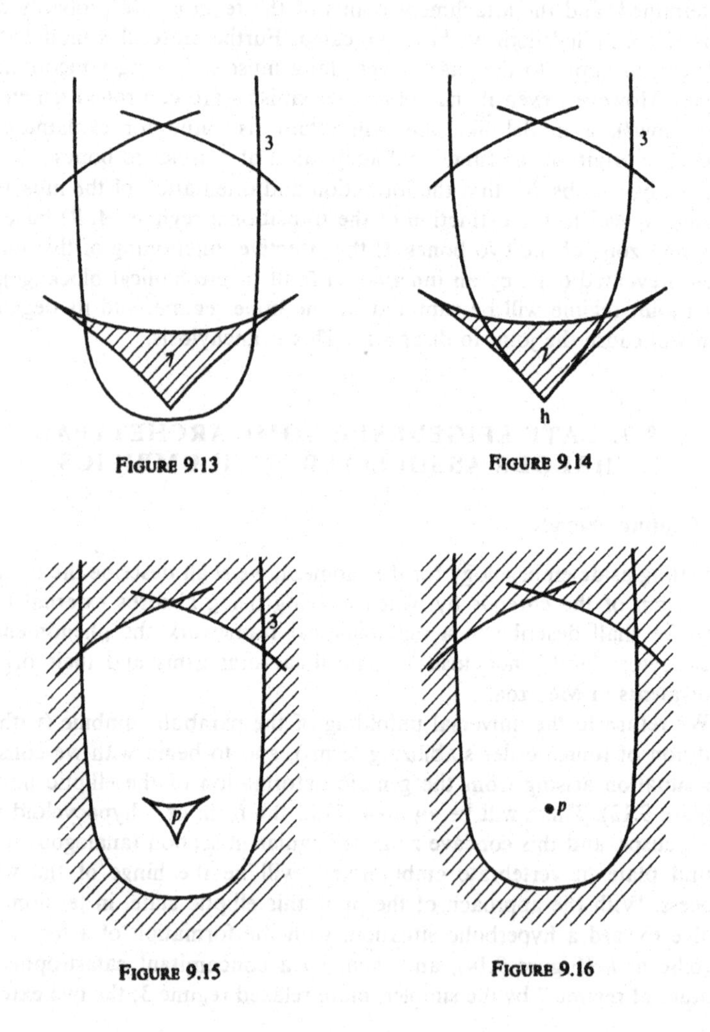

FIGURE 9.13

FIGURE 9.14

FIGURE 9.15

FIGURE 9.16

This general scheme seems to apply roughly to unicellular organisms like *Amoeba*. Sometimes phagocytosis is preceded by the emission of a pseudopod, when the process is the breaking of an elliptic jet, and the capture at the end of the jet is an evolution toward a locally hyperbolic state. In other cases, phagocytosis (or, according to some writers, pinocytosis) is connected with the folding of a sheet of hyaloplasm toward the cell body; here the same scheme can apply, perhaps with some asymmetry. There is also a third process, ropheocytosis, occurring on the scale of ultrastructure (300 Å in diameter), where the cellular cortex invaginates into deep depressions (1000 Å) like the fingers of a glove, which become occluded and create vacuoles. Probably this third phenomenon is a kind of inverse elliptic umbilic (dual, as is said in topology), a "hollow hair" (Figure 9.17).

We see from the description that the typical capture appears as a kind of inverse hyperbolic breaker; at the beginning there is an organically stable cusp, which then transforms into a hollow wave with the hollow representing the prey.

Returning now to the mouth chreod, we shall improve the previous mathematical description by applying the principle of threshold stabilization. Because there is a critical moment representing the transition between the elliptic and hyperbolic situations, it will be more economical if the organ is stabilized in the threshold situation, the parabolic state, provided that it can vary locally on either side of this threshold. We therefore consider the parabolic umbilic:

$$V = x^2y + \frac{y^4}{4} + tx^2 + wy^2 - ux - vy$$

with its critical set C:

$$2x^2 = (y + t)(3y^2 + 2w)$$

(see Section 5.4.C).

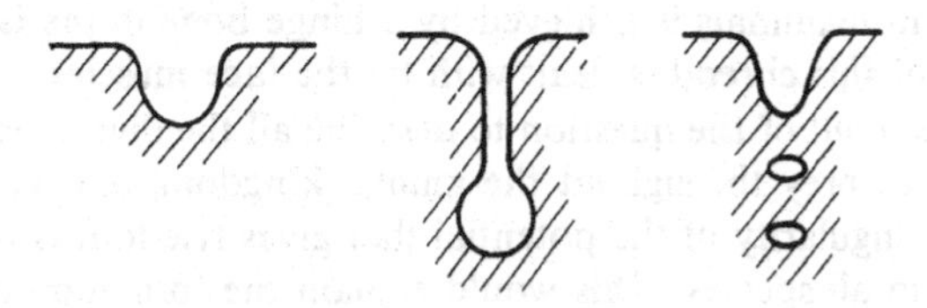

FIGURE 9.17. Ropheocytosis.

When $t > 0$ and w varies, we obtain (Figure 9.18) the following:

for $w = 0$, C has a double point at the origin with distinct tangents;

for $w < 0$, C has two connected components, one compact, lying between $-t$ and $-\sqrt{-2w/3}$; and

for $w > 0$ and very small, C is connected.

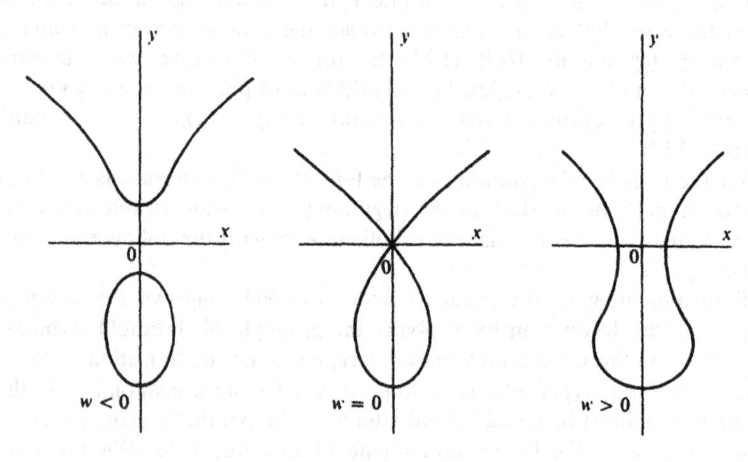

FIGURE 9.18

In the image space Ouv, for $w>0$, each branch of the critical curve has two cusps forming a swallow's tail, and as w tends to zero these points join up beak-to-beak on the symmetry axis Ov at the origin. This is the shutting of the mouth (Figure 9.19), and the external cusps that continue after the mouth closes might represent the lips, while we might reasonably interpret the internal cusps as organs like the teeth (as an effect of a later threshold stabilization which will intervene locally at the closure point). The closing of the mouth in mammals is achieved by a hinge bone in the jaw, and the variable part of this chreod is dealt with by the face muscles.

It is obviously out of the question to describe all the diverse morphology of the seizing of prey throughout the animal kingdom, but we may hope that the local singularity of the potential that gives rise to it is fundamentally the same in all species. This would explain the formation of chimera, as in the well-known experiment by Spemann, where the ectoderm of a *Triton* participates, in its own genetically determined way, in the construction of the mouth of a frog larva.

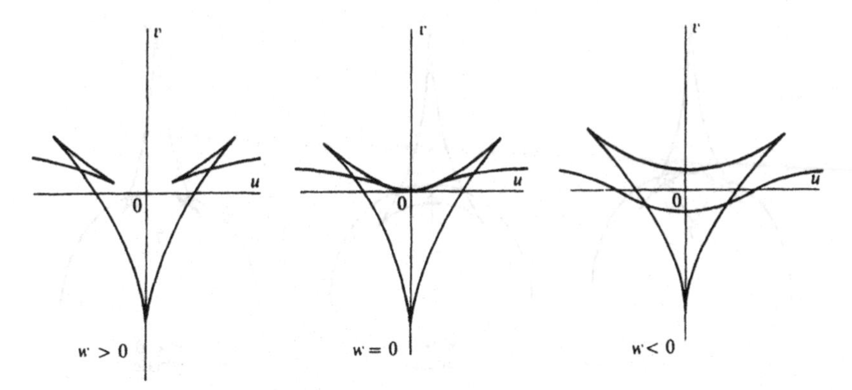

FIGURE 9.19. The mouth closing. (See also Figures 5.24 and 5.25.)

The closure of the neural plate, in vertebrate embryology, probably has a similar model, although here the external cusps become unstable after the closure (they are the neural crests) and become submerged in the epidermis. A detailed study of the formation of the embryonic axis in vertebrates would evidently be of the greatest interest. (See the appendix.)

B. Genital chreods

A typical example of breaking is the emission of gametes in Metazoa, but here the sexual act itself, consisting of the spatial ejaculation of the gametes from the male organism, is long preceded by the biochemical maturation of the gametes. This results in an inhibited or virtual breaking, and so we should not be surprised to find that the organizing center is the parabolic umbilic, the singularity by which a generated structure separates itself from the parent organism. We therefore consider the parabolic umbilic with $w = 0$ and $t > 0$, giving Figure 9.20 (where the axis Ov has been reversed), the hermaphrodite situation; region (5) represents the gonad, and the cusps (7) the gametes breaking in the gonad. Taking $w < 0$ gives the male deformation (Figure 9.21), in which the critical curve C decomposes into two connected components, and the gametes (7) separate topologically from the gonad wall and take up a regime (5) cutting into this wall; this amounts to saying that the gametes tend to leave the gonad (this being the medullary component). For $w > 0$ (Figure 9.22) the gametes remain bound to the wall (cortical), while an oviduct extends toward positive v up to the higher cusp a; by supposing that C is suitably displaced parallel to itself in Oxy, we can remove this cusp and get the male and female curves m and f (Figure 9.23). Here it is superfluous to point out the close resemblance of f to the global configuration of a pistil.[1]

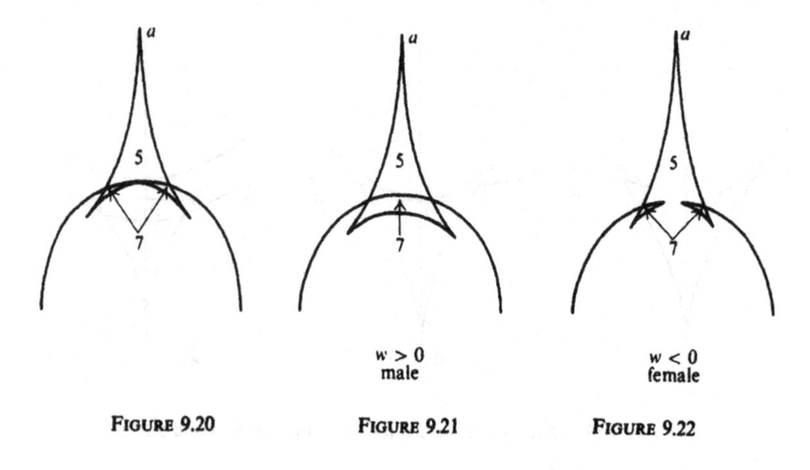

FIGURE 9.20 FIGURE 9.21 FIGURE 9.22

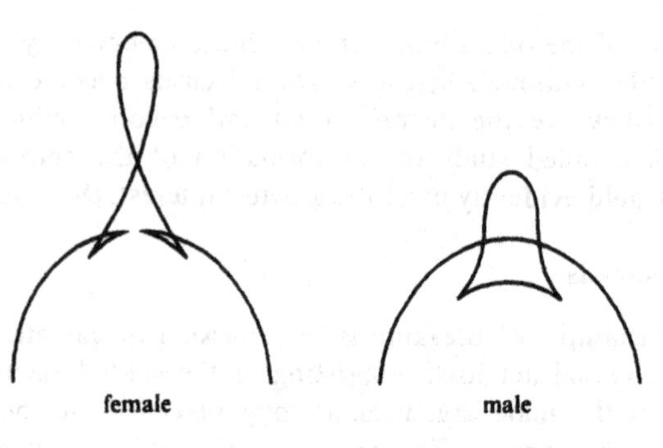

FIGURE 9.23

In fact, and for obvious physiological reasons, the cusp a indicating the end of the efferent canal must be open to the outside, but this cannot occur in this model which is too local. Another singularity, another organizing center, is necessary to bring about the junction of this gonad canal with the external world, and, curiously, this is again the parabolic umbilic but this time unfolded in two orthogonal, complementary directions. In the male (e.g., in man) there is an unfolding in the proximal-distal direction, but coupled with this is an evolution m: there is the beginning of an oviduct (Müller's canal) which then atrophies, while the closing and thickening of the wall d represents the closing of the urogenital sinus in the scrotum. Orthogonal to this, parallel to the body axis, and tangential to the skin, there is the other development f: the formation of a penis by a stretching of the genital tubercle, with free communication between this new organ and the new scrotum. The female undergoes a complementary evolution with development f, the formation of Müller's canal and the opening of the lips in the distal-proximal sense, and in the orthogonal direction the development m involving the regression of the genital tubercle into the clitoris. The formation of the glans at the end of the penis corresponds to the action of a third organizing center, derived from the second; this has as representation the hyperbolic umbilic with $w < 0$ and $t < 0$ (see Figures 5.25-6 and 9.24), giving the typical mushroom configuration in the Ouv-plane. A variation of t leads to the deformation given in Figure 9.25; here again it is superfluous to remark on the similarity to the mushroom *Phallus impudicus*. The analogy of Figure 9.24 with 9.23f can be explained if we view copulation as a reversible capture.

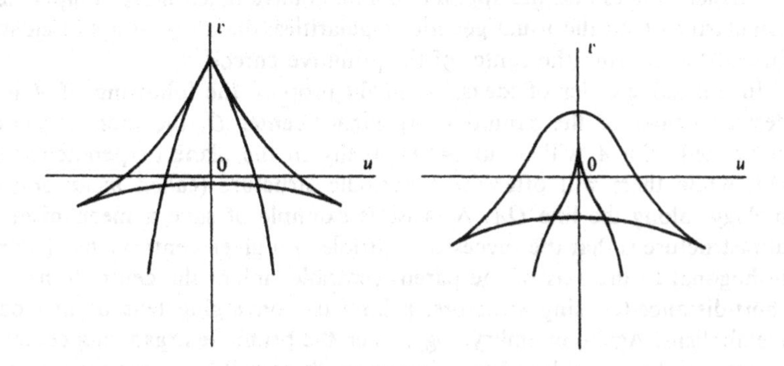

FIGURE 9.24 AND 9.25. The plan for the glans. (See also Figures 5.24 and 5.25.)

It might be argued that these analogies are largely arbitrary, and that any mathematician could construct simpler functions that were just as suggestive. To this I can only reply that I was led to these equations by the theory of structurally stable singularities with the additional hypotheses of bilateral symmetry and the stabilization of the umbilic into a parabolic situation. To what extent might one object to this last hypothesis? That the human embryo has a hermaphrodite structure until a late state is, without question, not the result of our having had distant hermaphrodite ancestors, as the law of recapitulation would have it, but rather occurs because epigenesis has found it more economical, before constructing either males or females, to construct first a threshold situation that can then direct the organization in one or the other direction after a short lapse of time. Hence this is yet another instance of threshold stabilization.

Another problem that arises is the appearance of secondary organizing centers in the chreod of the basic organizing center. In the example above, the secondary center appears at a singularity of the primitive chreod (here, the cusp *a*), and it is tempting to think that, in effect, this secondary orgainzer appears by a kind of focusing phenomenon, often at some distance from the primitive center. Similarly, the triple bang produced by the passage of a supersonic airplane can occur in a very localized way extremely far from the path of the airplane. Also we must explain how, in such cases, this derived singularity can be as complicated as, if not more complicated than, the primitive singularity, although this is, in principle, impossible in a global static model. It is tempting to think that the constraints imposed by the holonomy conditions $V_{xy} = V_{yx}$ (or the more subtle constraints of threshold stabilization), which, it is known, imply that the critical values and the apparent contours have much more complicated singularities than the usual generic singularities, die away at a sufficiently large distance from the center of the primitive chreod.

In this same order of ideas, we might propose the following: if A is a derived center of the primitive organizing center O, the morphogenesis associated with A will occur preferentially in the plane perpendicular to OA, while there will often be a periodic structure (called *metameric* in biology) along the line OA. A possible example of such a mechanism in ultrastructure is that the successor centriole of a given centriole has its axis orthogonal to the axis of the parent centriole; might the centriole have a short-distance-focusing structure, a kind of converging lens of the local metabolism? Again, in embryology, when the primitive organizing center is located in the dorsal lip of the blastopore, there will be a secondary center in the front end of the notochord. Then the associated morphogenesis, that of the neural plate, will occur essentially in the orthogonal (bilateral) direction, by the folding of the plate, and there will be the metameric structure of somites in the radial direction. Furthermore, in the case of Coelenterata, Tunicata, and the like, where the animal reproduces by splitting or by strobilation of a stolon from the abdomen, the morphogenesis occurs in the plane perpendicular to the axis of the stolon, while an approximately periodic structure is set up along the axis. Finally, a principle of this kind could account for the dendritic growth of crystals.

APPENDIX

Neurulation and the formation of the spinal cord. Let us return to the gastrulation of a chicken's egg at the stage of the primitive streak; we know that the head end of this line moves for some time toward the cephalic direction, stops, and gives rise to the formation of Hensen's node. This node then moves back in the caudal direction; meanwhile the notochord appears, extending along the axis below the epiblast, and induces the neuralization of the overlying epidermis. The somites form on both sides of the notochord, separated by lateral mesoderm, which exfoliates in two layers (somatopleure, splanchnopleure) surrounding the coelome. Meanwhile the epidermis neuralized into the neural plate folds up and creases, and the neural folds join up progressively from front to rear, forming the neural tube; later mesenchyme, emitted by the somites (sclerotome), and the neural crests form a periodic structure of rings around the neural tube. Finally the ossification of these rings gives the vertebrae, as the notochord regresses and disappears.

An interpretation of such a complicated morphology is obviously difficult. Also the algebraic models that we give here are unavoidably arbitrary, and they are presented with this warning; a more detailed description can be found in [5].

The first qualitative idea is that the notochord, by virtue of its rigidity, must be justified by a dynamical scheme characteristiic of elongated, pointed organs—hence the tapered cone bounded by an elliptic umbilic; in section, this organizing layer has a hypocycloid with three cusps. At the same time we can suppose that the local dynamic is approximately invariant under translation parallel to the notochord and also has bilateral symmetry with respect to its symmetry plane. This leads us to suppose that the local dynamic in the neighborhood of the axis is given by a static field which is an unfolding of the parabolic umbilic of the type considered in Section 6.A. The elongation of the germ at the time of neurulation must be considered, in this model, as a controlled breaking, held in a parabolic situation.

The beginning of the formation of Hensen's node involves a problem that makes it difficult to give a satisfactory model, despite the versatility of our methods. The most probable hypothesis is that the swallow's tail at the cephalic end of the primitive streak degenerates through flattening, and the lines defining the ectoderm and mesoderm become tangential while the swallow's tail shrinks to a point. Then, after an ill-determined catastrophe, probably involving the virtual layer representing the endoderm and perhaps also a virtual layer affecting the ectoderm (and involving, as well, a cusp representing the future neural fold), we have the situation of Figure 9.26-4, where the shaded region represents the notochord, which so far covers only a portion of the curvilinear triangle. The point penetrating into the ectoderm represents neurogenic induction, and on moving forward this point plunges under the ectoderm, where the notochord ends in an elliptic umbilic (Figure 9.26-5). When Hensen's node begins to regress in the caudal direction, the figure complicates. A swallow's tail develops at the point where the arc crosses the axis of symmetry and then pierces arcs *1-4* and *1-10* bounding the cord (Figure 9.27). Finally this produces a kind of five-pointed star crossed by a horizonatal line, and it is this that we might consider as the ultimate director plan of the mesoderm: the quadrilaterals, such as *23lm*, define the somites; the points plunged into the ectoderm, like *ml9*, represent the induction of the neural crests; the triangles, such as *345*, represent the lateral mesoderm and their interior the coelome; and at point 3, the milieu of the median layer, an intermediary piece develops, the support of the future gonad (in fact, point *5* has two symmetric representatives).

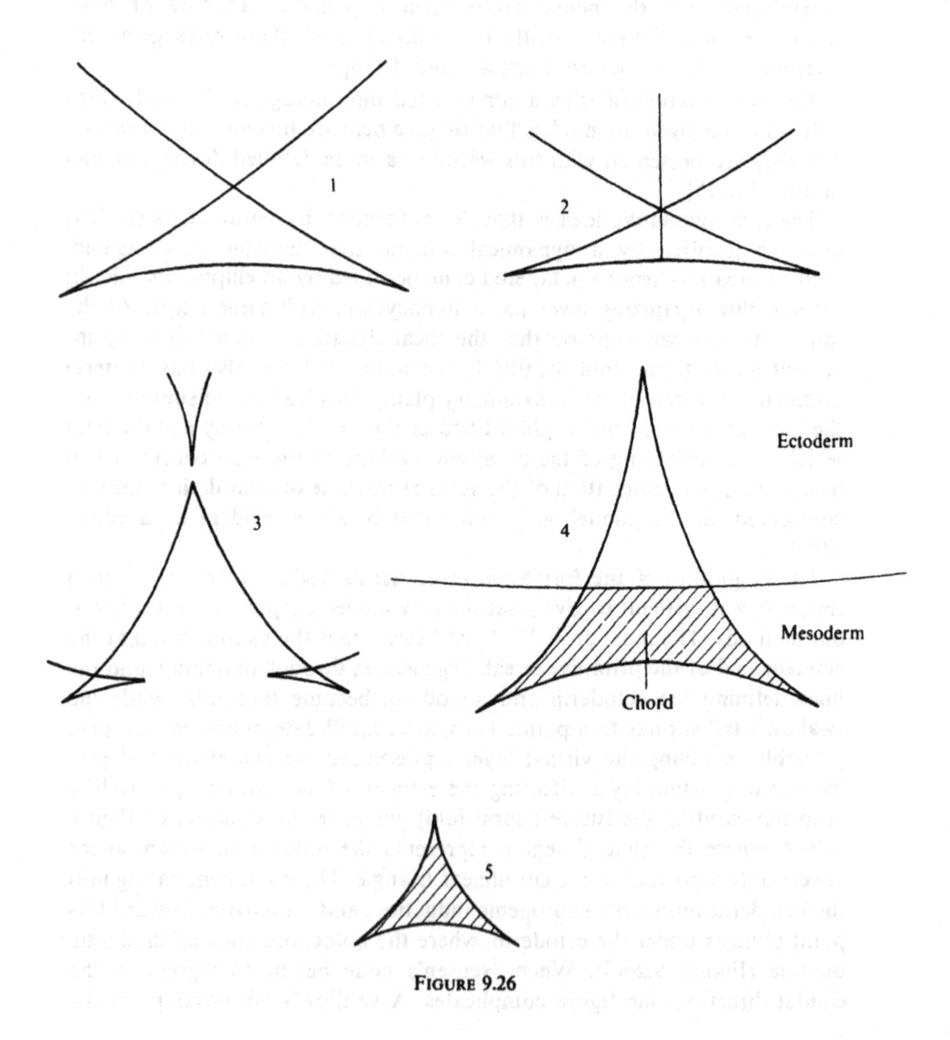

FIGURE 9.26

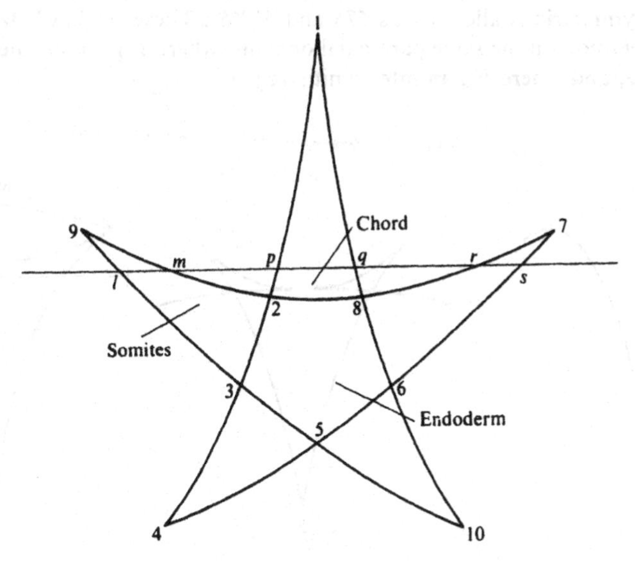

FIGURE 9.27

Now consider the ectoderm. Under the neurogenic influence of the mesoderm, it will take up a local dynamic defined by a potential V_1, coupled to the potential V of the mesoderm by a relation of the form $V + V_1 = \text{constant}$, and suppose further that V_1 and V have the same unfolding coordinates (u,v), except that u has opposite sign in V_1, representing the dorsoventral gradient. Under these conditions the organizing layer of the nervous tissue will be arranged symmetrically with respect to that arising from the unfolding of V in the mesoderm. However, since the ectoderm shows a certain delay, this configuration will only develop progressively, as in Figures 9.28-1, 2, and 3, and this will provide a description of the closure of the neural fold. Finally another swallow's tail, *5′7′9′*, develops, a symmetric to *579* and representing the formation of the anterior, motor horns of the spinal cord; later point *1′* retracts above arc *7′9′*, giving the *H*-shaped formation that represents the gray matter of the spinal cord, and the double points *4′* and *10′*, remains of neural folds, become the posterior, sensitive horns. The periodicity of the somites can be interpreted as follows. In principle the local dynamic of the ectoderm and the endoderm will be invariant under translation parallel to x, the coordinate defining the cephalocaudal gradient. Now we can suppose that this symmetry will be broken at the end of the catastrophe by a perturbation of the form $V + \cos x$, $V_1 - \cos x$, affecting precisely the unfolding of the symmetric swallow's tails *578* and *5′7′8′*. These swallow's tails are in competition on the same para-axial domain; where V governs, there will a somite, and where V_1, an intersomitic region.

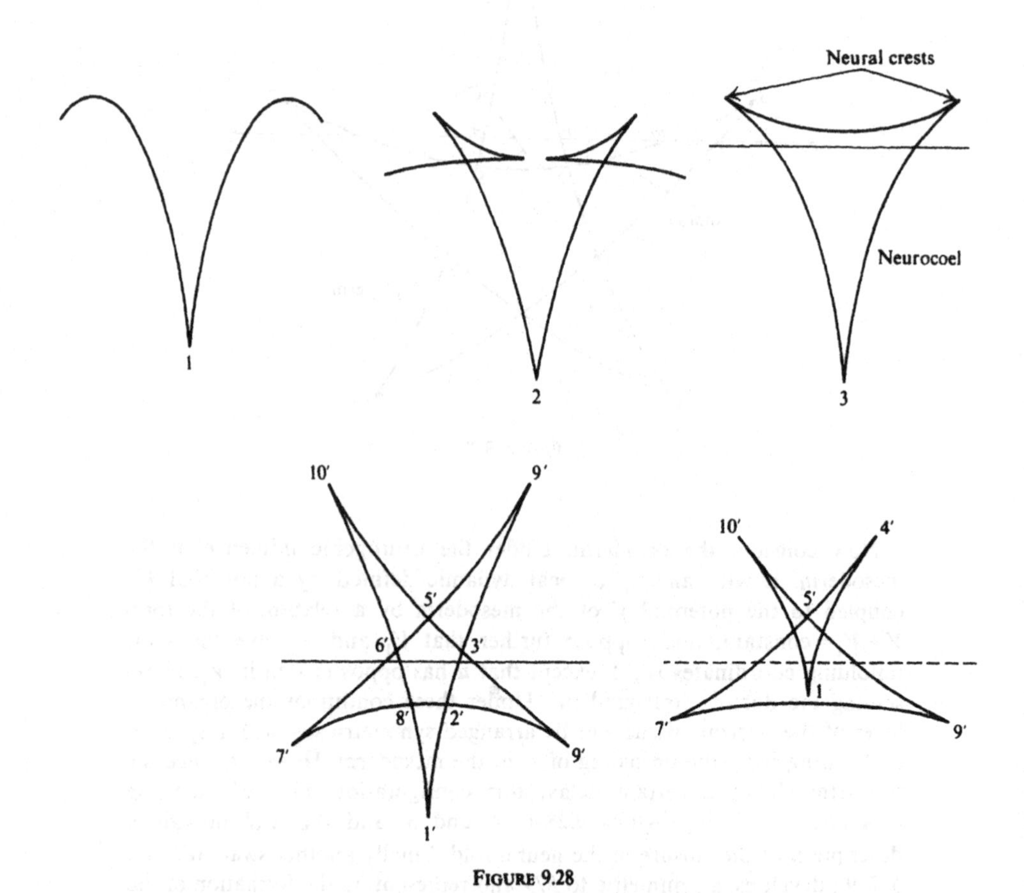

FIGURE 9.28

As the horizontal *mpqr* moves downward, it eliminates the cord, which necroses, and some of the somites, which disintegrate into mesenchyme (sclerotome); here the validity of the model ends. Considerations of the gradient defined on the regulation figure (Section 10.2.C) provide an interpretation of why the mesenchyme surrounds the neural tube (in the somitic zones); let ρ denote a function that increases from states r to states s, so that ρ will be maximum on the boundary of the neurocoele. (In fact, the neurocoele is the support of the representation of the animal's external world.) Since the epidermis has a lower value of ρ than the bony mesenchyme, the latter will infiltrate between the neural tube and the epidermis. We can then suppose that the orgainizing center of the mesoderm, defined by Figure 9.27, re-forms in this tissue, with the neural tube occupying the central polygon *23568*; the global plan of a vertebra, with its processes, is not unlike the five-pointed star of Figure 9.27.

It would clearly be interesting to pursue this kind of study of the anatomy of the nervous system further; let me give two examples. First, the retraction of swallow's tail *5′7′8′* (symmetric to *578*), defining the motor horns, must occur in the spinal cord at the level of its junction with the brain; might this be the origin of the interlacing of the motor and sensory fibers at the level of the pons? And, second, an interpretation of the neurocoele as the "support of infinity" of the external world is curiously confirmed in the cephalic extremity of this cavity. It is known to divide into two horns in the brain, the vestiges of the interior cavities of the peduncles of the optic vesicles during eye formation; therefore we can say that the neurocoele ends at its cephalic end in the retinas of the eyes, exactly that nervous zone specialized in a particularly precise simulation of distant phenomena.

NOTES

1. "... Un sourire lui plissa les lèvres, car subitement il se rappelait l'étrange comparaison du vieux Nicandre qui assimilait, au point de vue de la forme, le pistil des lys aux génitoires d'un âne ... ." J. K. Huysmans, *A rebours*, Fasquelle, Bibliotheque Charpentier, 1965, p. 221.

REFERENCES

1. K. Vogt, Gastrulation und Mesodermbildung bei Urodelen und Anuren, *Arch. Entwicklungsmech. Organ*, **120** (1929), 384–706.
2. L. Gallien, *Problèmes et Concepts de l'Embryologie Expérimentale*, Gallimard, 1958.
3. C. H. Waddington, *Principles of Embryology*, Allen and Unwin, 1956.
4. S. Ulam, Combinatorial analysis in infinite sets and some physical theories, *S.I.A.M. Rev.* **6** (1964), 343–355.
5. R. Thom, A global dynamical scheme for vertebrate embryology, *Lecture on Mathematics in the Life Sciences*, **4**, Amer. Math. Soc., 1971.

CHAPTER 10.

GLOBAL MODELS FOR A LIVING BEING (METAZOA)

Der Mechanismus jeder beliebigen Maschine, z. B. unserer Taschenuhr, wird immer zentripetal aufgebaut, d. h. die einzelnen Teile der Uhr, wie Zeiger, Feder und Räder müssen immer erst fertiggestellt werden, um dann einem gemeinsamen Mittelstück angesetzt zu werden.

Im Gegensatz dazu geht der Aufbau eines Tieres, z. B. eines Triton, immer zentrifugal von einen Keim aus, der sich erst zur Gastrula unformt und dann immer neue Organknospen ansetzt.

In beiden Fällen liegt der Umbildung ein Plan zugrunde, von denen wie gesagt der Uhrplan ein zentripetales Geschehen, der Tritonplan ein zentrifugales Geschehen beherrscht. Wie es scheint, werden die Teile nach gänzlich entgegengesetzten Prinzipien ineinandergefügt. [1]

J. VON UEXKÜLL,
Bedeutungslehre

10.1. THE STATIC MODEL

A. Preamble

I do not think that this description by the physiologist Uexküll of the essential difference between the dynamic of life and the anthropomorphic constructions with which it is often compared can be bettered. This is not to say that comparisons of the life dynamic with some manifestations of human technology (automata, electronic computers, etc.) are pointless, but ratner that these comparisons have validity only for partial mechanisms. fully developed with their complete functional activity[2], and they can never be applied to the global living structure or to its epigenesis and physiological maturation. I hope that I may here be excused a highly technical comparison borrowed from algebraic topology. The topologist has two radically different means with which to construct a given space. One, centripetal, is to make the space from a cellular subdivision, constructing it by successively adding cells or balls, and specifying the means by which each is attached; the other, much more subtle method consists of making the space out of successive fibrations in Eilenberg-MacLane complexes, where the nature of the fibration at each step is specified by an appropriate invariant. There is little doubt that the epigenesis of a living being as a metabolic form resembles the second method much more than the first, with the epigenetic memory, localized mainly in the genes, corresponding to the invariants of the successive fibrations. Of course the development is permanently restricted by biochemical and spatial constraints, and all these constraints have to be coordinated in such a way as to satisfy the a priori requirement of structural stability. Here I am going to present a model that, whatever its defects, gives a relatively accurate global picture of the working of Metazoa.

B. The static global model

Suppose, to begin with, that the genetic endowment of an animal species A can be characterized geometrically by a germ W of a differential manifold at the origin O of Euclidean space E, and by a manifold Q of linear projections p of the form $E \rightarrow E'$. The basic idea is intuitively as follows: at any moment t, and any point m of the animal, only a fragment (generally small) of the genetic information of the species will be used effectively, and this fact that most of the global genetic information will be neglected locally is best expressed by the geometrical operation of projection; hence at a given time t the local biochemical state at m is represented by the projection $p(m)$, and the qualitative properties of the genetic information used will be defined by the topological type of this projection map $p(m) : W \rightarrow E'$. As usual, there is, in the space of projections Q, a closed subspace K, the bifurcation subspace, such that the topological type does not change in each connected component of Q-K (in fact, only a stratified subset of K plays a part when we consider a static field of the type described in Chapter 4). We suppose that, among all the projections $p(m)$, there is one that is the most degenerate and so the most topologically complex, and this projection will correspond to the dynamic of the egg before fertilization; let $p(a)$ be the corresponding point in Q that will correspond to the most "repressed" local dynamic (to use a very fashionable word). The other projections in Q will then give rise to less complex projections than $p(a)$, and so we can think of Q as part or all of the universal unfolding of the singularity with organizing center $p(a)$. We can also suppose that the topological complexity of $p(m)$ is a continuous function of m, taking its maximum at a.

The egg polarizes immediately after fertilization; this state, after fertilization, will be represented by a map (we might say, without punning, a germ of a map) F of the three-dimensional ball B^3 into a neighborhood of a in Q, and we shall suppose this map to be differentiable and of maximum rank (even though it was the constant map a before fertilization). The map F will vary, as a function of time, during development according to a differentiable relation of the form

$$\frac{dF}{dt}(m, t) = X(m, t),$$

where the vector $X(m, t)$ depends both on the point $F(m, t)$ in Q and on the conditions prevailing in the local biochemical environment. Because this model is not precise, we shall not attempt to specify $X(m, t)$ quantitatively but will restrict our attention to describing the qualitative evolution of the image $F(B, t)$ in Q during normal conditions—this will be the *growth wave* of the individual. We might consider this growth wave F_t intuitively as a three-dimensional hypersurface in Q whose local evolution is governed by the metabolic wind defined by X. On the whole, there will be two different types of local regimes: catabolic regimes, characterized by a loss of energy (consumption of reserves) and information (decrease of topological complexity); and anabolic regimes, in which the energy and complexity increase.

At the beginning of development, when the embryo is living on its reserves, the evolution of $F(B, t)$ will be catabolic in the sense that the energy is decreasing; $F(B, t)$ will move in Q so as to cut certain bifurcation strata transversally, resulting in histological separation between the different tissues on the counterimages of these strata by F. Then the organs will form and will develop up to the time when they enter into functional activity; the transfer of energy and complexity with the external world will balance, and X will die down except in one region, the gonad, where it will be, in some sense, tangent to F and will cause some kind of breaker. Some isolated cells will detach, topologically, from the parent organism and be swept by a kind of cyclone toward the initial germinative point a; this is gametogenesis, the most anabolic of regimes. From this point of view, the thermodynamical structure of anabolic regimes raises some very interesting problems:[3] an anabolic regime seems, a priori, to defy all natural thermodynamics, which generally evolve toward decreasing complexity.

This evolution might be explained by analogy with a yacht that can sail against the wind by tacking on either side of it. Now, in this analogy, the sea will correspond precisely, in the cell, to the chromosomal DNA, the only element of the cell relatively independent of the metabolism in an environment of other molecules constantly swept along by the cellular metabolic wind. During this beating against the wind, the helmsman must cancel the component of the drift exerted by the wind on his sail by an appropriate orientation of the tiller, and this produces a large wake. This analogy may explain the strange chromosomal formations (lamp-brush) observed during oogenesis; if they are not the sign of a wake, they at least show the great tension undergone by the genetic material at this moment. We shall discuss sexuality and meiosis later, for they cannot be accommodated in so crude a model.

What should be the dimension of the space Q in which the growth wave of the animal evolves? For a situation of strict structural stability, it would be restricted to not more than four, the dimension of space-time. Such a restriction is only locally valid, however, and we might suppose Q to be of some finite-dimension, while the evolution of the growth wave is fairly strictly canalized by epigenetic and physiological regulating mechanisms into a tubular neighborhood of a manifold F, which itself cuts transversally the different strata of the bifurcation set determining the differentiation and so is structurally stable. We might note that this manifold F is not simply connected and has a noncontractible loop, which, starting from the germinative point a, may be realized as the evolution of the germ line; this loop returns to a after gametogenesis and completes the circuit from egg to egg in one generation (Figure 10.1 gives a model for F in three-dimensional space). Finally, this model reduces to considering the whole animal as a part of the universal unfolding of its egg, the gametogenesis being the inverse process that brings the local dynamic into its most degenerate, folded form.

In spite of the defects of this model that we will point out later, it provides a correct global insight into the development of an animal. Moreover, it lends itself to a reasonably good interpretation of the regeneration process in species like planarians, which have a broad regenerative capacity. We now consider this point in more detail.

C. The geometry of regeneration in planarians

In planarians, we might suppose that the growth wave $F_t(B)$ has values in a two-dimensional space Q, which contains a map of the adult animal in which one coordinate, x, corresponds to the cephalocaudal gradient, and the other, y, to the mediolateral gradient. We suppose also that $x = 1$ corresponds to the empty regime defining the boundary of the individual at its cehalic extremity, and $x = -1$ to the empty regime at the caudal extremity. On cutting the planarian from the head along the neck (normally mapped by F into the line $x = 0$, e.g.) the wave F will jump discontinuously from the neck regime to the empty (cephalic) regime, but as such a discontinuity cannot persist in this model the wave F must regularize by diffusion. The head chreod, situated in Q in $0 < x < 1$, will again be swept out by F, and the head will be reconstructed (Figure 10.2). If two decapitated planarians are grafted in opposite directions along their necks, the growth wave F of the double individual so constructed will have no discontinuity and there will be no regeneration (Figure 10.3).

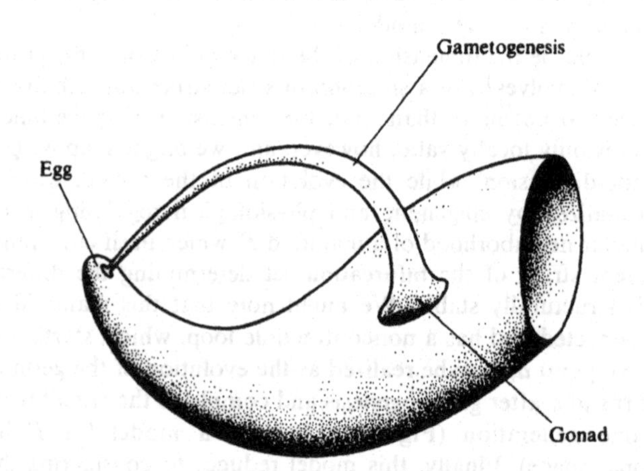

FIGURE 10.1. The trajectory of the growth wave.

Here again the model is too crude. The particular choice of empty regime at the amputation point is, no doubt, a function of the position in the body of the animal and might result in the regeneration of either head or tail for certain cuts. Finally, when considering the details of regeneration, it is necessary to consider chains of induction between the organs (e.g., the brain induces the eyes and the gonads), which seem appropriate to a more complicated genetic model that we will give later and, in any case, cannot be described by the propagation of a rectilinear wave in the plane.

D. A digression: preformation and epigenesis

The essential feature of the genetic equipment of an animal resides not in its morphology—many features of anatomy have, in the end, only a secondary importance—but in its mechanisms of homeostasis and

SPECIAL ILLUSTRATIONS

Nos. 1-27

1

2a

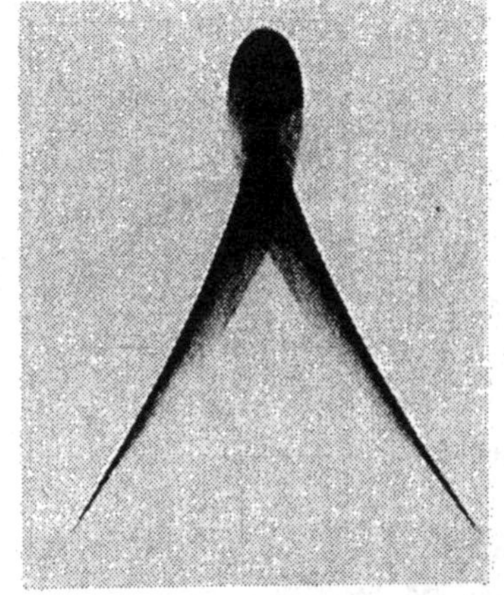

2b

2c

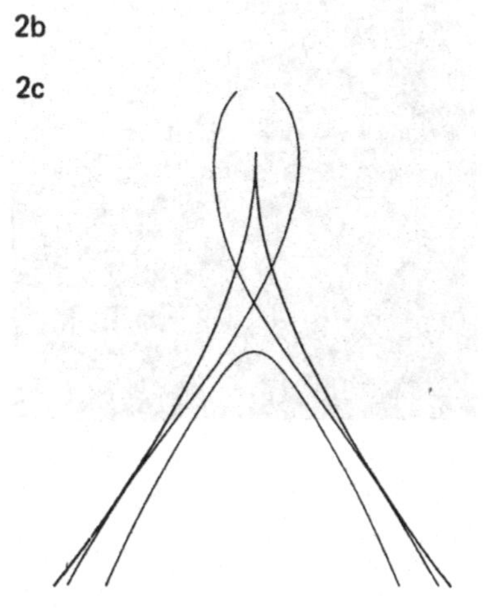

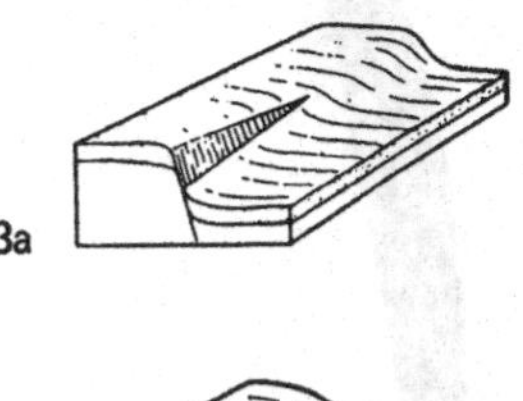

3a

3b

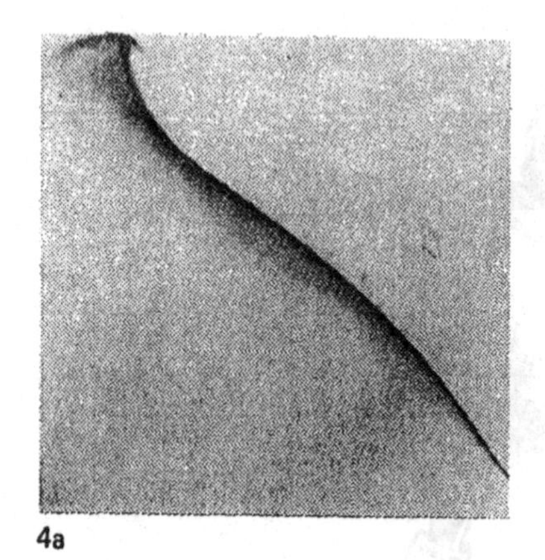
4a

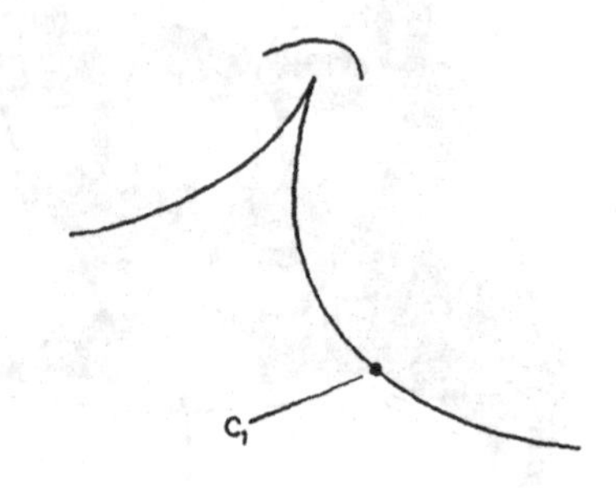

4c

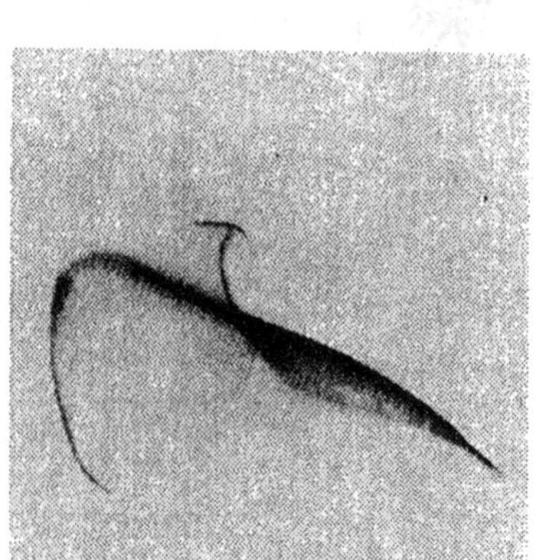
4b

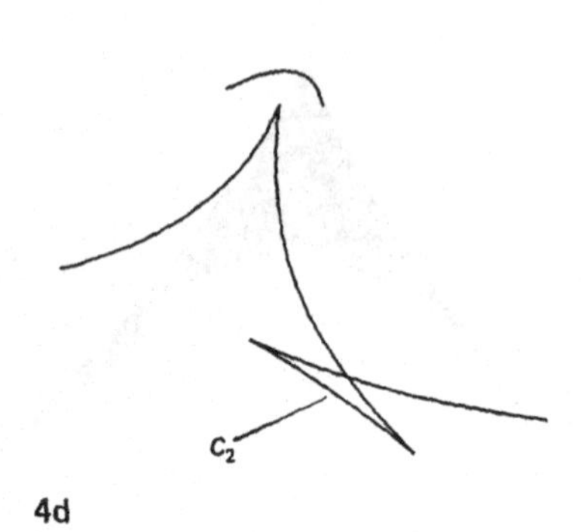

4d

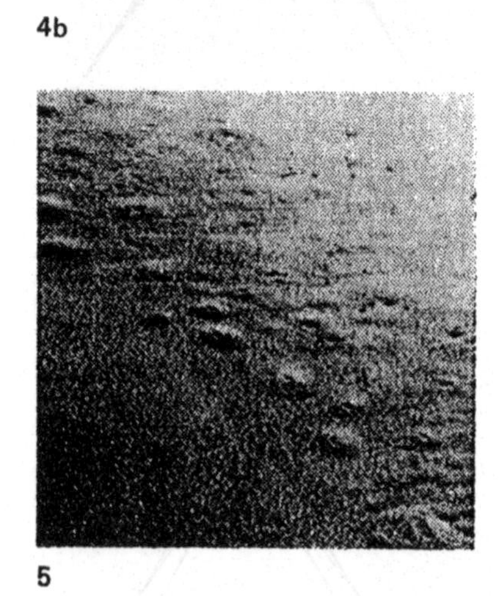
5

6

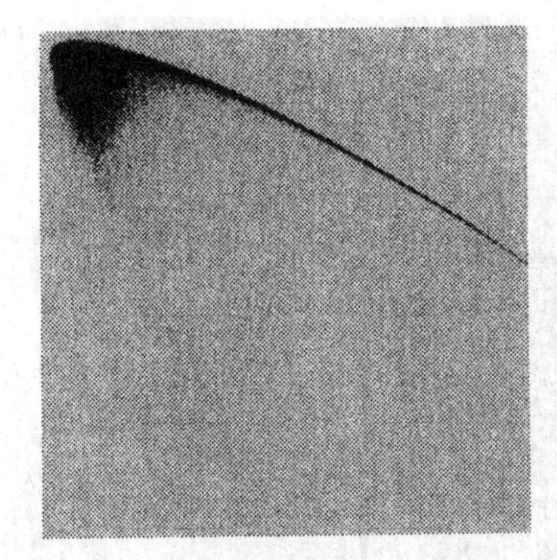

8a

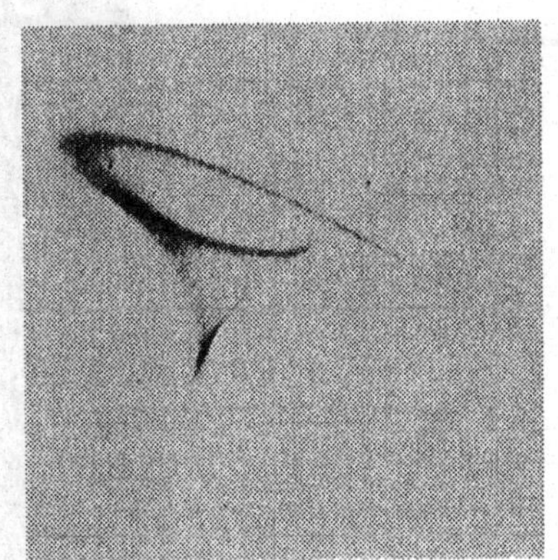

8b

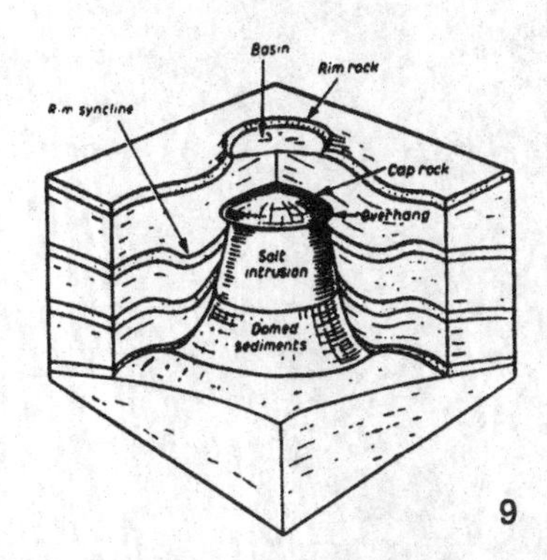
Basin
Rim rock
Rim syncline
Cap rock
Overhang
Salt intrusion
Domed sediments

9

10

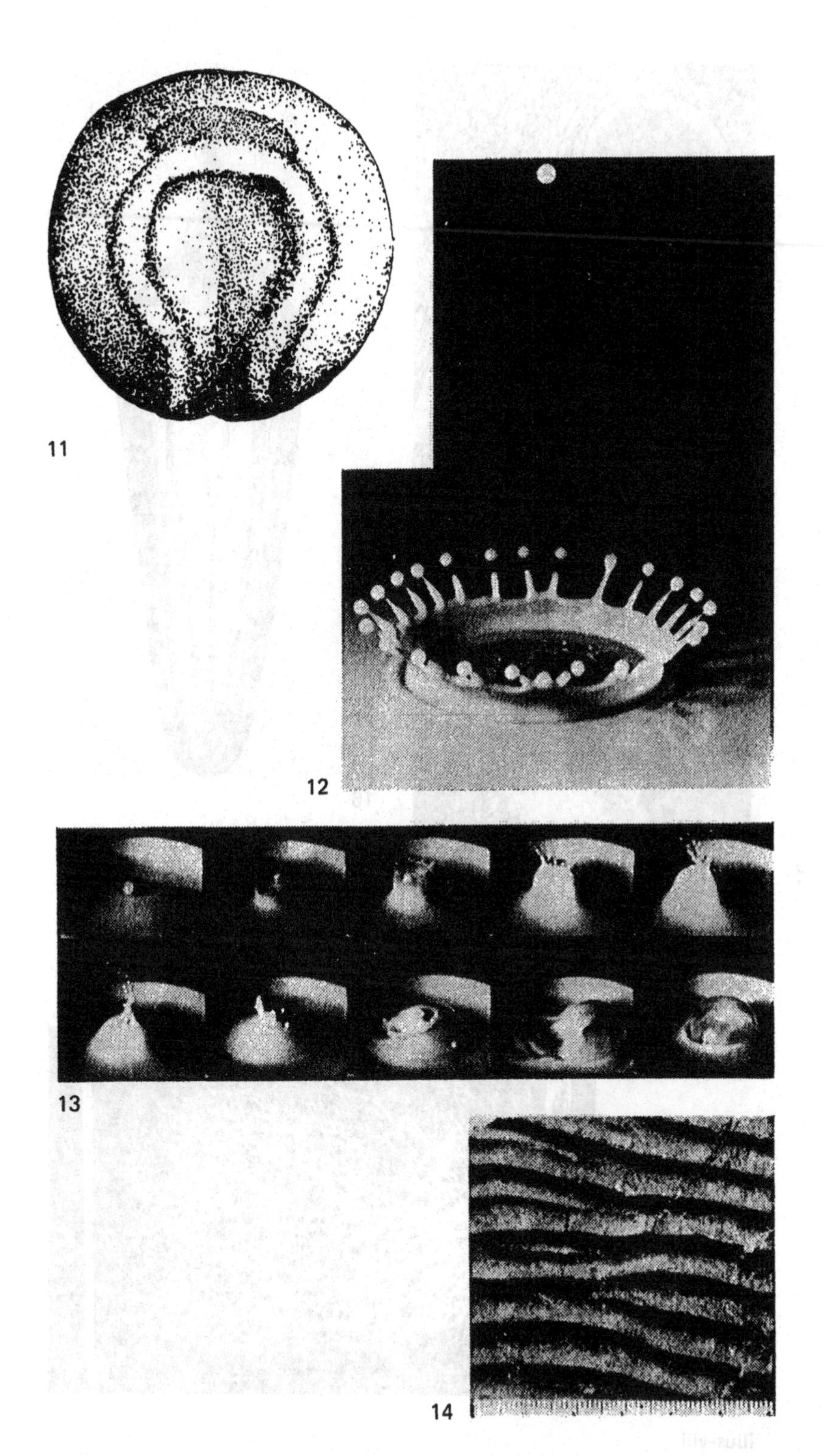
11
12
13
14

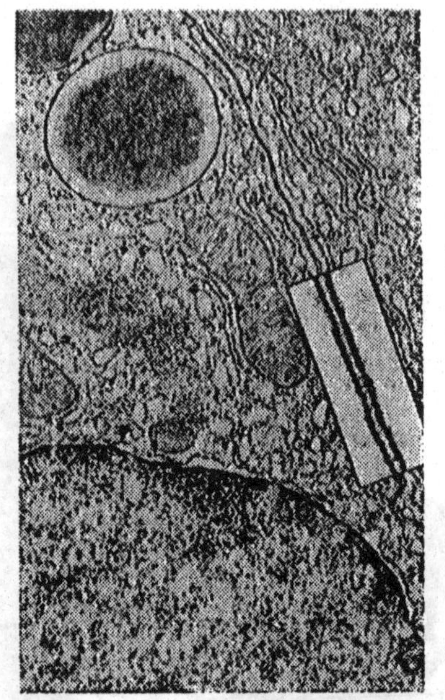

15

16

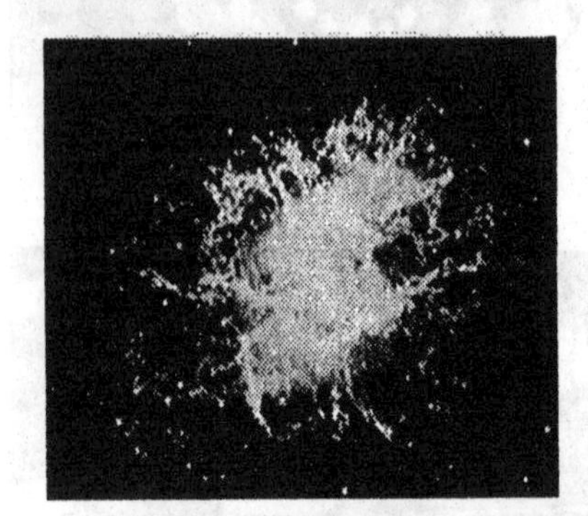

17

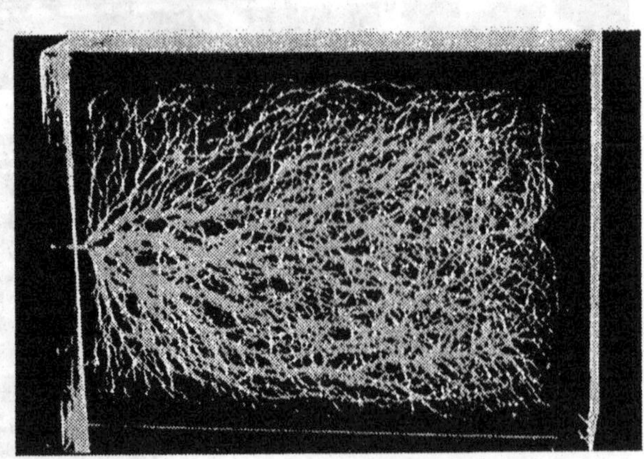

18

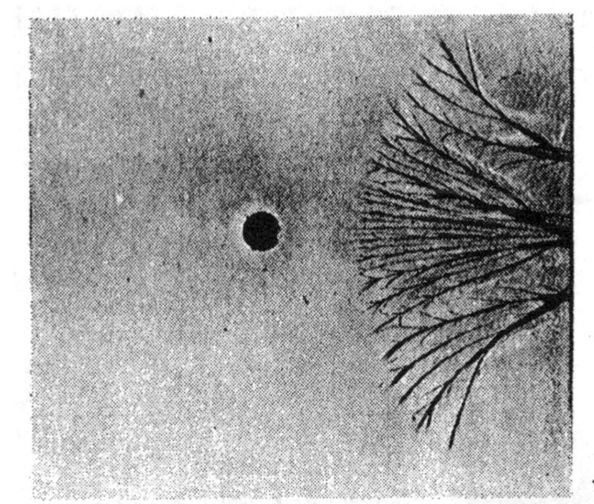

19

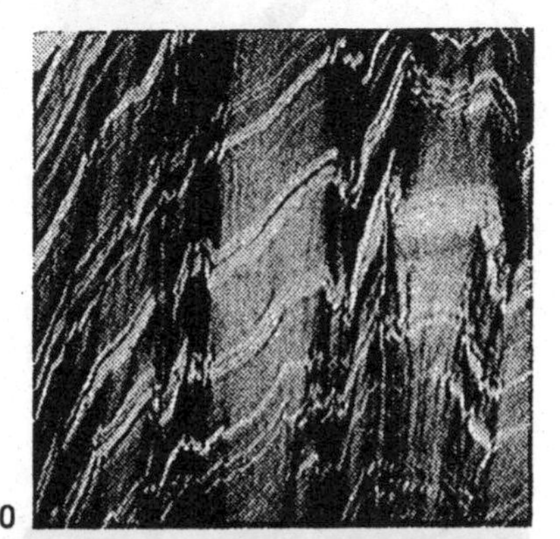

20

21

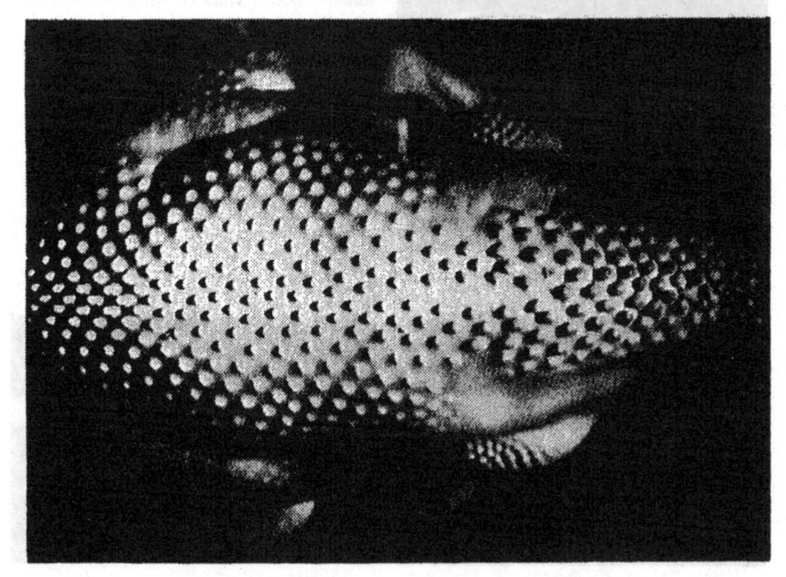

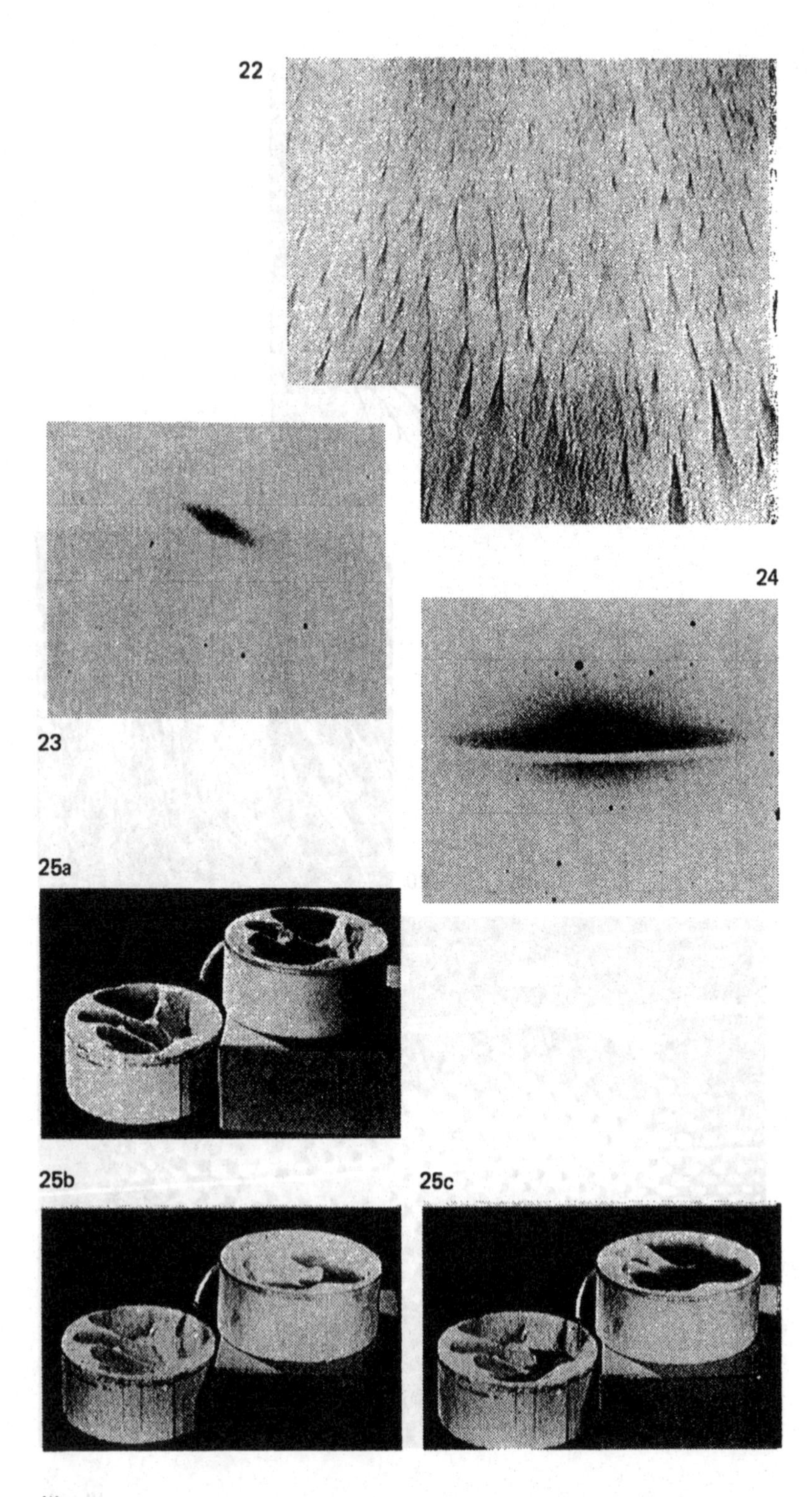
22
23
24
25a
25b
25c

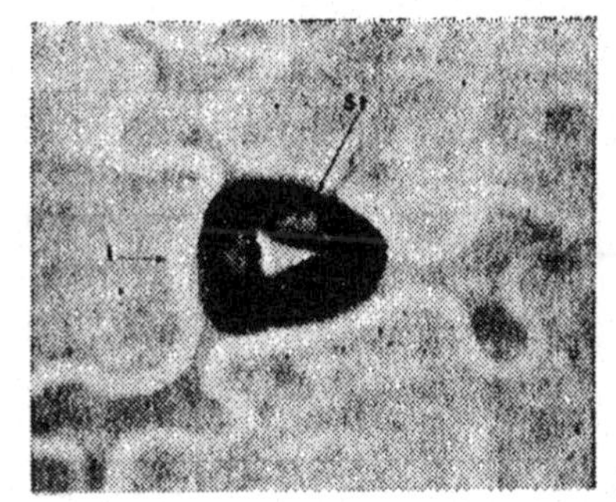

26a

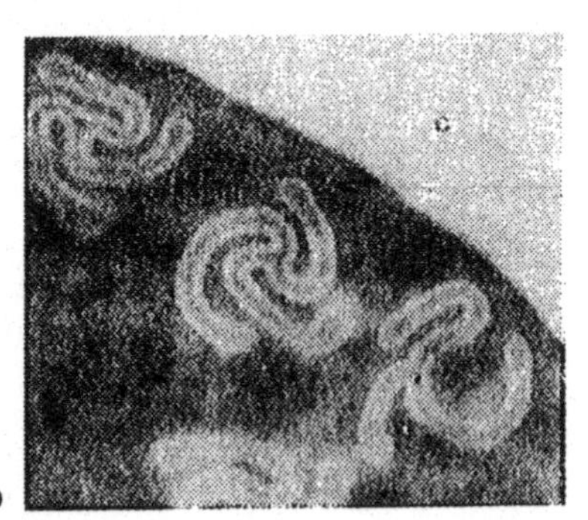

26b

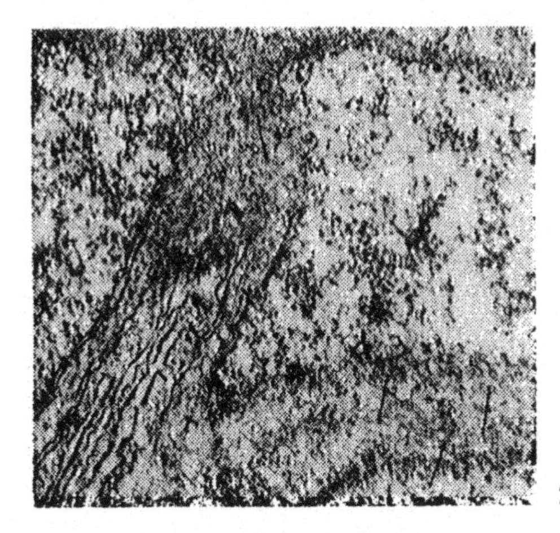

27

CAPTIONS OF PHOTOGRAPHS

1 "The effect is to upset or disquiet the observer." An American advertisement for a brand of shaving brush, late 19th century.

From Romi, *Histoire de l'Insolite,* Pont-Royal, 1964.

2a,b,c The cuspidal caustic associated with the Riemann-Hugoniot catastrophe. The two fine lines in 2c are the bounding curves of the pencil of light-rays, and the cusp is the section by the screen of the caustic surface of this pencil.

3a,b The Riemann-Hugoniot catastrophe in geology: the origin of a fault.

From E. Sherbon Hills, *Elements of Structural Geology,* Methuen, 1963.

4a,b,c,d The swallow's tail in geometric optics. The point singularity developing at c_1 in 4b is unfolded at c_2 in 4d (see Figure 5.7).

5 Domes on a beach (see Figure 5.13).

From W. H. Bascom, *Waves and Beaches,* Doubleday, 1964.

6 The universal unfolding of the hyperbolic umbilic in geometric optics.

7 The hyperbolic umbilic in hydrodynamics: a wave breaking.

8a,b The parabolic umbilic in geometric optics (see Figure 5.24).

9 The parabolic umbilic in geology: a mineral mushroom (see Figure 5.25-5).

From *Elements of Structural Geology,* Methuen, 1963.

10 The hyperbolic umbilic in geology: a fold on the point of breaking.

From *Elements of Structural Geology,* Methuen, 1963.

11 A biological fold on the pint of closing: neurulation in the toad *Xenopus Levis.*

From P. D. Nieuwkoop and P. A. Slorsthurst, Archives de Biologie, **61** (1950), p. 113.

12 A generalized catastrophe and example of symmetry in hydrodynamics: the diadem of a splash.

Photograph H. E. Edgerton, M.I.T.

13 The birth of a bubble. The falling drop leads to the formation of an open bubble with a delicately breaking boundary. This bubble then closes after a complicated catastrophe provoking a spray of elliptic breakers.

Photograph H. E. Edgerton, M.I.T.

14 A laminar catastrophe in geology: a riverbed.

From *Elements of Structural Geology,* Methuen, 1963.

15 A laminar catastrophe in biology: ergastoplasm in an exocrine cell of the pancreas.

From K. R. Porter and M. A. Bonneville, *Structure fine des cellules et tissus,* Ediscience, 1969.

16 A partially filament catastrophe in hydrodynamics: cavitation behind a sphere in water.

From G. Birkhoff and E. H. Zarantonello, *Jets, Wakes, and Cavities,* Academic Press, 1957.

17 A partially filament catastrophe in astrophycis: the Crab nebula, the remains of the explosion of a supernova.

18 A catastrophe with spatial parameter in physics: an electric spark in plastic.

Photograph kindly communicated by Mr. Robert Mouton.

19 A catastrophe with spatial parameter in physics: cracks in glass arising from the impact of a bullet. These cracks start from the edge and ramify towards the point of impact.

Photograph H. E. Edgerton, M.I.T.

20 The superposition of the two laminar catastrophes in geology: *Gleibretter* in slate, Donnelly's Creek, Victoria.

From *Elements of Structural Geology,* Methuen, 1963.

21 A generalized catastrophe and example of symmetry in biology: feather buds on a chicken embryo.

Photograph kindly communicated by Professor Philippe Sengel, Faculté des Sciences, Grenoble.

22 A generalized catastrophe and example of symmetry in hydrodynamics: the periodic formation due to the flow on a beach.

From *Waves and Beaches,* Doubleday, 1964.

23 A young nebula: NGC 4530.

From S. Chandrasekhar, *Principles of Stellar Dynamics,* Dover, 1960.

24 An older nebula with equatorial breaking: NGC 4594.

From *Principles of Stellar Dynamics,* Dover, 1960.

25a,b,c Hydraulic models.

These plaster models were kindly made by Mr. Marcel Froissart.

26a,b The section of a nematocyst of a cnidoblast (the urticant cell of hydra). The triangular character of this section is evidence of the elliptic origin of this organogenesis.

a. Sections of the central body.
b. Sections of the thread.

Photographs E. Van Tongeren.

27 Ganglion cell of hydra: the arrows indicate the microtubes associated with the nervous system.

Photograph E. Van Tongeren.

physiological regulation. It is due to these mechanisms that the animal can face the inevitable variations in its environment and proceed to the manufacture of the gametes that will assure the survival and multiplication of the species. This attitude is defiantly Lamarckian; it supposes that, on the whole, *the function creates the organ* or rather, more precisely, that the

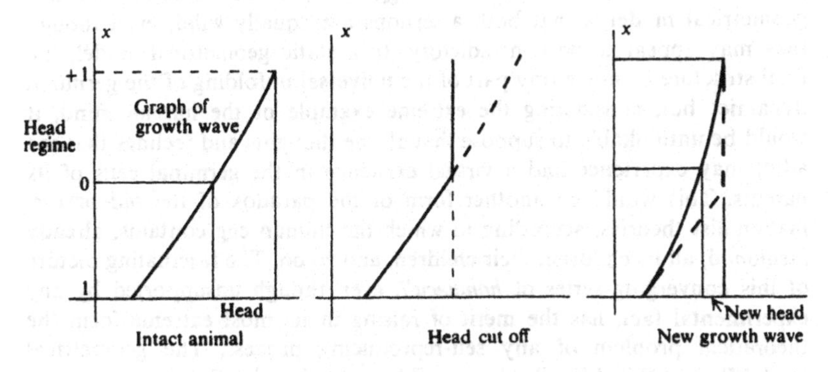

FIGURE 10.2. Regulation in planarians by diffusion.

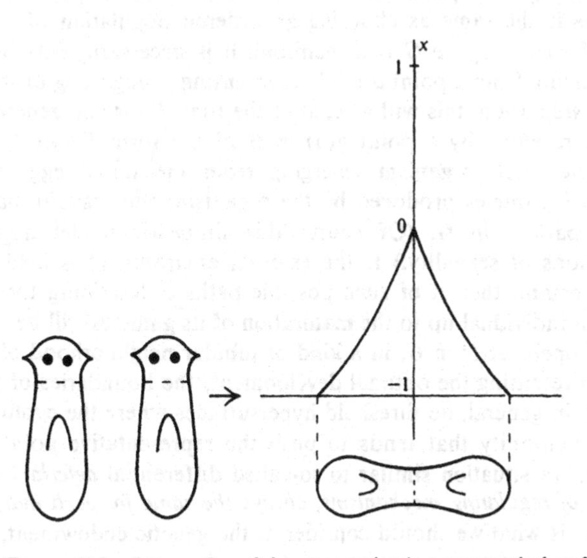

FIGURE 10.3. Interruption of the regeneration in a symmetrical graft.

formation of an organ is the result of a conflict between a primitive field with a functional aim (or significance) and an organic raw material resisting it and imposing on it certain genetically determined paths for its realization (chreods). This viewpoint raises the metaphysical question of whether the final functional field of the adult individual is, in the end, richer and more complex than the initial primitive factor transmitted by the egg, or whether this final realization is only an infinitesimal part of the potential or virtual content of the egg. One of the paradoxes of this geometrical model is that both assertions are equally valid, even though they may appear to be contradictory. In a static geometrical model, the final structure is only a tiny part of the universal unfolding of the germinal dynamic; but, considering the extreme example of the human mind, it would be unthinkable to suppose that all the thoughts and feelings that the adult may experience had a virtual existence in the germinal cells of its parents. This would be another form of the paradox of the old preformationalist theories, according to which the human egg contains, already fashioned, all its children, their children, and so on. The fascinating picture of this converging series of *homunculi*, even though unsupported by any experimental fact, has the merit of raising in its most extreme form the theoretical problem of any self-reproducing process. The geometrical model flushes this difficulty by a well-known trick: let G denote the space parameterizing all the physiological states of a finite population of the species considered; then there is a canonical embedding of $G \times G$ in G, since choosing a population of p individuals and a population of q individuals is the same as choosing an ordered population of $p+q$ individuals. If such a space G is a manifold, it is necessarily infinite dimensional. Starting from a point a in G, representing a single egg of the species under consideration, this will give, after the time T for one generation, the system represented by a point $g(a)$ in G of the form $S+k(a)$, where S denotes the adult organism emerging from the initial egg, and $k(a)$ represents k gametes produced by the organism; this transformation will define a path c in G. (Of course this simplistic model neglects the complications of sexuality.) If the external environment is held approximately constant, the set of such possible paths c describing the development of an individual up to the maturation of its gametes will be contained within an open set U in G, in a kind of tubular neighborhood of an ideal path c_0 representing the optimal development; the boundaries of this open set U will, in general, be threshold hypersurfaces where the evolution field has a discontinuity that tends to push the representative point into the open set U (a situation similar to so-called differential *deferlant* systems).

This set of regulating mechanisms, always the same for each individual of the species, is what we should consider as the genetic endowment, and it is this set alone that is transmitted in the egg. And this is how the paradox above is resolved: the heritage of the genetic equipment is *the set of bounds of the variations of the many physiological activities of the adult*, because for any physiological function there are maximum and minimum thresholds beyond which there occur irreversible catastrophes within the functioning of the regulating mechanisms themselves, whereas within these thresholds the variation of the fields can be relatively arbitrary. In this way the formation and maintenance of a spatially and physiologically stable form is not incompatible with the great variety of local fields of living organisms; in fact, this variability is beneficial to the global stability, allowing adaptation to different types of aggression. The role of the nervous system is to canalize this variability into a system of functional chreods, either inherited genetically or built up by education, and this explains why the physiological structure of the adult can be richer and more complex than the primitive germinal field that generated it. Even so, we can, to some extent, say that this complexity existed virtually in the egg; to return to a crude but striking mathematical analogy, to require that a differentiable function on a domain D of $\mathbf{R}^n$ be bounded by ϵ does not prevent this function from having a topological type of arbitrarily large complexity. Similarly the genetic constraints, essentially the constraints of thresholds, do not prevent the field of the adult individual from evolving in an infinite-dimensional space and exhibiting an infinity of forms.

10.2. THE METABOLIC MODEL

A. Limitations of the static model

As we have described it, the static model has an obvious defect: if it is supposed that the growth wave unfolds, in the universal unfolding space Q of the germinal dynamic, transversally to the universal catastrophe set, all the final structures of the organism must appear at the beginning of development. This will be a mosaic type of situation, without any possibility of secondary catastrophes during the course of development. A finer theory must take into account first the nature of the repression dynamic, which constrains the egg at the germinal point in gametogenesis, and then the de-repression, which quite visibly comes into play by a series of discontinuous steps during epigenesis. This occurs as if the growth wave F were first restricted to the strata of maximum codimension in the bifurcation set, and then spread out by discontinuous steps into strata of decreasing codimension. This leads us to improve the model by introducing a metabolic (recurrent) element into the dynamic, leading to the idea of the *epigenetic polyhedron* [1].

The same difficulty already occurred in the idea, put forward by Waddington, of the *epigenetic landscape*. Here the valleys descending from the mountain range can only meet, whereas development requires divergences and bifurcations. Divergences can occur only when descending from a plateau cut into by lateral valleys, and it is difficult to conceive how this situation could be generalized throughout development.

B. The epigenetic polyhedron

We shall describe the formalism of embryological development by a geometrical object, a polyhedron E defined as follows. There is a real, positive function, the time t, defined on E. The level hypersurface of this function at time $t = 0$ is reduced to a point, the germinal point. Each stable cellular or spatial differentiation of the embryo tissue corresponds to an edge A of E on which t is strictly increasing. Each morphogenetical field corresponds to a zero-, one-, or two-dimensional simplex according as one, two, or three tissues participate in the interaction giving rise to the field; and to simplify matters we shall suppose that time t is constant on each of these simplices. Then each simplex s associated with a field gives rise to new edges having as origin the barycenter of s and characterizing the new tissues produced by the morphogenetical field associated with s. We shall consider E as being embedded in a Euclidean space whose coordinates have biochemical significance (corresponding roughly to the important gradients of development, such as animal-vegetal, cephalocaudal, and dorsoventral).

The interpretation of the simplices associated with morphogenetical fields requires the use of a metabolic model and appeals to the idea of resonance; for example, if the one-dimensional simplex s has vertices p and q, and the corresponding tissues have local biochemical regimes described by attractors G and H of the corresponding local dynamics, then, when these tissues come into contact, they will produce a resonance that will cause the topological product $G \times H$ to degenerate into one or more attractors K of lower dimension. Let Q denote the space of all embeddings of K in $G \times H$; then the local morphogenetical field associated with s has Q as director space (universal unfolding), and the chreod defined by s is described by a map g of the support of the field in the space $Q \times t$, as in the case of a static model. The new tissues created by the field s, and associated with the new edges coming from the barycenter of s, will correspond to the new attractors K produced by the catastrophe $G \times H \rightarrow K$. (The polyhedron E is not strictly a simplicial complex because some faces of the two-dimensionl simplices may be missing, and some of the barycenters may also be vertices.)

As an illustration, Figure 10.4 shows the epigenetic polyhedron of the frog embryo up to the neurula stage. We might consider this polyhedron as defining a deterministic semantic model, the normal development being considered as a global chreod and each simplex of E defining a subchreod. Clearly the epigenetic polyhedron gives only a schematic description of the successive steps of development and does not in itself add anything new to the usual accounts in textbooks of embryology. However, we shall try to give it a dynamical justification, but first it will be helpful to develop a geometrical picture of the regulating mechanisms of a living being. This is the purpose of the next section.

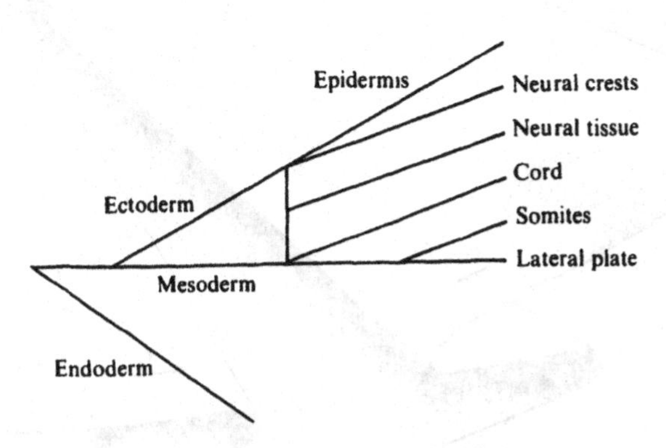

FIGURE 10.4. The epigenetic polyhedron.

C. The regulation figure

Every living being is characterized by its global stability, and after a shock s it will respond by producing a reflex $r(s)$ with, in principle, the effect of canceling any perturbation produced by the stimulus s; to be more exact, a stimulus is not the external physicochemical cause of the perturbation, but its immediate repercussion on the organism. It is permissible, in this case, to suppose that all the instantaneous states of the organism under study can be parameterized by the points of a locally Euclidean space W, the origin O of this space corresponding to the normal resting state of the organism. Applying a given external stimulus will cause the point representing the organic state to move first to a point s characterizing the stimulus (i.e., specifying its immediate effect on the organism) and then return to O, following a path characteristic of the reflex $r(s)$ excited by the stimulus (Figure 10.5). Now, although the stimuli obviously make up a many-dimensional geometrical continuum, if only from the fact that they originate in the ambient space, the corrector reflexes are, at least to the first approximation, only finite in number, and each corrector reflex is thus defined by an attracting curve leading to O. The space W is divided in this way into basins of attraction, each associated with a corrector reflex. This global configuration will be called the *regulation figure* of the organism under consideration.

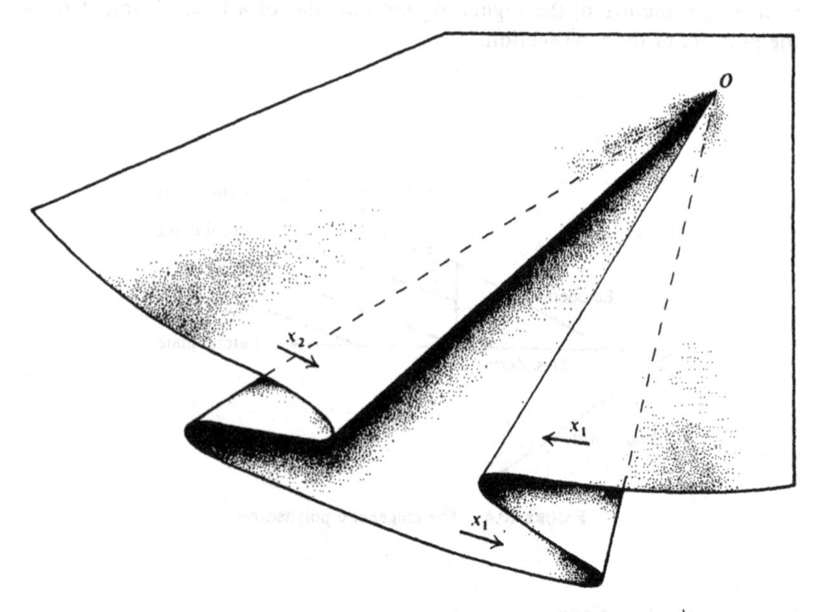

FIGURE 10.5. Isometric view of the regulation figure defined by two cusps.

This model raises many questions, of which the most important is to know which of the states s are susceptible of being corrected by the reflexes r_j. Leaving aside for the moment any consideration of excessive perturbations, which traumatize the organism to the extent of paralyzing its regulating reflexes, we shall probably be left with two main types of stimuli: the common stimuli, dealt with genetically, for which the association $s \longrightarrow r(s)$ is almost automatic, and the strange or unusual stimuli, for which the association is indeterministic or imprecise. We shall return to this question in the discussion of evolution in Chapter 12.

Now take a dynamical system (P,X); we say that it *simulates* the regulation figure T of an organism with state space W when the following conditions are satisifed: there is a fibration $p: P \longrightarrow W'$ over a locally Euclidean space isomorphic to W, defining a system of approximate first integrals of (P, X); this defines a mean coupling field Z on W' by integration along the fiber; and there is a homeomorphism of W' onto W, transforming the trajectories of the field Z into the curves $s \longrightarrow r(s)$ defined by the corrector reflexes r_j. In most cases, a typical reflex will be simulated by an attractor of the fiber dynamic $P \longrightarrow W'$.

D. A preliminary description of the global model

The basic idea of this model is as follows. Let A be an animal species; the local metabolism of a primordial germinal cell (oogonium or spermatogonium), shortly before meiosis, can be described by a differential system (M, X), *which simulates the global regulation figure T of species A*. Of course such a statement can be made precise only by exhibiting the isomorphism $W' \cong W$ relating the parameter space of the cellular metabolism W' to the space of global states W of the organism, which we are unable to do. The global intensity of the metabolism decreases abruptly in these germinal cells shortly before or shortly after meiosis, and this leads to a contraction of the topological structure defined by the system (M, X). The essential part of this structure crystallizes out in macromolecular structures (nuclear and cytoplasmic); and, as is often said, the "information" associated with T becomes coded in these structures.

An inverse process occurs after fertilization: with the freeing of many biochemical degrees of freedom and the starting up of many reaction cycles, the dynamic (M, X) re-establishes itself in approximately its original state and so to some extent simulates the regulation figure T. However, this figure is far too complex to be stable on a many-celled organism, and so certain cells specialize and retain only the states r: first the endoderm, which conserves only the reflexes relating to the performance of alimentary functions, and then the mesoderm, which specializes into the state r of (internal and external) spatial regulation and the reflexes of biochemical regulation (blood circulation, excretion). Conversely, other cells lose the states r and retain only the states s: these are the nerve cells which, in effect, having lost the capacity for regulating their metabolism, *are irreversibly affected by everything happening to them*, and the only regulation that they are capable of is the crude and undifferentiated discharge of nervous impulses. Other cells, in the epidermis, evolve through aging toward an attractor between s and r. The primitive functional fields $s \longrightarrow r(s)$, momentarily broken by the catastrophes of organogenesis, then reform as functional fields directed by neuronic activities canalized into chreods; for example, the typical alimentary field is represented on a blastula cell by a preferential oscillation of the metabolism $s \rightleftarrows r(s)$, and this field decomposes in the adult into the following:

1. A sense stimulus (e.g., seeing a prey).
2. An external motor field (capturing the prey and bringing it to the mouth).
3. Eating it.
4. A visceral field, directing the glandular and motor activities of the digestive system.

The field does not die away definitively until after the complete digestion of the prey. Later, we shall give a dynamical model of this apparently mysterious process, the global reconstitution of a field at the end of development (see Appendix 3 and Section 13.1A, The Predation Loop).

E. Self-reproducing singularities

Let U be an infinite-dimensional vector space and V a potential function $V: U \to \mathbf{R}$ having a singularity of codimension three at the origin. Denote the universal unfolding space of this singularity by $Ouvw$, so that, for example,

$$V_1 = V(x) + ug_1(x) + vg_2(x) + wg_3(x)$$

is the universal unfolding of V. There will generally be a direction, say Ou, along which the unfolding of the singularity is at its most complex, and along this direction all the critical points of V concentrated initially at O are real, with a maximum concentration of critical values. Let a be the point $u = 1$, $v = w = 0$ giving such an unfolding. Adjoining to V_1 a term like $v^2 + w^2$ (a *stopping potential*) and applying Maxwell's convention at each point of the plane $u = 1$ gives a potential V_2 defined continuously on this plane—this potential will be differentiable except along the conflict shock waves, where there will be discontinuities in the first derivatives (angular points). We then say that the singularity $V(x)$ is *self-reproducing* if the potential V on the (two-dimensional) space of internal variables x has the same topological type as the potential V_2 on the plane $u = 1$. For example, the Riemann-Hugoniot catastrophe $V = x^4/4$ has universal unfolding

$$V_1 = \frac{x^4}{4} - \frac{ux^2}{2} + vx + v^2 ;$$

the curve $x^4/4 - x^2/2$ is an even curve with two symmetric minima, and the curve defined by $V_2(v)$ on $u = 1$ has, after applying Maxwell's convention, the same topological type as V_1 (Figure 10.6), although with an angular maximum for $v = 0$; I do not know whether such a phenomenon occurs generally for a static singularity—probably not, because, as I said earlier, the static model is too rigid to be correct. In biological applications,

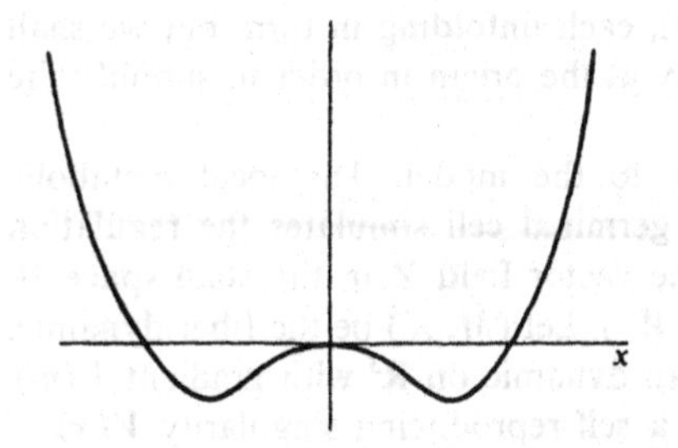

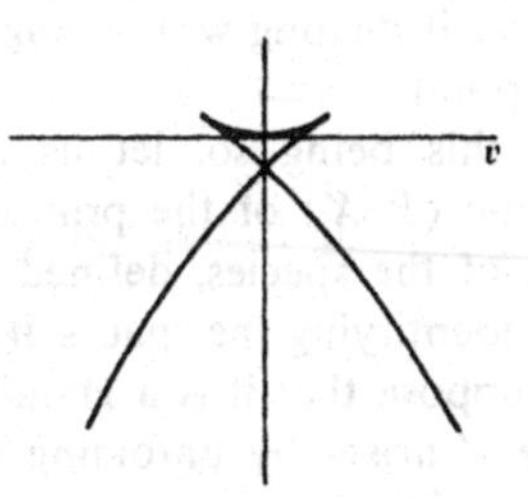

(a) The original potential $\frac{x^4}{4} - \frac{ux^2}{2} + vx$ at $u = 1$, $v = 0$, and its critical values when $u = 1$

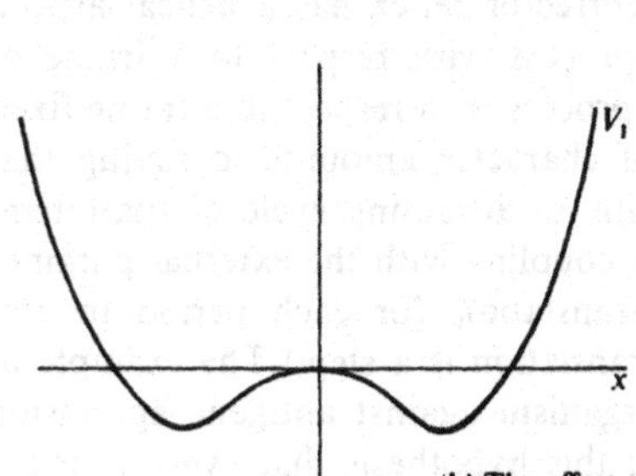

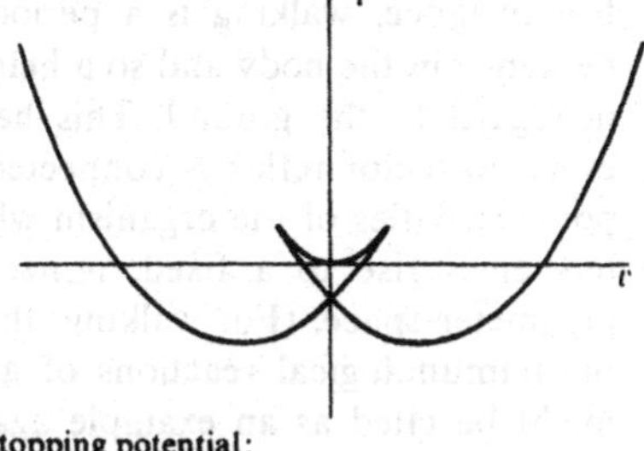

(b) The effect of a stopping potential:
$V_1 = \frac{x^4}{4} - \frac{ux^2}{2} + vx + v^2$ at $u = 1$, $v = 0$, and its critical values when $u = 1$

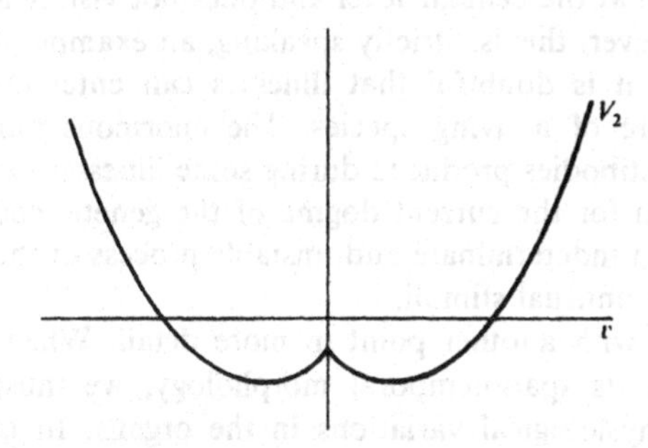

(c) Maxwell's convention applied to the critical values of V_1, leading to a function V_2 of the same topological type as V_1

FIGURE 10.6

there will be a string of singularities, each unfolding in turn, but we shall argue as if dealing with a singularity at the origin in order to simplify the description.

All this being so, let us return to the model. The local metabolic dynamic (P, X) of the primordial germinal cell simulates the regulation figure of the species, defined by the vector field Z in the state space W (after identifying the spaces W and W'). Let (M, X) be the fiber dynamic, and suppose that it is a Hamiltonian dynamic on $\mathbf{R}^3$ with gradient $V(m)$, where V arises by unfolding from a self-reproducing singularity $V(x)$ of codimension three.

These hypotheses might seem arbitrary a priori, but they can be justified as follows. First, observe that every corrector reflex has a helical aspect. For instance, walking is a periodic process with respect to a frame of reference in the body and so a helical process with respect to a frame fixed in regard to the ground. This helical character amounts to saying that every corrector reflex is connected with an attracting cycle of spatiotemporal activities of the organism whose coupling with the external parameters gives rise to a fixed, nonzero translation for each period in this parameter space. (For walking, this translation is a step.) The example of the immunological reactions of an organism against antigene aggression might be cited as an example against this hypothesis that every reflex is connected with a spatiotemporal field of the organism, because this reaction takes place at the cellular level and does not visibly affect the overall structure. However, this is, strictly speaking, an example not of reflex, but of illness, and it is doubtful that illnesses can enter explicitly into the regulation figure of a living species. The enormous number of protein structures of antibodies produced during some illnesses (which pose a very serious problem for the current dogma of the genetic code) suggests that we have here an indeterminate and unstable process of the type associated with strange or unusual stimuli.

Let me deal with another point in more detail. When we characterize each reflex by its spatiotemporal morphology, we must also take into account the physiological variations in the organs. In this way the discharge of nervous impulse and the distribution by the blood of a glandular secretion must also be considered as spatiotemporal reflexes of a special kind.

The helical nature of a reflex appears in our model $(P, X) \to W$ of the regulation figure. Every reflex is associated with an attracting cycle of the fiber dynamic (M, X) (this is the periodic component), and the mean coupling field Z in W gives the corrector or translation component of the reflex. In fact, the general situation is more complicated and this helical description works only for simple reflexes, for irreducible functional fields. The fundamental reflexes are generally made up of periodic chains of simple reflexes; also, in W itself (which contains, as we shall see, an image of the body), the global corrector field is often discontinuous and has a funneled character, rather like the mean coupling fields described in Chapter 6.

F. The mixed model

Now we are going to suppose that the fiber dynamic (M, X) is a Hamiltonian dynamic defined on a finite-dimensional Euclidean space, coordinates x, with Hamiltonian

$$H = \frac{x'^2}{2} + V_1(x),$$

where V_1 is the most complex unfolding of a self-reproducing singularity $V(x)$ at $x = 0$. We might then visualize this dynamic as that of a heavy ball rolling without friction on a surface whose height is given by $V_1(x)$. Suppose that the ergodic properties of the dynamic on the energy hypersurface $H = c$ are as follows: a large ergodic component whose coupling component Z on W is zero (representing the nonexcited, resting state of the organism), and vague attractors corresponding to simple reflexes for which the coupling field Z gives the corrector translations in W. In a relief map, these attractors would correspond to something like highways crossing the mountains of the landscape through carefully aligned passes, and on which the ball would oscillate indefinitely. Note also that the global configuration defined by $V_1(x)$ is of a potential well with an internal relief, and the highways abut orthogonally against the potential walls and so reflect back along the road at each end. In the most degenerate possible case all the sinks will be at the same height, as will be the saddles and the summits, and this is the most favorable situation for the ergodic component; since the system depends on the parameters of the space W', when moving away from the origin of W' and going toward a very distant point s, the landscape $V_1(x)$ will change, adjusting the heights of the summits and the saddles so that, in principle, a corrector path will be favored and its basin of attraction will be increased. This can be arranged by lowering the altitude of the passes traversed by the path, and increasing the heights of the hills and the passes on either side into the form of a drainage channel.

After development, the same situation will occur for the dynamic defined by the potential $V_2(x)$ on the universal unfolding space with parameter v. The potential well defined by V_2 will be isomorphic to that defined by V_1 on the x-space; the Hamiltonian dynamic defined by $H' = v'^2 + V_2(v) = k$ will therefore have isomorphic properties to the dynamic defined by $V_1(x)$, so that the deformations, as functions of points of W, will be homologous to those for the corresponding points of W'.

This being so, suppose that, during gametogenesis, the parameters c and u tend to zero in such a way that the local dynamic on the ripened gamete is represented by the singular dynamic $V(x)$; this will be the germ of potential representing the genetic endowment of the species, that is, all the information coded in the genes. The organism will unfold, after fertilization, in the universal unfolding space $Ouvw$ of the self-reproducing dynamic $V(x)$, and we may suppose the development to be halted at the plane $u = 1$, as the addition of a potential of the form $v^2 + w^2$ will stop the spatial development. At the same time, the Hamiltonian H will increase, so that the spatiotemporal dynamic defined by V_2 will simulate the regulation figure of the species. We can present a rather more intuitive model for this entire process.

10.3. THE HYDRAULIC MODEL

A. Description of the model

We suppose that the terrain described by the relief map defined by the height function $V_1(x)$ is inundated by a flood of water at height c; this will make a twisting archipelago. The main reflexes give rise to straits, long passages crossed from end to end by navigation routes. Decreasing the height c of the flood will result in the passages being blocked, one after the other, by the saddles as they emerge in succession from the water. The water will move down into the valleys and create sinuous fjords; as c decreases even further, it will retreat until it leaves only a circular pool of water covering the origin, the germinal point, at the lowest sink (Figure 10.7). Development corresponds to the opposite process; the manifold of constant energy $H' = k$ projects onto the base space $Ouvw$, and it has a level hypersurface $V_2 = k$, the shore of the sea. Because the density (in terms of the Liouville measure) above a neighborhood of this critical level is very large, we can think of this surface $V_2 = k$ as defining the spatiotemporal development of the embryo as k increases. Since V_2 is defined by a static model (at least as long as the stopping potential $v^2 + w^2$ is negligible), we should not be surprised to see on this surface the singularities of shock waves discussed in Chapter 9; the topological type of the embryo will vary every time the Hamiltonian H has a singular point, that is, at each singular point of V_2 corresponding to a saddle point in the relief map. Of course there are two different types of saddle points in three dimensions: those of index one and those of index two; the first characterize the interaction of two tissues, and the second, three tissues. In practice, each saddle is associated with the formation of an organ, and the saddles are arranged along the long straight lines characterizing the main functional reflexes operating these organs.

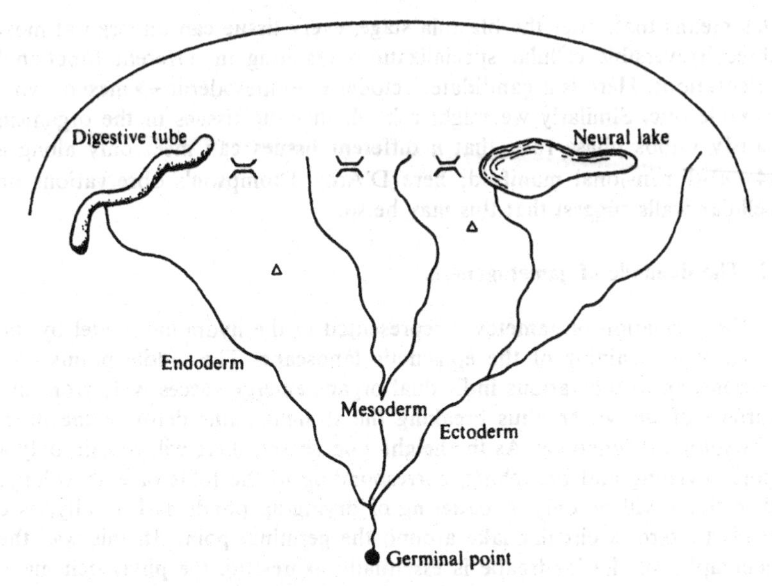

FIGURE 10.7. The hydraulic model.

B. Relationship between the hydraulic model and the metabolic model

Consider two connected components of the manifold $H' = k_1$ which join at a saddle point of index one at $k_1 = s$, and suppose, for simplicity, that x is a one-dimensional space. The representative point will describe periodic cycles, with periods T_1 and T_2, respectively, on each of these connected components; as soon as the connection is established, it will describe these cycles successively. If the quotient T_1/T_2 is rational, a resonance will be set up on the product torus of the two cycles, and this resonance will have the same asymptotic behavior (rotation number) as the movement considered. It seems reasonable to believe that the saddle will develop precisely at height s such that the ratio T_1/T_2 is a simple rational number, and in this respect both static and hydraulic models have the defect of ignoring the kinetic nature of the reactions occurring during development, even though this aspect is certainly very important. Similarly, a metabolic model is necessary to explain phenomena like induction and loss of competence through aging.

In a strictly static model, we might interpret all irreversible cellular specialization (where the tissue acquires a new functional adaptation) by the representative point passing from a stratum of codimension r to one of codimension $r - 1$. Since the germinal dynamic is of codimension three, this means that, after the blastula stage, every tissue can undergo at most three irreversible cellular specializations resulting in different functional orientations. Here is a candidate: ectoderm ⟶ mesoderm ⟶ mesenchyme ⟶ sclerotic. Similarly we might ask whether the tissues in the organism satisfy Gibbs phase rule, that n different tissues can meet only along a $(4-n)$-dimensional manifold; here D'Arcy Thompson's observations on cellular walls suggest that this may be so.

C. The dynamic of gametogenesis

The formation of gametes is represented in the hydraulic model by the progressive draining of the epigenetic landscape. The saddle points corresponding to the various individual organs emerge successively from the surface of the water, thus breaking the straight paths defining the main physiological functions. As the height c decreases, there will remain only a fjord, twisting and branching, corresponding to the folds of embryology; then there will be only a scattering of drying-up ponds and, finally, as c tends to zero, a circular lake around the germinal point. In this way the geography of the landscape is essentially expressing the physicochemical constraints imposed on the realization of the different catastrophes of organogenesis, and so it expresses also the burden of past evolution, as required by Haeckel's law of recapitulation. Because of this, the branches of the fjord take on a more and more abstract functional significance; the ingestion catastrophes of food or oxygen are taken care of by the endodermic arms, while the excretion catastrophes are handled by a branch of the mesoderm waters, and the locomotion reflexes are concentrated in another branch of the mesoderm river that flows into the previous one. Finally, the stimuli are represented by the neural lake, a kind of large, flat swamp with only one ill-defined outlet into the ectoderm valley.

In a strictly metabolic interpretation, the appearance of a saddle point can be seen as the extinction of a resonance between reaction cycles, and the draining of an isolated pond as the extinction of an independent reaction cycle. At the point where no saddle remains, the local dynamic will be almost a direct product of independent oscillators and so nearly linear. This fact explains why embryonic tissues of different species are so readily able to fuse together and join in the construction of chimeras, as in the classic experiment of Spemann on the tadpole-newt chimera. If tissues of different origin, coming from phylogenetically distant species, can develop harmoniously together, the reason must be that their regulation figures and the epigenetic landscapes of the different species are isomorphic in respect to their main features. Insofar as embryological induction is a resonance phenomenon, it must be normal for it not to have any zoological specificity. Nonetheless, this lability of primitive tissues is short-lived, and with aging and loss of competence the local attractors degenerate through resonance into highly attracting, nonlinear attractors, incapable of entering into any pronounced interactions. Nevertheless, the general isomorphism between the early stages of development in different animals is a striking and well-known fact.

D. Reproduction in the hydraulic model

To incorporate in the model the process of reproduction, we start with the simple observation that this is a morphologically periodic process, because the descendant is isomorphic to the parent. Let us neglect the complications of sexuality and consider only a simple model of the type of vegetative reproduction. The periodic character can be represented by using a linear oscillator with conjugate variables U and v; and suppose that the movement is maintained by a push at each oscillation, rather like the effect of the escapement on the pendulum of a clock, so that, for example, at $U = 0$, the coordinate $v = v_0$ has a discontinuity, a jump to the value $v_0 + \Delta v$, while the rest of the trajectory is a contracting spiral as a result of friction, joining the point $(0, v + \Delta v)$ to $(0, v)$. Suppose further that this jump Δv is due to the passing of a shock wave of a function $G(\theta)$ of the angle θ, having a Riemann-Hugoniot cusp at the origin, three-sheeted in the first quadrant; such a function acts as a potential for maintaining the oscillations, here relaxation oscillations (Figure 10.8).

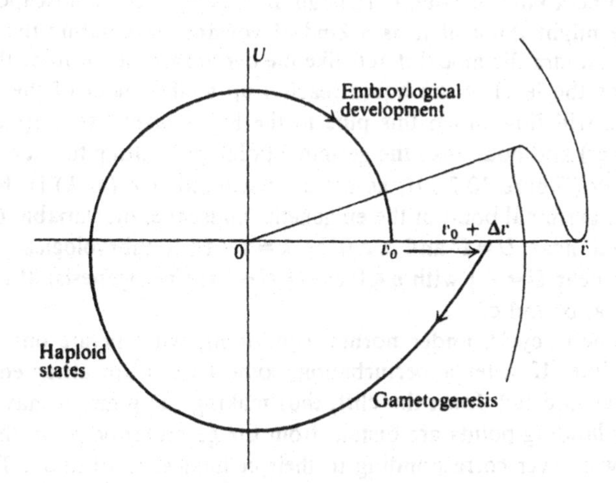

FIGURE 10.8. Relaxation jump.

Here is the main assumption of the model: the variable v, on $U < 0$, can be identified with the variable c of the hydraulic model, this magnitude expressing in some sense the distance from the germinal situation defined by $c = 0$. When $U < 0$, the unicellular states (eggs) are the only stables states; before the genetic jump, $U = 0$, $v = v_0$, the coordinate v is identified with k; after this jump and the corresponding separation of the germinal cell from the parent organism, v will be identified with c. In this cycle, only the quadrant $U > 0$, $v > 0$ corresponds to a many-celled state, the state of embryonic development; the second quadrant corresponds to oogenesis, and the half-cycle $v < 0$ to the spatial transport of the gametes through a large distance from the parents.

With this arrangement, when the organism approaches the point $U = 0$, $v = v_0$, the potential G will enter an unstable situation as a result of the presence of a branch of lower potential. As often occurs in similar cases, this overhanging potential cliff will tend to cave in, and the least stable part of the organism will be the most sensitive to such an attraction. Therefore the (central) junction zone of the organism, where the local dynamic is the most topologically complex, will be the most sensitive, and, in this geological analogy, the erosion will leave intact the large blocks defined by simple attractors and attack preferentially the interstitial zones, where it will create fissures. This central zone will finally produce a kind of funnel leading down to the lower layer; this will correspond to the gonad. Mathematically, we might suppose that the v coordinate of the genetic pendulum is coupled to the V potential of the hydraulic model through a relation of the kind $V + v =$ constant, and so the abrupt increase Δv of v will occur at the expense of the potential V; at this stage the gonad will appear to be a kind of potential trough. In the epigenetic landscape defined by V, we might think of it as a kind of volcano dominating the scenery, and in the hydraulic model it acts like the overflow of a bathtub; therefore, as soon as the level of the liquid reaches up to the crater of the volcano, the liquid will flow down this pipe to the lower layer (isomorphic to the upper layer) and pour onto the germinal point, generating the development of progeny (Figure 10.7). If, as before, the number c (or k) is the height above the germinal point in the epigenetic landscape, the variable U can be interpreted near $U = 1$ and $v \geqslant 0$ as $k = 1 - U$ (embryological development), or near $U = -1$ with $v \leqslant 0$ as $c = U + 1$ (gametogenesis). (See Photographs 25a, b, and c.)

This genetic cycle, under normal conditions, will operate only for the germinal line. If, after a perturbation, some tissues are attracted by the lower layer and fall down the cliff, thus making the jump, it may happen that their landing points are distant from the germinal point, in the valley of the lower layer corresponding to their cellular specialization. Then the genetic cycle can restart with a reduced radius and a much shorter period;

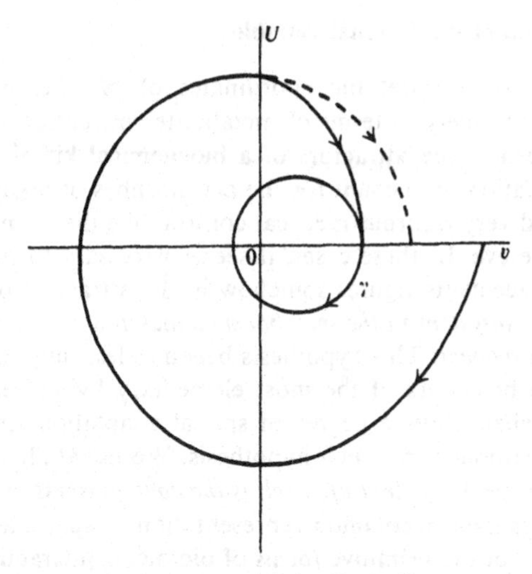

Figure 10.9. Periodic oscillations of the regulation figure. Cancerous states are represented by the circle γ.

this might be the interpretation of the mechanism of cancer (Figure 10.9).

E. Interpretation of the animal-vegetal gradient

The radius $\sqrt{U^2 + v^2}$ of the genetic cycle can be interpreted as the animal-vegetal gradient. At the moment of the genetic jump, the egg undergoes a positive discontinuity of Δv; insofar as the egg remains attached to the maternal organism, which itself does not have any discontinuity, it is natural to suppose that the pole of the egg situated at the ovarian wall has a minimum of value v, while the opposite pole (the animal pole) has maximum v. This polarization of the egg could be conserved throughout the rotation of the genetic cycle, with the u coordinate, at the moment of fertilization, acting as the animal-vegetal gradient during development.

This model exhibits only one polarizing gradient in the egg, but we could introduce both one more gradient and sexuality by elaborating the model, for instance, by taking a loop in the universal unfolding of the parabolic umbilic. Then there might be a point on such a loop where two isomorphic attractors compete through bifurcation and later fuse together, and this would provide an interpretation of the haploid-diploid alternation that is characteristic of sexual reproduction (see Section 10.4.C). It might also explain why the plane of fusion of the gametes usually determines the bilateral symmetry plane of the future germ.

F. Interpretation of the internal variables

If we were to interpret the coordinates of M, the manifold of the dynamic (M, X), solely in terms of metabolite concentrations, it would be difficult to see how the attractors of a biochemical kinetic could contain enough information to account for the catastrophes of organs, particularly those that need very rigorous metrical control, like the formation of bones, muscles, or the eye. In these cases, it seems necessary to suppose that the group of displacements figures somehow in the attractors or, equivalently, that *there is a subsystem in the metabolism simulating the metrical configuration of the environment*. This hypothesis becomes less outrageous on further reflection; the behaviors of the most elementary living beings, like paramecia or amoebae, show a degree of spatial adaptation that can scarcely be explained without some such hypothesis. We might almost say, anthropomorphically speaking, *that life itself is the consciousness of space and time*; all living beings have a common representation of space, and competition for space is one of the primitive forms of biological interaction. It would be sufficient to suppose, biochemically, that the cytoplasm contains, in addition to small molecules, macromolecular populations having locally a kind of symmetry or homogeneity so that their position with respect to their surroundings can be described by a Lie group G. If the position and orientation of the macromolecules relative to the smaller molecules have an enzymatic effect on the metabolism of the small molecules, depending on G (this coupling being reciprocal, with the metabolism of the small molecules influencing the position of the large ones), this would give a sufficiently rich structure to account for all the groups needed in organogenesis. Furthermore, it would explain how certain catastrophes can already be realized at the macromolecular level, as in the duplication of chromosomes or the contraction of myofibrils. Of course these are just speculations, but the existence of such mechanisms certainly seems to be a logical necessity.

10.4. THE FORMAL ANALYSES OF ORGANOGENESIS

A. Origins of organogenesis

To understand organogenesis, we must move up the scale of living beings, starting with the unicellular organisms like amoebae. At the lower level, the basic functions of nutrition, respiration, and excretion are carried out not by fixed organs, but by reversible chreods of limited duration and labile implantation: pinocytosis, phagocytosis, the formation of excretory vacuoles, and so forth; in more highly evolved unicellulars (e.g., the stentor) some of these functions already have irreversibly fixed cytoplasmic localizations, the other parts of the organism having lost their ability to exercise these functions. How can these chreods, initially labile on an everywhere-competent tissue, become localized in fixed organs whose structure will be transmitted hereditarily? This is one of the major problems of biology, since it is also the problem of the change from an essentially differentiable thermodynamical regulation to a discontinuously acting genetic regulation by biochemical mechanisms of control, genetically transmitted through specialized molecular structures. We shall now give a formal scheme for this transformation, although its biochemical realization is clearly a matter of speculation.

It is essential to understand that every function corresponds to the *catastrophic correction of a state of metabolic disequilibrium*. Consider the example of the function of excretion in an unicellular organism such as *Amoeba*. When the cell is not too large, so that the surface/volume ratio is sufficiently high, elimination of the waste products of metabolism can be carried out continuously along the cortex, by diffusion; but when the size of the cell exceeds a certain threshold, these waste products will accumulate inside to the point of creating a local phase of reduced metabolism at the point where they are accumulating. This phase will then form an excretory vacuole, which will finally evolve toward the outside. In this way the effect of the small surface/volume ratio will be corrected locally; the global morphology of the process will behave as if it were controlled by an organizing center which, in this case, is located outside the chreod that it defines (the codimension of the singularity being more than four). Thus the realization of any function is subordinate to the action of a specific organizing center in the local metabolism entering into a critical state, and the unfolding of the corresponding singularity generates the correcting catastrophe. Let Y denote the critical set (bifurcation stratum) defining the organizing center of this function in a local metabolism represented by a metabolic field (M, X). In the case of a function with a labile implantation, like excretion in amoebae, any point whatsoever of the cytoplasm can enter into the critical state Y and trigger the corrector catastrophe. How, then, did it happen that the majority of these functions have become localized, in the course of evolution, in fixed, genetically determined positions? Here is our model.

B. Localization of functions

First observe that the localization of a function corresponds to a phenomenon of generalized catastrophe: an initially homogeneous situation changes into a situation where the function can take place only in a narrowly specialized zone of tissue. We must therefore expect the phenomena of competition characteristic of generalized catastrophes; in fact, it frequently happens that a function is performed by organs distributed periodically throughout the organism, as in the well-known metamerism and pseudometamerism. However, particularly in vertebrates, most functions are localized in a single organ except for those, like locomotion, whose expression imposes mechanical or physiological constraints and requires a cluster of coordinated organs. Incidentally, we should not suppose that the localization of a function is always favorable; since phagocytosis is not localized in *Amoeba*, the organism has not had to create prehensile organs to carry its prey to its mouth, an indispensable corollary of having a digestive tube. The concentration of one function in one organ in vertebrates is probably a result of the perfectioning of the circulatory system of the blood and the consequent increase in the competitive power of each isolated organ.

We shall suppose that the localization of a function at a point of an individual is the result of a facilitation process; each corrector catastrophe of the organizing center Y frees substances into the cytoplasm with the effect of facilitating the return of the singularity Y. If, for some reason, one location seems to be more favorable for the triggering of the catastrophe Y, this advantage will be developed, often to the point of preventing the appearance of Y at any other part of the organism. We now give an idealized model of how such a monopoly might be established.

To simplify matters, suppose that the bifurcation stratum Y is that associated with a Riemann-Hugoniot catastrophe of a static model of local dynamics, and let Ouv be the universal unfolding space of this catastrophe, and P the semicubical parabola of the fold points with equation $u^3 - v^2 = 0$. As in Section 6.2A(2), we suppose that two stable regimes compete within P: an indifferent one, a, defined on $v < 0$, and a corrector one, b, defined on $v > 0$, and also that the correcting effect of regime b is defined by a vector field with components $U = 0$, $V = -A^2$. Then, if the representative point (u, v) of the local state leaves the region $v < 0$, $u > 0$, and so the regime a, under the effect of an external perturbation, and moves into the region $v > 0$, regime a will first persist, under the effect of delay, and then, as the point (u, v) meets the upper branch of P (or slightly before), this regime will be captured by regime b. The corrector field $V = -A^2$ will then operate moving the representative point back toward a neighborhood of the lower branch, when regime b will finish by being recaptured by the indifferent regime a. In this way the catastrophe associated with the singularity Y will completely correct the perturbation.

Let M be the space of local biochemical states, and s a substance having a catalytic effect on the environment; this amounts to saying that the s molecules bind themselves temporarily to the molecules of the environment and provoke a sequence of reactions that can be represented by a path c in M. In this way the catalytic action of each substance s can be characterized by a function associating with each point m of M a path $c_s(m)$ from m to $m' = s(m)$, and s will, in principle, be unaffected during the reaction cycle represented by $c_s(m)$. Now suppose that, among all the substances that can be synthesized by the environment at m, there is one, s', with the property that the reaction cycle leading to it is represented in M by a path c' inverse to $c_{s'}(m)$. In this case s' will be self-catalytic, since it provokes reactions in the environment leading to its own synthesis. Then, if there is only the single substance s' at some point m of M, there cannot be equilibrium at m, since s' will be synthesized massively. Hence equilibrium can occur at m only if there are several such substances in competition; each of these substances will exert a tension at m in the direction tangential to $c(m)$, and equilibrium will occur because the sum of all these tensions is zero. This reminds us of Waddingtons's picture of the genes as cords pulling at the epigenetic landscape.

Let p be a point on the upper branch of the semicubical parabola P. Suppose that among the substances s governing at p (or, more precisely, at the point $\hat{p}$ of M on the fold that projects to p on P) there is one, s', such that the path $c_{s'}(\hat{p})$ projects on Ouv into a path looping the origin O, and so transforms regime a into a regime b of lower potential. Since $\hat{p}$ is on a shock wave leading to the capture of regime a by regime b, the reaction cycle leading to a synthesis of s' will be preferred to other cycles, because it realizes in M a transformation imposed by the local decrease in potential defining the catastrophe Y; the effect will be a massive formation of s' at p. Then, by an analogous argument, we can suppose that another substance S, whose characteristic path $c_S(\hat{q})$ loops O in the opposite sense to $c_{s'}$, will form at a point q of the lower branch, where the $b \rightarrow a$ capture takes place.

The situation will, in this way, produce two substances associated with the catastrophe Y: a *sensitizing* substance s', with the effect of starting the corrector mechanism, and a *repressing* substance S that stops this mechanism when its action is no longer necessary. The global effect of substances s' and S on the environment will be as follows.

1. They will smooth out the shock waves defined by P and will facilitate the $a \rightarrow b$ and $b \rightarrow a$ transitions. In this sense the synthesis of these substances will appear as a viscosity effect of the ambient metabolism. From this point of view, the synthesis cycle of s' might be thought of as analogous to the mechanical situation of a turbine being moved by the flow of energy breaking along the shock wave $a \rightarrow b$, forming a kind of potential cascade.

2. Suppose that regime a is no longer undifferentiated but has a mean coupling field of the form $u = 0$, $v = a^2 > 0$. Then the representative point of the state will be trapped between the two branches of P. After the

shock waves have been smoothed by the substances s' and S, we might suppose that the representative point comes under the influence of a local potential W, which has a crest along the parabola P and so restrains this point to a basin inside P. The simplest equation for such a potential W is

$$\frac{dW}{dv} = v(u^3 - v^2),$$

so that, if $W(u, 0) = 0$,

$$W = -\frac{v^4}{4} + u^3\left(\frac{u^3v^2}{2}\right),$$

for which we introduce new coordinates by the transformation $(u, v) \longrightarrow (U, v)$ where $U = u^3$. It is easy to see that all the trajectories of grad W converge toward the origin in the half-plane $u < 0$ (Figure 10.10). This leads to a threshold stabilization. Any field of the form grad $W + u_0$, $u_0 > 0$, has a funneling effect around the origin on the side $u < 0$; this means that the chreod Y will have an attracting effect on the whole of the zone close to the origin O in this half-plane $u < 0$. Then the appearance of catastrophe Y will be facilitated at places where substances s' and S occur, and this pair of substances can act as the germ triggering Y. Furthermore, once the catastrophe occurs in the zone considered, correction of the disequilibrium will take place first locally and then globally through

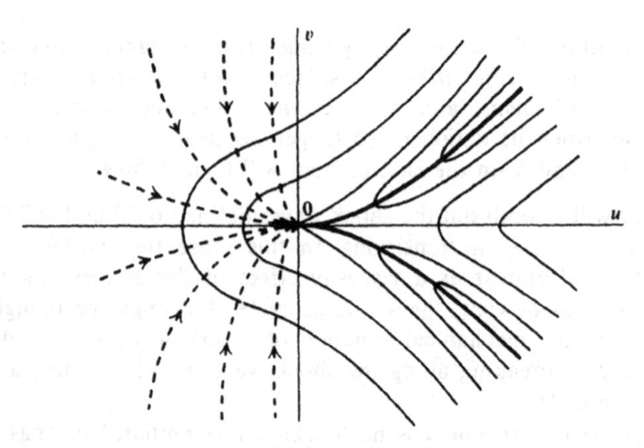

FIGURE 10.10. Threshold stabilization.

diffusion, and hence the probability of Y occurring at another part of the tissue will diminish. In this way the zone of substances s' and S will end up with the monopoly of Y-appearances, while, at the same time, the appearance of W can create a secondary morphology, attracting in the half-plane $u<0$.

Of course this explanation can have value only if the substances s' and S are stable and relatively fixed in the organism. Now it is known that most, if not all, of the organelles in unicellular organisms are related to organizing centers with hereditary properties; these organelles divide before the division of the individual in such a way that each descendant will have its own stock of organelles. Therefore let us now consider the central problem of the genetic material.

C. Formalism of reproduction, and the genetic material

Consider a gradient dynamic defined by a potential $V : U \longrightarrow \mathbf{R}$, and suppose that this dynamic has a projection $p : U \longrightarrow \mathbf{R}^4$ as a system of approximate first integrals; then a mean vector field X, a result of the coupling, will form on the base space $\mathbf{R}^4$. The dynamic V will be called *self-reproducing* if, when $\mathbf{R}^4$ is interpreted as space-time, the field X has a closed attracting trajectory g such that, if g is described in the sense defined by X, the associated attractor of V will bifurcate into two or more attractors, each of which will take the part of the initial attractor after g has been described. More precisely, we require that g be, not a closed cyclical trajectory, but only a system of arcs having a bifurcation at some given point; such paths g occur, or may be constructed, in the universal unfolding of the pocket singularity (the exfoliation of a shock wave) and the parabolic umbilic.

We shall suppose that the global dynamic of a living being has a structure of this type. This amounts to saying that *reproduction of a living being must be considered as a catastrophe whose formalism is similar in every respect to the catastrophes of physiological regulation* that we have just been considering. The scission catastrophe has an organizing center O in $\mathbf{R}^4$, which we can consider as the center of the cycle g; in the case of the mitotic cycle (to which we return in more detail in the next chapter), the figure will be of the type where g is a loop of strict duplication. For Metazoa, it is better to consider g as defined in the universal unfolding of the parabolic umbilic, and this leads to breaking phenomena in the gonad and allows an interpretation of the haploid-diploid alternation characteristic of sexual reproduction. This way of looking at things gives a solution to the old problem of the chicken and the egg: both chicken and egg are nothing but time sections of a global configuration whose organizing center never becomes manifest, but is circled indefinitely by the growth wave.

If we suppose this, it becomes natural to postulate that the division catastrophe of a cell itself can have germ structures; such a substance s will have an enzymatic effect c_s, defined by a path looping the organizing center O of the duplication catastrophe—that is to say, s will start multiplying itself as a preliminary to the scission catastrophe. Such a substance will behave like a chromosome in interphase.

Now it remains to explain how a germ structure of a catastrophe can have genetic properties. Vaguely speaking, we can probably say that the organizing centers of the metabolic and scission catastrophes are close in the space U of local states, and so it would not be surprising that, if one of the paths $c(s)$ loops one of these centers, it will loop the other. We shall propose, in the next chapter, an interpretation of the coupling between the metabolic effect and the genetic transmission of the structure.

Returning to the Riemann-Hugoniot catastrophe considered above, let us suppose that, in addition to the substances s' and S favoring the transitions $A \to B$ and $B \to A$, respectively, there are also substances, called a and b, whose enzymatic effect in the local environment is represented by vector fields $U = 0$, $V = \pm a^2$, characteristic of the regimes A and B. Then we might suppose that the system $as'bS$ of the four substances might form a complex whose enzymatic effect favors oscillations inside the semicubical parabola P. If there is a moment when these oscillations stop, under the effect of a global reduction of the ambient metabolism, the complex might duplicate, as with chromosomes, or multiply under the effect of coupling. The dying away of these oscillations, particularly in gametogenesis, results in the excitation of the portions of the chromosome, of which the corresponding enzymatic loops resonate with this cycle of small oscillations. Then the germ substance, diffusing into the cytoplasm from the loops of the lamp-brush chromosome, will synthesize massively. From this point of view, it is reasonable to suppose that cytoplasmic structures of the ooplasm are more important during development than the nucleic material. Conversely, the DNA plays the major part in the repression phase of gametogenesis.

D. Formal effects of localization: the reversibility of transitions and threshold stabilization

Let us consider a functional catastrophe like the formation of a digestive vacuole in *Amoeba*. This can be represented by the variation of a growth wave $F : \mathbf{R}^3 \times T \to U$, the universal unfolding of the catastrophe, where the local organizing center O corresponds to a stratum S of the bifurcation set. Then $F^{-1}(O)$ will give the critical point p where the digestive vacuole closes. The localization of this function will be represented as follows: the growth wave F can cross the bifurcation set S only at a well-defined point O of S, and it is natural to suppose that this localization spreads out into a local section J transverse to S at O, originated by the facilitation phenomena described earlier, with attracting and funneling mechanisms capturing the wave F and directing it toward this obligatory pathway. When the function is not in a state of activity, the permanent organic structure associated with it corresponds to the inverse image by F of J in $\mathbf{R}^3$, containing the critical section $F(\mathbf{R}^3, O)$ of the stratum S. This means that the threshold situation (here, when the vacuole closes at O) will always be realized organically; the organ can then trace out the transverse path J reversibly and periodically, giving a periodic cycle defined by the damping cycle of the singularity at O (like the chattering of teeth). In fact, this gives a spatiotemporal realization of a competition cycle analogous to that considered earlier in the parabola of the Riemann-Hugoniot catastrophe. However, it must be remarked that this reversibility applies only to the organic structure itself and not to the total field, which normally is oriented and irreversible; the mouth is made to take in food, and the heart to pump the blood, always in the same direction.

To summarize, the collection of transformations leading to organogenesis can be set out as follows (see Figure 10.11). The primitive functional catastrophe corresponds to a path c in the universal unfolding V of an organizing center, of which only the beginning and end points are significant, the actual path between them being relatively unimportant. This path c meets the strata $S_1, S_2, \ldots, S_k$ corresponding to critical topological situations, thresholds; each of these thresholds stabilizes in a well-defined point C_i on its bifurcation stratum and gives rise to a transition path J_i,

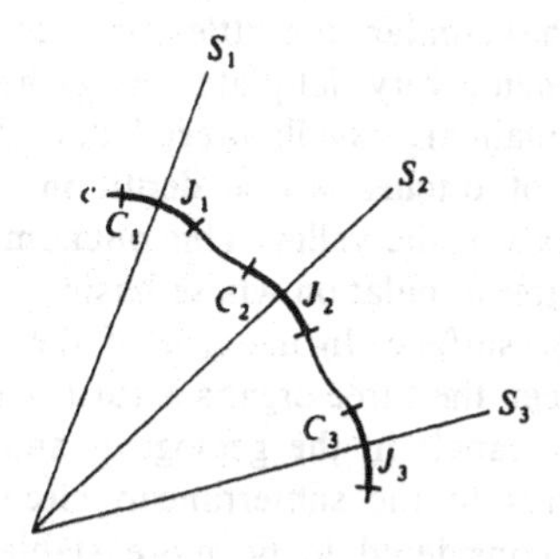

FIGURE 10.11. Scheme of organogenesis by successive threshold stabilizations.

also stabilized. Embryologically speaking, each of these transitions J_i will be created by a specialized morphological field, which will construct the corresponding threshold situation as well as the movable organs that ensure the transition section J_i; the functional transition between these primitive fields of codimension one is taken care of by organs created by secondary fields arising from the stabilization of the primitive fields and their interactions through resonance (like the digestive tube, which connects the mouth to the intestine). In this way, we again find the formal structure of the epigenetic polyhedron, at least as far as the construction of the organs associated with the main functions of physiology (feeding. breathing, etc.) is concerned.

E. Organs of the embryo

This raises a natural question: if organs are indispensable to life, how can the embryo, at the first stages of its development, manage so well without them? We might first reply that the conditions of the embryo's life are not natural. Even setting aside the extreme case of mammals, where the embryo is parasitic on the maternal organism, it usually has the reserves of the egg (yolk, etc.), allowing it to delay the primordial necessity of seeking food; and, as far as the indispensable functions of breathing, excretion, and circulation are concerned, these can be carried out by diffusion across the skin as long as the surface/volume ratio does not fall below a certain threshold, that is, as long as the organism is not too big. But then, as soon as it achieves a certain size, the creation of organs becomes necessary.

In this context, it is striking to observe how the definitive implantation of a function might necessitate several successive catastrophes of which the last and final catastrophe undoes the earlier ones: the example of the successive kidneys (pronephros, mesonephros, and metanephros) is well known, and the notochord in vertebrates might be considered as a protoskeleton which disappears with the formation of the vertebral column. This, too, is somewhat similar to a little-observed but indubitable geological phenomenon: when a very flat plateau is gashed by erosion, the valleys that descend to the plain are usually preceded, at the upper reaches of their source, by a zone of ditches whose depth increases toward the plain, situated along the axis of the valley. This phenomenon is doubtless due to the subterranean water circulation whose basin is more extended than that of the stream on the surface. In the case of the three vertebrate kidneys, the continuity between the three organs is furnished by the epithelium that forms the excretory canal; in the geological analogy, the epithelium catastrophe corresponds to the subterranean circulation, so that this catastrophe must be considered to be more stable than the mesenchyme catastrophe that it induces (a phenomenon called a *virtual catastrophe*).

10.5. A THEORETICAL PLAN OF A DIFFERENTIATION CATASTROPHE

Let H be a morphogenetical field defined by a chreod C with support U. Suppose, to fix the idea, that the field H arises from the interaction of two parent tissues E_1 and E_2, and that it gives birth to two new tissues t_1 and t_2; this can be represented by a one-dimensional simplex of the epigenetic polyhedron (Figure 10.12). Each of the tissues E_1 and E_2 is defined, in the metabolic model, by an attractor h_1 or h_2 of the local biochemical kinetic, and the catastrophe is provoked by the two oscillators h_1 and h_2 entering into resonance; the product dynamic $h_1 \times h_2$ forms on U, and this local dynamic will be structurally unstable and so will degenerate into a lower-dimensional attractor H. It will be convenient to suppose that there is a fibration of the form $H \longrightarrow h_1 \times h_2 \longrightarrow Q$, where the quotient space Q parameterizes all the possible positions of the final resonance—this will be the case when, for example, h_1 and h_2 are closed simple trajectories and H is therefore a circle. This space Q will be the director space of the catastrophe associated with C; in effect, the final morphology after catastrophe will be defined by an auxiliary static or metabolic field Y with internal space Q which characterizes the competition between the resonances. This field Y can be defined as a mean coupling field; initially the field X in $h_1 \times h_2$ is tangent to the fiber of the fibration $h_1 \times h_2 \longrightarrow Q$; it will cease to be so, however, after a slight perturbation, and the integration along the fiber of the horizontal component will give rise, by projection, to the field Y on Q. The later evolution of the catastrophe will be described by the auxiliary metabolic field (Q, Y), and the attractors of this field will define the stabilized resonances c_1 and c_2, which, in our case, generate the new tissues t_1 and t_2. We can suppose that the field (Q, Y) depends itself

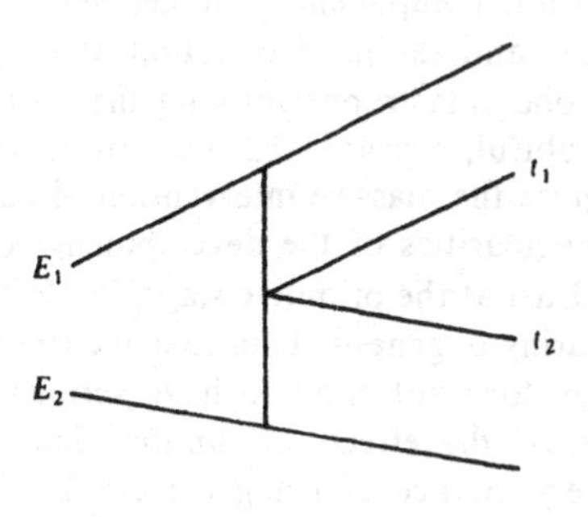

FIGURE 10.12

on a static model which, at the central point O of U, admits a finite-dimensional universal unfolding W; then the morphology of the chreod will be defined by a map F of U into W, the local growth wave, transversal to the universal catastrophe set K of W. All the catastrophes studied in Chapter 9 follow this pattern.

What mechanism controls the mean field Y? We may suppose that there is a germ structure G at the beginning of the catastrophe with the following properties: G has an enzymatic effect on the ambient metabolism, described by a mean field Y_1 with attractor H_1, but this germ structure becomes unstable in the region H_1 that it catalyzes; G then evolves to form complexes GA and GB, for example, catalyzing the formation of regimes c_1 and c_2 in (Q, Y), and such that these complexes are stable for the regimes that they catalyze. In these circumstances regimes c_1 and c_2 will compete; and, if the different phases t_1 and t_2 ever have different properties with respect to the main gradients of the organism, the final configuration of the division between these two phases will be well determined. In very refined organogenesis, like that of the eye and of limbs, it is doubtful that such a simple process can account for the elaboration of so complex a local structure, and here there are perhaps other mechanisms, probably requiring an anisotropic macromolecular structure of G, which we shall later be calling a polarization.

This model raises the following questions.

1. The origin of the germ structure. When there is only one initial tissue of the chreod, the germ structure G will be a component deposited in the cytoplasm of the egg during oogenesis and will essentially be of protein type, although the presence of the DNA and RNA nucleic acids is not excluded. When the chreod is the result of the interactions of several tissued $h_1, h_2, \ldots$, we might expect that G would arise by intercombination between the G-structures associated with the regimes $h_1, h_2, \ldots$ and possibly other ooplasmic components. The geometry and the chemistry of this intercombination, and the later evolution of G, provide a molecular model of the metabolic catastrophe forming the organizing center of the chreod. It seems doubtful, a priori, that these transformations imply that the G-structures require the massive intervention of chromosomal DNA in the embryo. The irregularities of the development connected with mutations are caused, at least at the primitive stage, by defects in the ooplasmic structures created during oogenesis. Thus massive irradiation of an embryo at the gastrula stage does not seem to have serious effects on the later morphogenetical fields; the effects of random genetical irregularities in many cells have every chance of being corrected. The situation is quite different, howevei, for the introduction of chemical substances into the environment, and the teratogenetic effects of many drugs, particularly pharmaceutical ones, are well known.

2. The frequent disappearance of organizing centers in embryology. We have seen how, in principle, G is unstable. Also the support of G generally divides into several distinct differentiations, if it does not completely disappear through necrosis. Examples of this type of evolution are well known in embryology (notochord).

3. Polarization. Consider a morphogenetical field, defined by the degeneration $H_{i-1} \rightarrow H_i$ of attractors of the local biochemical kinetic. We have seen that the corresponding chreod is defined by a map F of the support U of the chreod into a universal unfolding space W of the singularity of the metabolism defined by the field (Q, Y) of the coupling. Now, if the morphogenesis is to be well defined, the map F must be an embedding into W (transversal to the universal catastrophe set of W); equivalently, different points of the tissue must have different local biochemical properties. How might such a polarity be created and, once created maintained? This is an essential problem of biological morphogenesis, and it will need a fairly long discussion.

Classically, it is said that every tissue has stable gradients. For example the animal-vegetal gradient of the egg gives rise, roughly speaking, to the cephalocaudal gradient of the animal, and no doubt there are other gradients (left-right and dorsoventral), although much less fixed than this one. In geometrical terms, we say that each tissue A admits a pseudogroup G_A of local equivalences, a subgroup of the local displacements. When this pseudogroup G_A is the whole set of local displacements, the tissue is said to be homogeneous; at the other extreme, if G_A is reduced to the local identity displacements, A is totally polarized, and this amounts to saying that A contains local biochemical coordinates distinguishing all of its points. Now, as we saw in Chapter 6, each metabolic field like H_i admits a pseudogroup of local equivalences $G(H_i)$, and a catastrophe $H_{i-1} \rightarrow H_i$ is possible only if $G(H_i)$ is a sub-pseudogroup of $G(H_{i-1})$; therefore a totally polarized tissue in a catabolic catastrophe can give rise only to a tissue that itself is totally polarized. We have here one of the dynamical justifications of phenomena of induction; in the neuralization of the gastrula ectoderm, presumed homogeneous, by the underlying notochord, the mesoderm provokes, by contact, not only the appearance of the neural regime but also its polarization (in the cephalocaudal direction), and this polarization occurs by a continuous extension from the polarization of the mesoderm itself. In order for an initially homogeneous tissue to give birth to an even partially polarized tissue, it is necessary that the former undergo the action of an inductor tissue itself polarized (although of course the statement that the ectoderm of the gastrula is homogeneous must be qualified). In a catastrophe where $G(H_i)$ is a proper sub-pseudogroup of $G(H_{i-1})$ or, equivalently, in the case where an asymmetry is created from the void (violating Curie's principle), we should expect to have an effect of critical masses; the catastrophe can begin only when there is a sufficiently large mass of tissue at its boundary. In this way a solid crystal can provoke the solidification of a supercooled liquid only when its volume exceeds a certain threshold defined by the surface/volume ratio. One formal reason for these phenomena can be expressed as follows: the realization of the catastrophe $H_{i-1} \rightarrow H_i$ requires a polarization of the local sections $G(H_{i-1})/G(H_i)$; and, if the polarization is the result of gradients of the concentrations of coordinate substances, the gradient must form and maintain itself despite the diffusion across the boundary surface—hence an area/volume condition.

If the necessity of this polarization phenomenon in tissues prior to any morphogenesis is admitted, then what might be its biochemical support? The process of palisading of cells is often spoken of as a prelude to the epigenesis of an organ. Indeed we might suppose each cell wall in a tissue of cubical cells whose edges are parallel to the *Oxyz*-axes to have an asymmetrical behavior with respect to some coordinate substances; so the walls normal to *Ox* might allow preferentially the passage of a substance q_x in the sense of increasing x, and inhibit it in the opposite direction—phenomena of this type are known for a neuron wall and potassium ion. However, this scheme would require that the physiological properties of a cell wall depend on its spatial orientation, and this seems a rash assumption in the present state of physiological knowledge. But if it were so, three substances, q_x, q_y, and q_z, could provide enough concentration gradients to separate the points.

This scheme is notoriously no less insufficient, because morphogenesis very often has an infracellular, purely cytoplasmic support; for example, the spicules of holothurians, with their highly refined elaboration, form themselves in a syncytium. We are reduced to conjectures about the biochemical support of these polarities; I want only to emphasize the fact that certain biochemical parameters may have a purely kinetic character that will render them completely beyond investigation by chemical analysis. For instance, we cannot exclude the hypothesis of a cytoplasmic infrastructure with a circulatory character, like the decomposition of a fluid, under the influence of two opposed gradients, into convection cells (Bénard's phenomenon). We might also conjecture, as we did earlier, that each local regime is connected with the presence of a periodic macromolecular configuration of such lability that any attempt at its observation and fixation destroys it. The interaction of two tissues will then present itself by an interpenetration of the corresponding configurations, followed by a recombination whose geometry simulates the catastrophe $H_{i-1} \to H_i$ characterizing the morphogenetical field. In the case where a catastrophe creates order [so that $G(H_i)$ is strictly smaller than $G(H_{i-1})$], there is rarely any doubt that the polarization of the tissue procedes the catastrophe itself, because only this can allow the catastrophe to take place. We should probably assume that, before a new regime H_i can be set up, the tissue must enter a state of virtual catastrophe; many regions of the new regime will form randomly in the tissue but, being too small, will be unstable and will disappear—this fluctuating situation may persist until a stable growth regime of the new regime H_i can be achieved. Thus the polarization of the virtual catastrophe will precede the starting of the real catastrophe. A mathematical theory of this type of phenomenon has still to be made, and I should like now to give a reason why, in such an order-creating (or, more precisely, asymmetry-creating) catastrophe we should expect the support space to be polarized.

Suppose that the original attractor H_{i-1} is fibered over a quotient space Q, the fiber being the new attractor H_i, and the original regime is homogeneous on its support U and ergodic on the manifold H_{i-1}—this means that the probability that the representative point of the system is in a volume element dH_{i-1} of the manifold H_{i-1} above the volume dU of the support U is proportional to the product $dH_{i-1} \wedge dU$ of these volumes. Such a situation has every chance of remaining structurally stable, at least as far as the average over large volumes U is concerned. This means that the projection of the representative point onto the quotient space Q will itself have its probability of being in volume dQ of Q proportional to volume dQ, at least for large volumes of U. The final catastrophe will be defined by a map $h : U \to Q$ (here Q is the director space of the catastrophe), which must be surjective and of homogeneous measure on U; the volume $h^{-1}(dQ)$ must be independent of the position of the element dQ. Consequently h must be surjective and periodic in character, and this defines a polarization on U. In fact, in most cases the type of the function h is well defined, and only for a function h of this type will the new regime have a thermodynamical advantage over the old one and supplant the old.

This requirement of the transmission of a polarity in order to free (*à la mode*, to "de-repress") a potential regime in a piece of tissue has been observed by many biologists, but I do not think that they have given it the consideration that it merits. The typical example is the fertilization of the ovum, where it is well known that the penetration point of the spermatozoon often determines the future symmetry plane of the germ. Of course this action is not indispensable for the starting of the catastrophe; artificial fertilization by chemical agents, as in artificial inductions (particularly by an exogenous inductor on a tissue culture), can lead to the starting of the new regime and the setting up of a morphogenetical field almost as well controlled as that obtained by the action of a natural inductor. But here the polarity is set up by a sequence of ill-determined, structurally unstable processes, and minute changes in the initial conditions may alter entirely the orientation of the morphogenesis and often its qualitative form. The dynamic of life cannot tolerate such instability; hence we should distinguish two distinct, though thermodynamically connected, features in every inductive action.

1. The setting up of a new regime to the detriment of the pre-existing one.
2. The provision of a polarity to the induced tissue.

When the catastrophe creates asymmetry, the second feature will be indispensable to coherent development. Moreover, it greatly facilitates the progress of the first.

In most natural inductions, the polarization of the induced tissue occurs parallel (by a continuous extension) to that of the inductor tissue; the orientation of the induced organs prolongs the orientations of the inductor organs. The extension of a polarity from the inductor to the induced can easily be explained by the thermodynamical diffusion of a polarized regime, as in the following dynamical scheme: let A be the inductor regime (supposed polarized), B the induced (homogeneous) regime, and $A \times B \rightarrow C$ the catabolic catastrophe defining the induction; after the contact between inductor and induced, the induced will adopt the product regime $A \times B$ and undergo the catastrophe leading to C. Then it is normal to suppose that it will be the polarization of regime A that will polarize the tissue supporting the catastrophe $A \times B \rightarrow C$.

*4. **Embryological induction***. This approach may explain the setbacks suffered by experimental embryology in its search for the specific inductor agent; if this agent has never allowed itself to be isolated, this may be for the good reason that it does not exist. Of course there are inductor agents; in fact, any chemical substance acting on a competent tissue can provoke a morphogenetical catastrophe in it. When the tissue is capable of many qualitatively distinct catastrophes, like the gastrula ectoderm, the choice of catastrophe is certainly a function of the chemical nature of the agent. It is said that the realization of late and very elaborate inductions requires complex inductors, like proteins extracted from living tissues; cruder and less specific inductors, such as heat, pH variation, or methylene blue, can provoke only the most immediate chreod in the ectoderm, that of cephalic neurulization. But it seems to me that induction appeals essentially to the biochemical kinetic, and that the catabolic catastrophe is the biochemical analogy of resonance in vibration dynamics. Of course some substances can exert, by their enzymatic effects, a catalytic or inhibitory effect on certain biochemical reaction cycles, and therefore on certain local biochemical regimes; but to what extent this action is specific, and how far there are specific catalysts or inhibitors for every local regime, is much less sure. At the most we might concede, as we did earlier, that every local regime is connected with the presence of a periodic metastable macromolecular configuration playing the part of the germ structure. Induction then corresponds to a recombination between these configurations.

A scheme of this type is compatible with the classical experiments in which a microfilter is interposed between inductor and induced. The diffusion of the "inductor agent" has, in normal development, a limited, oriented, and controlled aspect, a process fundamentally different from simple diffusion.

*5. **Abnormal evolutions.*** Each differentiation catastrophe $H_i \times H_j \rightarrow H_k$ has its associated chreod and director space W_k; such a chreod will be in the zone of influence of the chreods giving rise to H_i and H_j, while it will have in its zone of influence the chreods connected with catastrophes of the form $H_k \times H_s \rightarrow H_m$. It would be difficult to specify the structure of the model in more detail, particularly in regard to the choice of the successor chreod in the umbilical zone. Thus one might ask whether aberrant organogenesis caused by genes with teratogenic effect arose from a modification of the chreod atlas, or only from a change in the evolution of the growth wave. Probably the distinction between the two types of monstrosity is purely verbal; we might, by introducing sufficiently many parameters, be able to make a chreod atlas, a universal epigenetic polyhedron for the species under consideration, valid whatever the genotype of the individual. Any teratogenic effect would then be due to a perturbation of the evolution of the growth wave in this universal atlas, and it could imply a modification in the number and type of those chreods swept out by the growth wave.

10.6. EXAMPLES OF ORGANOGENESIS

A. Respiration and blood circulation

*1. **The origin of the blood circulation.*** We shall now apply this formalism to give a hypothetical explanation of the evolutionary origins of the systems of blood circulation and respiration. To this end, we first study a fundamental physiological process, the variation in the intensity of the metabolism j (measured, e.g., by the quantity of oxygen absorbed in unit time in 1 cubic centimeter of tissue) as a function of the local oxygen concentration q. It is natural to suppose that, for large values of q, j

increases practically linearly with q and then tends to a limiting value given by the amount of supply (combustible metabolites) provided for the tissue; while, as q decreases, we suppose that there is a threshold value a below which a catastrophe occurs in which the metabolism suddenly decreases to a smaller value j_0 and then tends to zero with q, corresponding to a state of slowed-down life or virtual death—then, if we wish to reactivate the tissue by increasing the oxygen pressure q, it will be necessary to go well above the threshold a to return to a normal metabolism. All of this behavior can be symbolized by representing the function $j(q)$ by a Riemann-Hugoniot folding (Figure 10.13); if the temperature also decreases, as well as the supply of combustible metabolites, the folding will stretch out to give j as a single-valued increasing function of q. In all that follows, we shall write j_0 for the lower branch (slowed-down life) and j_1 for the upper branch (normal metabolism), and the intermediate decreasing branch will represent an unstable and therefore nonexistent regime.

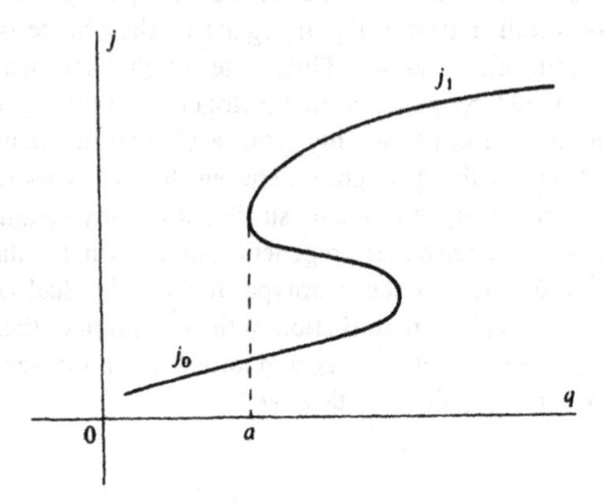

FIGURE 10.13. Graph expressing the intensity of the metabolism against concentration q of oxygen.

Now we consider an organism O, and let x denote the distance of a point of the organism from the periphery (the epidermis in Metazoa, the cortex in unicellular organisms), and u be the speed of oxygen diffusion in the direction of increasing x. Then we have the differential equations

$$\frac{dq}{dx} = -ku, \qquad \frac{du}{dx} = -j,$$

and so q is given as a function of x by the solution of

$$\frac{1}{k}\frac{d^2q}{dx^2} = j_1(q), \tag{E}$$

for which $q(A)$ (where A is a positive constant) is the oxygen pressure q_0 in the outside air. To simplify, we suppose that the organism is one-dimensional, defined by the segment $|x| < A$. Since the right-hand side of equation (E) can be majorized by a linear function, $j_1(q) < mq$, the solution $q = v(x)$ can be majorized by solution of the simpler equation

$$\frac{1}{k}\frac{d^2q}{dx^2} = mq.$$

Now the solution of this equation is

$$W = B \exp(\alpha x) + B \exp(-\alpha x),$$

where for symmetry reasons, $\alpha^2 = mk$. The value of this function at the center of the organism is $W(0) = 2B$, and the ratio $W(0)/W(A)$ is $2[\exp(\alpha A) + \exp(-\alpha A)]^{-1}$. This ratio tends to zero as A tends to infinity, so that diffusion alone will be insufficient to provide an adequate oxygen supply at the center of the organism when the size exceeds a certain threshold; $q(0)$ will then be lower than the catastrophe threshold, and a region of reduced metabolism will form at the center of the organism. This region might be like an excretory vacuole filled with carbon dioxide and other waste products, later to be conveyed to the exterior.

We could clearly dispense with these quantitative considerations and consider an obvious argument about the surface/volume ratio. However, it is interesting to plot on the same axes the solutions of both the equation

$$\frac{1}{k}\frac{d^2q}{dx^2} = j_0(x), \tag{E$_0$}$$

corresponding to the slowed-down conditions of life (since j_0 is small, these solutions will be almost straight lines), and equation (E). Considering the solutions passing through ($x = -A$, $q = q_0$) and $(0, a)$, we obtain two branches C_0 and C_1 with obvious interpretations: C_0, joining the line $x = -A$ to $(0, a)$, represents the circulation of an excretory vacuole to the outside, while C_1, leaving from $(-A, q_0)$, represents the ingestion by ropheocytosis of a vacuole of the outside environment, rich in oxygen (Figure 10.14). It is clear that the simultaneous use of these curves C_1 and C_2 instead of the initial solution $q = v(x)$ assures an oxygen supply at the center much higher than the threshold value, and eliminates the excess CO_2. How did this solution come about?

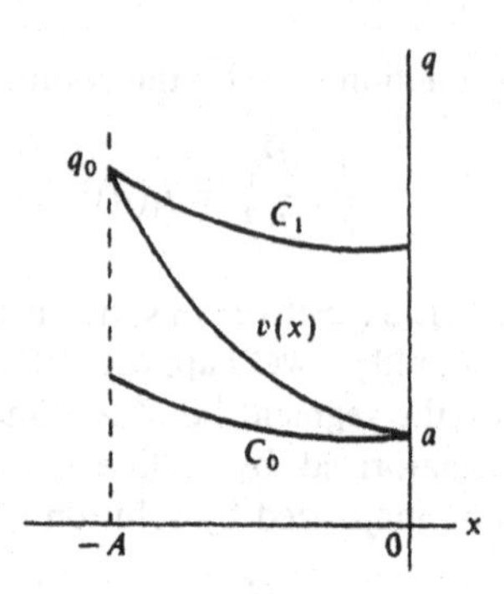

FIGURE 10.14. Hysteresis cycle for oxygen concentration.

2. *The blood.* Here the problem of the creation of hemoglobin occurs. Suppose, conforming to our global model, that the metabolism of a primordial germinal cell (ovogonium or spermatogonium) realizes an image of the global regulation figure of the species. There will be coordinates homologous to q, x, and j in the configuration space containing this figure, and we shall again write these coordinates as q, x, and j, even though they do not necessarily again have the same interpretation (particularly the spatial variable x).

Suppose then that the representative point of the local state describes a figure at first close to that given by the differential equation (E), and then by (E_0) as the size of the organism approaches the critical threshold. Under these conditions, a cycle can be set up consisting of the curves C_0 and C_1, completed by the lines $x = -A$ and $x = 0$, forming a closed trajectory of the spectrum of the local metabolism, a cycle whose description means the supply of oxygen and the elimination of CO_2. The vertical lines $x = -A$ and $x = 0$ define chemical catastrophes on the cycle that can be described as the rapid transformation of a phase rich in CO_2 into a phase rich in oxygen, and conversely (Figure 10.14). Then hemoglobin appears as a germ substance of these two catastrophes; and the two forms of the protein complex, as oxyhemoglobin and reduced hemoglobin due to a chemical threshold stabilization process, form a molecular model of these two catastrophes. I shall not here consider the problem of the origin of the structural gene of hemoglobin, apart from saying that it might well have arisen from a macromolecular combination of the genes stabilizing the regime j_0 of slowed-down life, but at a sufficiently high metabolic level to avoid the irreversible degradations of the machinery of life.

As soon as the hemoglobin is formed, a continuous phase in the organism can describe the cycle C_0C_1, and the blood circulation is thus set up; the critical threshold in the size of the organism can then be crossed. To begin with, the flow of blood can be achieved by simple diffusion, with the hemoglobin molecules being dragged along by the gradient of decreasing O_2 concentration, and increasing CO_2 concentration in the outside-inside direction.

3. The heart. Then a new threshold quickly appears. If, in equation (E_0), we have $j_0 = 0$, the speed u is a (positive) constant, and so q will become a linear decreasing function of x:

$$q = q_0 - kux.$$

Hence, for sufficiently large x, q will fall below the critical threshold a and asphyxia will again threaten the central tissues. Furthermore, the blood flow will become discontinuous, broken up into drops that move around in a periodic way. Therefore the tissue bounding the blood phase in the organism has to be flexible, but why does this tissue become muscle? Here again we must appeal to the metabolic picture of the process in the regulation figure, realized in the primordial germinal cells. We might suppose that this figure is refined during the course of evolution so that it contains more and more details; in particular, if it contains a representation of the outside world, it wil also have one of the internal world of the organism itself. Now a curve such as the arc C_1 represents at first the passage of a vacuole rich in oxygen from the exterior toward the interior, and so this curve has a representation as a spatial catastrophe defined by a loop of the growth wave $F(\mathbf{R}^3, t)$ in a universal space W representing the organ space. Then, after projection along trajectories of the form $F(m, t)$, $m \in \mathbf{R}^3$, the universal model of this catastrophe will become a flexible cylinder in $\mathbf{R}^3$ containing a bulb that passes from one end to the other. Now the only flexible and mobile tissues are either connective or, when work is required of the tissue, muscular. The loop $F(\mathbf{R}^3, t)$ will be stabilized in W through threshold stabilization, and its frequency will be determined by a correcting law which might be represented as follows: the intensity j of the metabolism of the tissues being supplied, as a function of the blood circulatory speed u, also has a folding with a vertical tangent at the critical speed u_0; the beginning of the heart's functioning might be represented by a smoothing of this cliff, this physiological shock wave. This will give a considerable increase Δv in the speed u at the cost of a local decrease Δj in j (Figure 10.15) and a transfer of this energy to the heart and artery muscles.

This too is a particular case of the fact discussed in Section 10.4.B, that the smoothing of a catastrophe (a discontinuity) implies the stabilization of its organizing center, this, in its turn, implying a secondary morphology. It is the later importance accorded to this cardiac function that has enabled the largest known animals to reach their size. Note that the basic plan of the heart (Figure 10.16), inspired by the scheme of Figure 10.15, has the arrows in parallel. The cardiac tissue is genetically characterized by a variable attractor whose universal unfolding is a contracting cylinder; this, it seems, is the only known case of cellular specialization in which the final result varies in a well-defined, periodic way in time. These considerations may illuminate the main problem of cardiology: why does the heart beat?

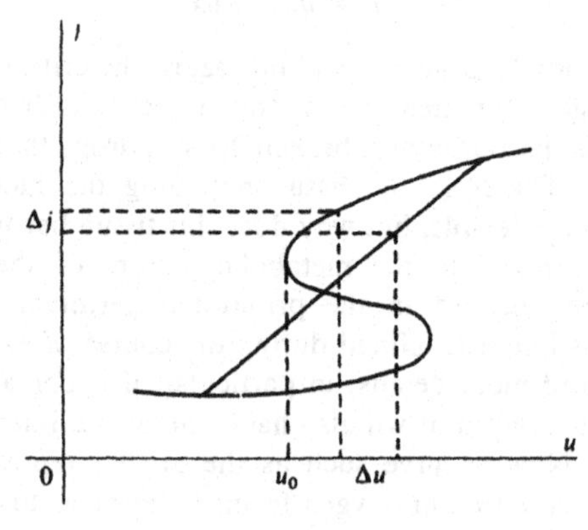

FIGURE 10.15

*4. **Breathing***. This is the place to study the morphology of the lungs and the capillary system, and we shall obviously consider a generalized catastrophe with a spatial parameter. The case of the lungs is particularly interesting, because here there is a first catastrophe consisting of the formation by invagination of vacuoles of the outside environment, the air. Then this catastrophe is stabilized in the threshold situation with the vacuoles maintaining the connection with the external air, and with a reversible functional path allowing them to be periodically filled and emptied. The fact that this is an ingestion catastrophe explains why, embryologically, the pulmonary epithelium has an endodermic origin. Then there is a capillary catastrophe with a vaguely periodic character on the surface of the vacuoles (the pulmonary alveoli), with the same type of cellular configuration as is observed in Bénard's phenomenon (pulmonary vesicles).

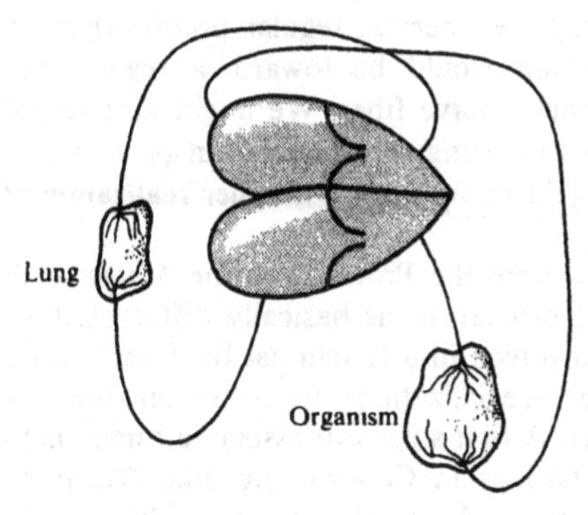

FIGURE 10.16

B. The nervous system

1. The origin of the nervous function. The evolutionary origins of the nervous system lie very far back; as soon as an organ is formed, in practice a specialized system will be created to control its activities. So we obviously should suppose that those unicellular organisms with fixed organelles such as cilia already have a cytoplasmic structure acting as a nervous system; otherwise how can we explain, without a structure of this type, how the beating of the different cilia are so efficiently synchronized into global functional fields? But although the existence of such a coordinating function is indispensable, neither its support nor its origin is known; on the latter point, we might postulate a genetic origin. We know that the ciliature of a unicellular organism derives from a basal body, a kind of organizing center itself derived from a centriole; in effect, it is more economical to achieve the division of the individual by doubling a unique organizing center than by doubling in isolation the germ structures of each of the organs of the field. The uniqueness of the organizing center then probably corresponds to the fact that the global structure that it generates is connected, so that the different organelles that are generated belong to a unique functional structure. It is not unreasonable to think that nerves and muscles have as a common ancestor the cytoplasmic structure represented by the microtubes 200 Å in diameter found in the infrastructure of cilia and in the primitive neurones of the nerve cells of *Hydra*. This structure, initially associated with all the movement catastrophes, might have undergone a later differentiation: where there might have been a mechanical resistance, there would be modification of the topology of the system toward muscle fiber; at regular points where there was no resistance, the evolution would be toward a regime favoring rapidity of propagation, namely, nerve fiber. We might also regard the fibers of the mitotic apparatus realizing the spatial transport of biochemically significant substances (chromosomes) as another realization of this hypothetical ancestor.

When we pass from the Protozoa to the Metazoa, the development of the coordinating mechanism is basically different; this is so because the various organs involved in a functional field have too diverse histological and biochemical specializations to be under the control of a unique organizing center. A first step is to associate functional specialization with cellular differentiation; in Ceolenterata, like *Hydra*, this step is not yet completed because, on the one hand, some cells, like the musculoepithelial cells, have a mixed functional vocation, whereas, on the other, there are some very complicated fields with a uniquely unicellular support [e.g., the stinging cells (cnidoblasts), which have a most improbably refined cytoplasmic apparatus]. In such cases we cannot yet use the full cellular scale, but must allow the fields a cytoplasmic support; hence we must suppose that differentiation of the omnipotent interstitial cells in *Hydra* is controlled, in generalized catastrophe, by nonlocalized functional fields which are fixed at the beginning of an ill-determined competition process. Thus there is the group of sense cell, nervous cell, and cnidoblast that forms an autonomous functional unit and of which many examples can be found in the epithelium of the animal, but with a very imprecise geometry.

In more evolved Metazoa, all of the cells with functional vocation seem to derive from the same layer of the embryo. This is the abstract (geometrical or biochemical) character of the functional catastrophe determining the origin of the cell. The global physiological fields, are formed only after this, by combinations of the several constituent organ fields and the nervous system guarantees the coordination of these fields. Let me give an idea of this process.

2. The structure and role of the nervous system. Let us return to the central idea of our model: that the metabolism of a germinal cell is the realization of a simulation of the global regulation figure of the species. Such a figure F is an unfolding of a degenerate germinal dynamic G; if then G unfolds in a system of spatiotemporal parameters, this will give a description of development and, at maturity, the dynamic Z defined by the mean coupling on this space will give another realization of F. This means that the construction and operation of an organ O will correspond to the presence of an attracting circuit in the macroscopic dynamic Z; this circuit will describe not only the spatiotemporal and biochemical field which has the organ O as support, but also the collection of nervous activities connected with the operation of the organ. Now we can surely suppose that the operation of every organ is completely described or, more precisely, simulated by the neural activities of all the nerve cells coupled to it; under this hypothesis we obtain a complete simulation of the figure F by considering only the set N of nervous activities of the animal. Then we may suppose (see Chapter 13, Note 2) that the set of nervous activities provides a description of the outside space $\mathbf{R}^3$ and of the position of the organism B in this space, or, equivalently, that the pair $(B, \mathbf{R}^3)$ is a system of approximate first integrals of the nerve dynamic N. The main correcting fields in this figure will be the movement fields (of walking, running, etc.) which guarantee the synchronization of the muscular activities of the limbs. The image B of the body will itself contain regulating fields associated with the main physiological functions (blood circulation, breathing, excretion, etc.), although, in man, these fields enter conscious thought only exceptionally.

We can say that, to a first approximation, the cerebral cortex is the support of the external spatial field $\mathbf{R}^3$, and that the spinal chord (with its sympathetic and associated systems) is the support of the field B; hence the initiative for the construction of the essential sensory organs (eyes, ears) comes from the cerebral tissue. The nervous representation of the skin must be situated on the boundary of B in $\mathbf{R}^3$, but in its organic realization the skin must be on the boundary of the organism; therefore the whole of the neural plate must invaginate into the interior while the mesoderm remains fixed—this precisely is the effect of neurulation. After all, the neurocoele is the support of a representation of the space at infinity, and its formation during neurulation symbolically represents the absorption by the animal of the ambient space in which it will live.

3. The epigenesis of the nervous system. We can finally explain the organogenesis of the nervous system as follows: the local metabolic regime of the nervous tissue can be characterized by the almost complete abolition of states r, leaving, in the form of approximate first integrals, only states s. The physiology of a fully differentiated neuron might be described as follows. Its configuration space will be a high-dimensional Euclidean space S, and there will be a hyperplane H, the excitation threshold, in this space defined, say, by $x_n = 0$; then, as soon as the representative point crosses this hyperplane to the side $x_n > 0$, there will be a flux of nervous energy represented by a sudden translation normal to H, $\Delta x_n = -1$. Each time that a neuron undergoes an excitation coming from an afferent neuron N_j, the representative point will undergo a translation Y_j characteristic of this neuron and of the instantaneous state of the synapse of the junction. In some sense, a neuron is a "stupid" cell whose stereotyped reactions are far removed from the subtlety and finesse of the other cells, particularly those, like leucocytes, that are concerned with the struggle against antigenes; the neuron is like a sick man, pampered, protected, and fed by its Schwann sheath, and allowed to keep the traces of all the influences that affect it, of all its past history.

But this degeneration of the metabolism—said in no pejorative sense, because it is accompanied by a large increase in the topological complexity—can only come about progressively. At first the local metabolism can regulate itself in many varied ways, according to a curve C whose orientation depends both on primitive epigenetical gradients and on the kind of excitations received by the cell. The regularization of this corrector curve toward the final direction x_n is accompanied by a compensatory sensitizing of the cytoplasm with respect to the horizontal component Z of the function field or fields passing through this cell; equivalently, the qualitative specificity of the functional fields acting on the cell disappears progressively, but, in localizing themselves in the cytoplasm, these fields sensitize it to epigenetical gradients (recall that these are the coordinates of the universal unfolding of the germinal dynamic G defining the configuration space of the dynamic Z). Hence *the biochemical specificity is transformed into a spatial localization*; the cell unfurls a long filament, the axon, whose extremity ramifies as the director gradient weakens. The connections thus made with innervated organs or other neurons will be as stable as the degree to which the contacts so set up allow the realization, through resonance, of the global regulating fields of the dynamic Z. The definitive structure of the connections between the neurons will be established only after a long process of successive approximations whose limit will be the best possible realization of the regulation dynamic Z (see Appendix 3).

This scheme contains many obscurities, even on a strictly theoretical plane. One of the primordial questions is to what extent the nervous system is capable of structuring itself through its own primitive epigenetical polarizations, without the influence of the peripheral innervated organs. Because of the high degree of functional lability of nerve tissue, even at adult age, we might think that the only parts of such tissue capable of their own internal structuralization are those ending in the formation and innervation of sensory organs. Most of the cerebral matter is a kind of exploded organizing center in which the regulating fields, whose source and support are the peripheral organs, have modeled the main structurally defined traits.

From this point of view, there is a certain homology between the brain and the gonad. It is as if the initial lability of the epigenetical fields finally becomes concentrated in these two organs: on the one hand, the gonad, manufacturing the omnipotent gametes, and, on the other, the brain, a functional field capable, at least in man, of an almost boundless variability. Nervous activity can then be represented as controlling an information flux; the functional fields of epigenesis, decomposed by the catastrophes of organogenesis, reconstruct as fields of stationary neuron activity in the ganglions corresponding to organs, while the functional interactions between the fields gives rise to the connections between the neurons of corresponding ganglia. This might explain how the embryo, at birth, has all the reflexes necessary for subsistence. However, in addition to directed connection, there is a mass of connections set up without any valid reason, that is, where no director gradient can have intervened; this is particularly the case in the cerebral cortex, where the attracting gradients of the sensory and motor organs have the least effect. The stationary nerve activities, supported by these more or less randomly set up circuits, constitute a large attracting manifold W which they fill out ergodically, and these activities are only weakly coupled with the flux of sensory-motor information which guarantees the physiological regulation.

Education operates by acting on this *tabula rasa*; by presenting to a young subject a system of pre-established forms acting as stimuli, this process of education excites by induction (resonance) analogous forms in W, and this can there cause a highly varied catastrophe (in principle, reversible because of the phenomenon of forgetfulness). If, however, these stimuli have not been presented to the subject at the age when the field W still has all of its lability and competence, W will evolve through aging toward a degenerate attracting state and any subsequent attempt at education will come to a halt; the mental capacities will be insignificant (this is the case with deaf-mutes from birth). From this point of view, education appears as the normal successor to embryonic epigenesis, and the inductor action exercised by the forms of behavior of parents or adult educators is formally the same as that acting in embryological induction.

APPENDIX 1

Plant morphology. Unlike the situation between animals, there does not exist any, even approximate, global isomorphism between two adult individuals of the same plant species; isomorphisms exist only at the level of the few individual organs. like leaf, flower, or root. This makes it normal to describe plant morphology as consisting of a small number of morphogenetical fields (chreods) associated with these elementary organs grafted onto the stem chreod, which plays the central part. The static model previously described for an animal also applies, in principle, to each of these organs taken in isolation. Each of the fields has its territory (not necessarily connected) on the stem where it is capable of manifesting itself. The starting mechanism of these chreods has been well studied in the case of the apex of the stalk, on which the leaf buds appear.

We might formalize the result as follows. A bud will appear, in principle, at a fixed period A (the *plastachron*) on a circle C with fixed radius and center at the apex; then the radius of the circle increases as a result of the growth of the meristematic tissue. Suppose that the kth bud appears at the point with argument x_k, and that the catastrophe connected with the birth of a bud leaves a definite trace on the circle, defined by a function of the form $g_k = 1 + \cos(x - x_k)$, zero at the point diametrically opposite to x_k and maximum at x_k. Then the $(k + 1)$th bud will appear on the circle at the minimum of the trace function of all the previous catastrophes, that is, at the minimum of a function of the form

$$F_k(x) = \sum_{j=1}^{k} e^{-b(k-j)} g_j(x)$$

where b is a positive constant; the exponentials $e^{-b(k-j)}$ are weighting factors, decreasing with the age of each catastrophe. This process very probably has an attracting cycle such that $(x_{k+1} - x_k)/2p$ gives rise to the golden number a satisfying $a^2 + a = 1$ [3].

We might interpret this mechanism by saying that there is a tangential gradient around the apex which polarizes a neighborhood of the apex, and a bud is formed whenever this tangential gradient exceeds a certain threshold. The catastrophe thus produced then has a depolarizing effect on the gradient and inhibits its reformation in this region until a certain time has passed.

This scheme is typical of the essentially repetitive morphology of plants. The polarization necessary for any morphogenesis is reduced here to the strict minimum, namely, to the embryonic tissue, the meristem, and once the chreod has spread and the structure has formed, the polarization disappears; injured organs are not regenerated. In its place, regulation is based on a competition mechanism: as long as the organs (e.g., leaves) are metabolically active, the formation threshold of the analogous virtual chreods will remain fairly high and no new organ (leaf) will form. If, however, the metabolic activity of the leaves decreases, for example, as the result of some internal or external accident, the morphogenetical gradient, no longer inhibited, will reform and built up until it exceeds the formation threshold; new leaves will then appear, compensating for the deficiencies of the old.

In this respect, it seems that the specialization of a plant cell has an aspect that is more geometrical than biochemical, and this should be interpreted as a relatively easily reversible catabolic catastrophe (Stewart's experiment).

Of course what I am saying in no way implies that plant morphogenesis is incapable of complicated constructions; floral structures, in particular, are evidence of a great richness of configuration. The well-known adaptation of certain flowers to the bodies of the Hymenoptera which visit them poses the full problem of the interactions and couplings between living species in functional relation. Here we must not draw back in the face of criticisms of finalism; I have no hesitation in asserting that the genetic endowment of the snapdragon contains a virtual bee and that this virtual bee determines the structure of the flower chreod. Only a global vision of the mechanisms underlying evolution will permit us to imagine how such a situation might arise (see Chapter 12).

We finally propose the following model for plant morphogenesis. The germinal regime will be defined by a metabolic attractor H (to fix the idea, we suppose this to be a finite-dimensional attracting manifold)—in fact, this regime will be the characteristic regime of the stem; H will contain submanifolds r, l, f, etc., characteristic of the regimes of the root, leaf, flower, etc., and the various possibilities for the position of these attracting submanifolds in H will define spaces U_r, U_l, U_f, etc., which are precisely the model spaces of the chreods corresponding to the root, leaf, flower, etc. Some of these spaces may not be simply connected, and when the growth wave F_t traces out a loop in these spaces a repetition morphology will appear; many examples of this kind among the infinite inflorescences are known (e.g., the biparous cyme of the small centaury). The most complicated chreod is that associated with the flower, and it is in this space U_f that the virtual bee of the snapdragon occurs. Without wishing to embark on the subject of sexuality, let me say only that the male and female gametes are represented by submanifolds h_1 and h_2 of f, and f/h_1 and f/h_2 are essentially functional (not morphological) chreods which anticipate the spatial reunion of the gametes of the two sexes. The germinal regime H is a submanifold of the product $h_1 \times h_2$, and the corresponding chreod $h_1 \times h_2/H$ directs fertilization (the formation of the pollen tube).

All the differentiations in a scheme of this kind are catabolic, involving a decrease in the dimension of the attractor, and, except for fertilization itself, there is no place for inductive phenomena of one tissue on another. Some writers have spoken of induction in plant morphogenesis, evoking, for example, the inductive action of procambium on the overlying meristem in the leaf bud. But in the absence of experiments on the separation of the tissues, it is difficult to see this formulation as more than a rhetorical device.

As we said earlier, the global morphology of plants is not fixed, and the only uniting principle lies in the regulating mechanism of the starting threshold of the chreods. Consequently very many global forms are possible, and the choice between these forms arises from very slight internal or external factors. The absence of a regenerative mechanism at the level of an elementary chreod such as a leaf may seem incompatible with the relative ease with which certain plant cells can "dedifferentiate." Perhaps it is a consequence of the relatively slow nature of plant metabolism: the maintenance of a polarized structure in a tissue requires a constant struggle against the leveling effects of diffusion, with its increase in entropy; thus the only possible source of a permanent polarization will be of metabolic origin. The maintenance of a polarization thus requires an active metabolism (this is certainly one of the origins of Child's theory, even when the intensity of the metabolism is not necessarily one of the polarization gradients), while in plants it may be that the destruction of part of an organ (e.g., a leaf) only produces a perturbation of the metabolism of the remaining part that is not sufficiently strong to provoke dedifferentiation. In fact, normal plant morphogenesis depends to a large extent on external gradients, like those of weight and light, almost as if the internal metabolism were incapable of maintaining its own gradient except during the embryonic period and in certain localized zones like the apex.

This short description may suffice to provide some idea of the plant dynamic. For a more complete description, the model must account for regulation by adjusting the thresholds initiating the chreods associated with fundamental organs. For this, we must introduce coordinates representing the local biochemical states in the space in which we have the threshold hypersurfaces; stability will be a result of the presence of an attracting cycle in a system with discontinuities of *deferlant* type. In this respect the situation is not fundamentally different from that of animals (see the idea of the regulation figure, Section 10.2.C).

APPENDIX 2

Physiological applications of the model: sickness and death. As far as the spatial structure is concerned (except in the gonad during gametogenesis, and during regeneration) the growth wave will stabilize at the end of somatic development but will continue to vary in the space of internal parameters, where it will describe the physiological states of the various organs; as far as the nervous system is concerned, it will describe the mental state of the subject at each moment. If we are interested only in physiological homeostasis, we might consider the space W of global states of the organism to be the product ΠW_j of the state spaces of the organs. The representative point $w \in W$ will vary in a region bounded by a number of hypersurfaces X_j in W, on which the evolution field will undergo breaking discontinuities defined by vector fields Y_j tending to force the point w toward the optimal state; each such field Y_j will be connected with the operation of one or several organs. However, these hypersurfaces X_j and the correcting discontinuities Y_j will themselves depend on the global state w via a coupling term $A(w)$, which is weak at first but increases as w moves away from the optimal state. If w moves too far away from the optimal state (suppose this to be the origin), then each organ taken by itself will no longer be able to act effectively and so the correcting effect of the corrector discontinuity Y_j will tend to diminish; of course, the points for which this effect will occur lie beyond the barriers formed by the hypersurfaces X_j and hence outside the normal regime defined by these barriers. But if w lies on or beyond one of these barriers, as a result of a shock from outside or inside (illness), the evolution dynamic will enter a zone of qualitative indeterminacy; some of the barriers will only have a weak correcting effect or may be displaced in the space W; the immediate qualitative effect will be to displace the barriers toward the outside and so to allow larger variations in the global physiological state, but the longer-term result will be that the corrector effect can weaken and each partial field W_i can tend to excite itself more and more. In the case of a fortunate evolution, the corrector effect can finally triumph; and, as soon as the most strongly affected organs have stabilized, the other organs that were excited by coupling can return to their less excited states, tending, in their turn, to stabilize the sick organ. This produces the healing effect, even though the positions of the barriers and the corrector forces of the final state may not coincide with those of the original state, this being the after effects of the sickness. Although this is still a very vague model, the number of main parameters involved in physiological equilibrium cannot be so large that we may not one day be able to construct reasonably simple standard models; then the job of a doctor, facing a patient, will be to set up his model and evaluate the effect of each medicine on the global evolution. This will be much better than the purely symptomatic treatments of present-day medicine, treatments whose effects on the global stability of the patient are often poorly appreciated.

In the case of an unsatisfactory evolution toward more serious illness, the representative point will finally cross the correcting barriers and their effect will weaken. The perturbation of the local biochemical states will be so great that the metabolic structure of the field will deform; the underlying metabolism will be stopped through lack of essential metabolites (oxygen), and the metabolic field will give way to a static field through a generalized catastrophe. As with all such generalized catastrophes, the evolution will first be very indeterminate, consisting of a mass of small, initially reversible phenomena; then the catastrophe will simplify topologically and enter on an irreversible phase governed by the new chreods of the body's decomposition. Our everyday life, on the physiological plane, may be a tissue of ordinary catastrophes, but our death is a generalized catastrophe.

APPENDIX 3

The epigenesis of the nervous system: a theoretical scheme. In the space V of the regulation figure, consider the following regulating reflex: suppose that V is one-dimensional with coordinate v, and the reflex is defined between the points $v = -1$ and $v = +1$ by a Riemann-Hugoniot folding in the vx-plane with equation

$$v = \frac{1}{3}x^3 - c^2x,$$

where $-1 = \frac{1}{3}c^3 - c^3$, that is, $c = (\frac{3}{2})^{1/3}$. Suppose, as usual, that the branch defined by $x < -c$ defines a mean coupling field $V_1 > 0$ on the axis Ov, while the branch $x > c$ defines a field with the opposite sign (Figure 10.17). Now suppose that this corrector reflex is the effect of the action of a single organ J, and also suppose that the coordinate v is an epigenetical gradient entering into the universal unfolding of the germinal dynamic; the number v will then have significance as a spatial coordinate. It is normal to suppose that the organ J is localized at O, the mid-point of the segment $(-1, +1)$, and also that J remains fixed at O during epigenesis, as the result of a localization of function. We then might express the epigenesis of the organ J by saying that the folding shrinks so as to form a cusp at O_1; for this to happen, it is sufficient to consider $c^2 = y$ as a time coordinate. The smoothing of this cusp will give rise to a local potential W (see Section 10.4.B) whose gradient lines converge toward O_1 in the sense of negative increasing y (Figure 10.18).

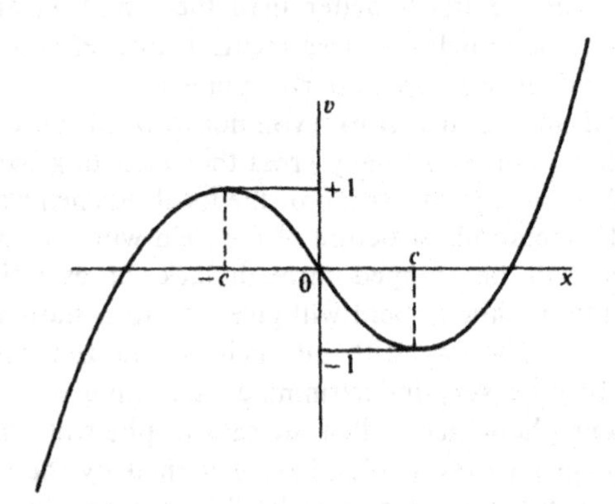

FIGURE 10.17

Now we can suppose that the metabolic unfolding of the germinal dynamic slightly precedes its spatiotemporal unfolding as a result of the delay inherent in all phase change and all morphogenesis. The primitive neurons have a metabolic regime characterized by a special affinity for rapid changes of regime, for physiological catastrophes, and their local regimes will therefore preferentially occupy the catastrophe set of the regulation figure; in our idealized model, the points $v = \pm 1$ will be filled with neurons long before the organ J has effectively come into being. The regulation direction of the metabolism will be $\Delta x < 0$ for $v = -1$ and $\Delta x > 0$ for $v = +1$, the signs of the jumps that take place on the folding. At the moment of the formation of the organ J, and for y small and negative, these neurons will be acted on much more forcibly by the vector grad W, and the regulation vector of the metabolism will therefore undergo a variation defined by the horizontal component of grad W. Nevertheless, since the local metabolism of a neuron has only one manner or direction in which it can regulate itself, namely, the discharge of nervous impulses, the cell will compensate for this difference by unfurling along the direction defined by grad W in the system of epigenetical gradients. This assures the innervation of the organ J; the axons coming from neurons with regime $v = -1$ will have an excitory effect, and those from $v = +1$, an inhibitory effect.

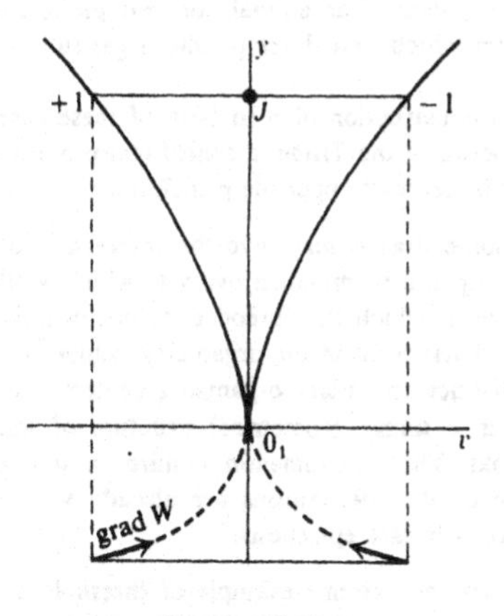

FIGURE 10.18

This specific sensibility of the expansion of the neurons to epigenetical gradients is probably a reciprocal phenomenon; when a nerve spreads out into its terminal ramifications, it can create epigenetical gradients in this zone. This may explain the part of nerves in regeneration: we have seen how nerves "pump," in some sense, epigenetical polarizations from a forming tissue and then use them in the construction of the central nervous system. This process might be at least partially reversible in the case of amputation, and the chemical markers of the local polarization of a field might return at the ramified extremities of the amputated axon and provide the regeneration blasteme with its indispensable polarization elements. From this point of view, the part of the nerves in regeneration would not be trophic, as is often believed, but polarizing, a geometrical role analogous to that of the apical cap. This would explain how the aneurogenic limbs of the salamander regenerate without innervation after amputation, although the normally innervated limbs refuse to regenerate if innervation of the blastemes is prevented: this occurs because the innervation has deprived the limb of the epigenetical polarization using it to construct the connections of the nervous system [4].

NOTES

1. The mechanism of any machine—consider for example a watch—is always built up centripetally, meaning that the various parts, the cogs, springs, and fingers, must first be completed and then mounted on a common support.

As opposed to this, the growth of an animal, for example of a *Triton*, is always organized centrifugally from its germ which first develops into a gastrula, and then constantly sprouts organ buds.

There is an underlying construction plan in both of these cases; for the watch, the plan controls a centripetal process, in the *Triton*, a centrifugal process. It seems that the parts are assembled according to diametrically opposite principles.

2. The formation of some organs may give the impression of a following a centripetal process; consider, for example, a mammalian ovary to which Müller's canal is adjoined from the outside. This is a case in which the chreod does not contain its organizing center; the growth wave passes at a distance from the organizing center and encounters only a certain number of apparently distinct secondary organizing centers, although they are, in fact, all connected algebraically in a larger geometrical structure of which the growth wave only sweeps out a submanifold. This phenomenon requires a sufficiently developed threshold stabilization, whose homeostatic mechanisms are already very efficient. Therefore such a phenomenon can occur only in late epigenesis.

3. Gametogenesis affords the extreme example of threshold stabilization, since it gives a structurally stable reconstruction of the primitive organizing center. We saw in Chapter 6 a formal example of a mean coupling field providing a funneled reconstruction of the organizing center.

REFERENCES

1. C. H. Waddington, *Introduction to Modern Genetics*, Allen and Unwin, 1939 (Epigenetic landscape and polyhedron, pp. 181–183).

2. C. H. Waddington, *The Strategy of the Genes*, Allen and Unwin, 1957 (see Figure 5).
3. F. J. Richards, Phyllotaxis, its quantitative expression and relation to growth in the apex, *Philos. Trans. Roy. Soc. London* Ser. A **235** (1951) 509–564.
4. E. Wolff, Le rôle du système nerveux dans la régénération des amphibiens, *Ann. Biol.* **4** (1965) (Yntema's experiment).

CHAPTER 11

MODELS IN ULTRASTRUCTURE

In this chapter we propose some simple dynamical models representing the global evolution of a cell, and then discuss to what extent these models can account for known phenomena.

11.1. THE DIVISION OF A CELL

A. The optimum size

Consider a ball b of radius r in local dynamical regime g, taking in energy from the external environment e, where there is a continuous flux of energy in the process of thermal degradation, and suppose that the consumption of energy is proportional to the volume; then the capture of energy arising from exchanges with the environment is proportional only to the surface area. If the radius is at first sufficiently small, there will be an excess of captured over consumed energy; and, if this excess is given over to the synthesis of the substances supporting the regime g, the volume V (and radius r) will increase. Were this increase strictly proportional to the increase of energy, the radius r would increase asymptotically toward a value r_0, the value at which the captured energy is equal to the consumed energy. Suppose now that the mechanism of synthesis has inertia, or delay; the volume can then increase above the optimal value to the detriment of the internal consumption of energy, and the ball, in the regime g with radius $r > r_0$, will enter into an unstable state that can be resolved only by division into two balls of radius $r_1 < r_0$, after which the process can repeat with increase and later multiplication.

B. Energy flux

What, then, can be a mechanism of synthesis to explain this delay? One of the most plausible hypotheses is that it is caused by the inertia of a metabolic configuration, by the stability of a circulation structure in the cell. Suppose first that there is an energy circulation; there might be an energy reception zone in the cell (playing the part of the vegetal hemisphere in an egg) and a consuming zone (analogous to the animal hemisphere), and if there are molecules specialized in carrying energy (like the ADP-ATP couple) these molecules would tend to set up a relatively stable circulation between these zones. Here is the topologically simplest configuration representing such a circulation in the cell. We start with a flow in B^3, the unit ball of $Oxyz$, along the circles

$$x^2+y^2=k, \quad z=k';$$

the fixed point set of the flow joins the two points $z=\pm 1$ of the cell (Figure 11.1). By applying a homeomorphism to B^3, we can bring these two points arbitrarily close together to give the situation of Figure 11.2; the fixed point set is now essentially a circle C attached to the boundary of the cell at a point, and the field is tangential to the boundary and null at the point where C is attached. Then we consider the following idealized situation: there is a surface W on which the synthesis takes place, a disc bounded by C and such that, when a trajectory of the field pierces W, the corresponding molecule is discharged of energy. The surface W will then be like a shock wave separating a zone rich in energy from a poorer, exhausted zone. We propose to view the circle C as the ancestor of the present-day chromosome.

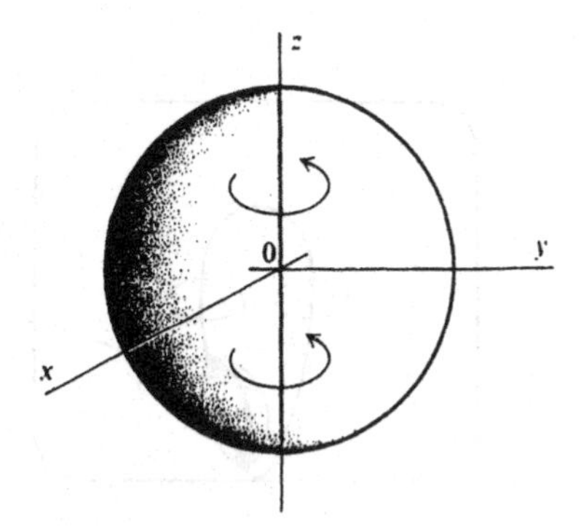

FIGURE 11.1

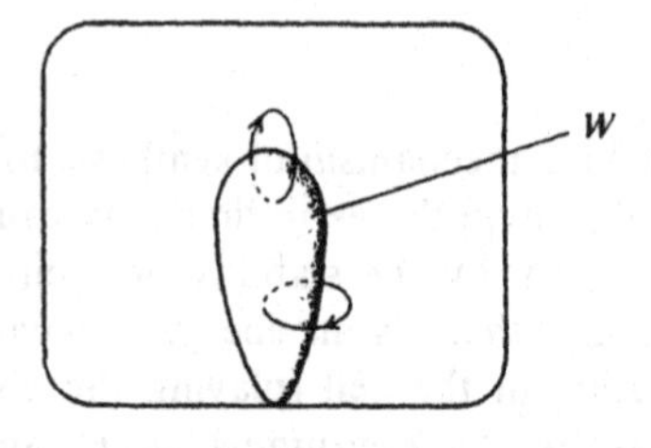

FIGURE 11.2

C. Duplication of chromosomes

In any case we must consider the surface W and its boundary C as connected in some way with supramolecular structures which have inertia arising from their internal energy. The intake of energy, after the growth of the cell and the decrease of the surface/volume ratio, will become insufficient to maintain a permanent flow across the whole surface W and so will give rise to a dead-water zone, starting in the neighborhood of the boundary C of W. The shock wave W will have an exfoliation singularity, and a new regime will develop in a neighborhood of a boundary point; the surface W will double itself there and develop a triple edge. The new fragment W_1 of surface will have as boundary an arc C_1 duplicating the corresponding arc of C (Figure 11.3), and W_1 will separate the dead-water regime from the regime of normal circulation.

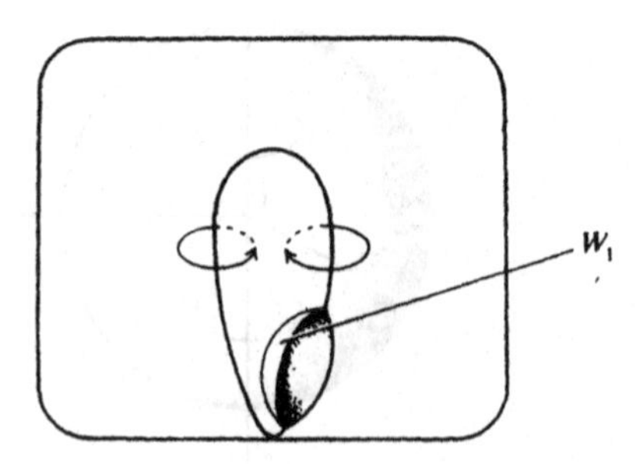

FIGURE 11.3

The surfaces W and W_1 are oriented in opposite senses, and so the energy effect on trajectories cutting both is zero. This duplication process will then have a braking effect on the synthesis of substances supporting the regime g, with the effect of decreasing the rate of growth; when W is wholly duplicated, the growth (and the synthesis) will stop completely, but the final state will be unstable (Figure 11.4). The abrupt separation of W and W_1 into two parallel, sufficiently distant planes will reorganize the circulation into two convection cells with axes C and C_1 (Figure 11.5), and the separatrix between the two convection cells will be realized as an intercellular partition in which the cortex, sucked in somehow by the internal flux, provokes a constriction leading to a spatial separation between the two cells and the division of the cell into two successor cells.

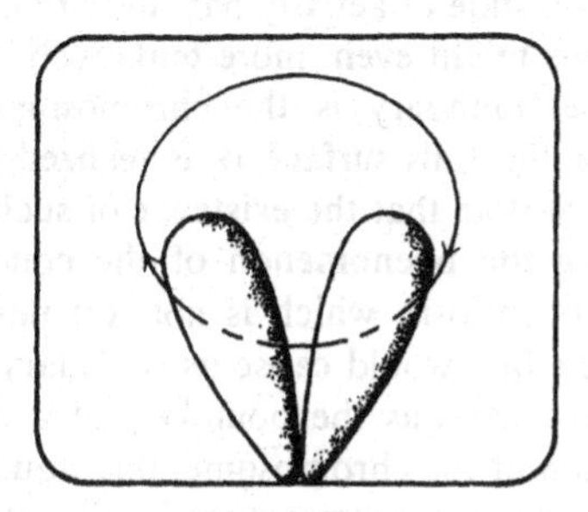

FIGURE 11.4

This model is open to obvious objections arising from its overschematic nature; very probably an information circulation is superimposed on the energy circulation early in the evolution, filtering it and subjecting it to a reaction by coupling. Actually, the elaboration of energy is the task of specialized organs, mitochondria, whose fine structure is not unlike that of the elementary cell envisaged above; they have a DNA chromosome (perhaps circular?) attached to a point of the cell wall, and the wall is a double membrane whose internal sheet has exfoliations (cristae). Furthermore, at a given time, the cell may adopt several diverse metabolic regimes according to the external conditions, each of these regimes corresponding to the activation of specialized information units of the type of the elementary cell considered earlier. These activities correspond geometrically to the unfolding of the corresponding chromosome, allowing a larger flux of information across the circuit.

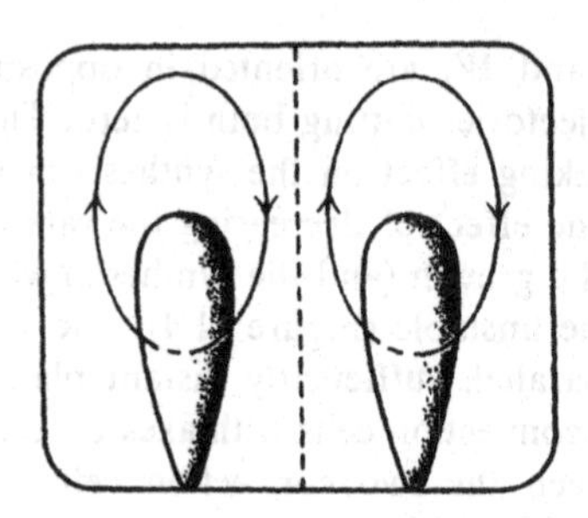

FIGURE 11.5

This raises a prejudicial question: to what extent has it been proved that every nucleic acid in a state of activity has this circular character? Here we have postulated it; we might even, more tentatively, suppose the existence of a disc W whose boundary is the chromosome, although it is not necessary to suppose that this surface W is realized by a stable macromolecular structure. First note that the existence of such a structure would be useful for explaining the phenomenon of the condensation of chromosomes in prophase of mitosis, which is not yet understood: the disc W, folding on itself like a fan, would cause its boundary to spiral, and then, if the bacterial chromosome was the boundary of a disc whose exfoliation caused the duplication of the chromosome, this would more or less explain why the two successor chromosomes were not intertwined. Next we might suppose that the surface W had only a local and purely kinetic existence around C; in the case where the chromosome is in a state of heterocatalytic activity, it seems probable that the filaments of messenger RNA which become detached (and perhaps also the flow of precursor molecules) have comparatively fixed trajectories in a neighborhood of C, and these filaments will sweep out a fixed surface W. Then the ribosomes, situated on this surface, collect together on the corresponding filament of RNA and synthesize the required protein, which leaves the surface W from one of its faces, giving rise to the information flux associated with C.

Whatever the nature of the activity centered on the chromosome C, it seems difficult to deny that only a circulation around C can account for the biological role of DNA, and we represent this situation geometrically by saying that the chromosome C is the zero set of a complex differentiable function $V : U \rightarrow \mathbf{R}^2$, defined on a domain U. Then, to obtain a closed curve, we postulate that, in general, the function V has a maximum rank at each point of $V^{-1}(0)$. The essential advantage of this point of view is that it provides an elementary model for crossing over between two DNA molecules.

D. A model for crossing over at the molecular level

Suppose then, that the chromosome is the zero set of a complex differentiable function V, regular on the counterimage $V^{-1}(0)$; what will happen at a point of $V^{-1}(0)$ where the function stops being regular, of rank two? Observe first that such a situation can only happen structurally stably during a deformation parameterized by time. Write (u, v) for the coordinates of the complex-plane, target space of V, and (x, y, z) for local coordinates of the source space with origin O; then $V(m, t)$ will be as follows:

$$u = x^2 \pm y^2 + t; \qquad v = z.$$

When the quadratic form defined on Oxy by the second-order Taylor development of u is positive definite or negative definite, then, as t changes from negative to positive values, a small circular chromosome will either disappear, or be created, at the origin. Such a phenomenon would be considered impossible from the normal viewpoint of biology, because a chromosome is in principle realized by a linear molecule of DNA, given once and for all or created by copying an already existing model. But in some circumstances (e.g., perhaps in virus infection) it is not impossible that certain chromosomes stop being functional through a breakdown in the associated circulation; then the associated molecule will become a phantom form and will lose its stability. The preceding model then has perhaps some value in describing this loss of viability of a chromosome.

When the quadratic form in (x, y) has a saddle point, $u = x^2 - y^2 + t$, then, as t changes from $-a$ to $+a$, the hyperbola $x^2 - y^2 = a$ changes to $x^2 - y^2 = -a$; the two branches of the hyperbola change, and this may perhaps be considered as a crossing over on the chromosomes, which exchange two threads of chromosome (Figure 11.6 and 11.7). The only accepted model of this phenomenon in molecular biology is the Campbell model for the joining of a lambda phage to the *Escherichia coli* chromosome, but the dynamical simplicity of the scheme given above indicates that it probably accounts for a very general mechanism. If bacterial cytoplasm is considered as being structured in energids U_i, which are convection cells directed by chromosomes C_i, it might happen that an energid U_1 can be captured by an energid U_2; then chromosome C_2 of U_2 will capture C_1 by crossing over (Figure 11.8). From the usual viewpoint of biochemistry, such a junction can happen only along certain segments, called homologues, of the chromosomes C_1 and C_2, and it is usually postulated that these homologous segments have a similar composition of nucleotides. The dynamic above is not incompatible with this point of view because a singularity of the complex potential V can probably occur only when the local enzymatic effects of C_1 and C_2 are very similar; but, even in

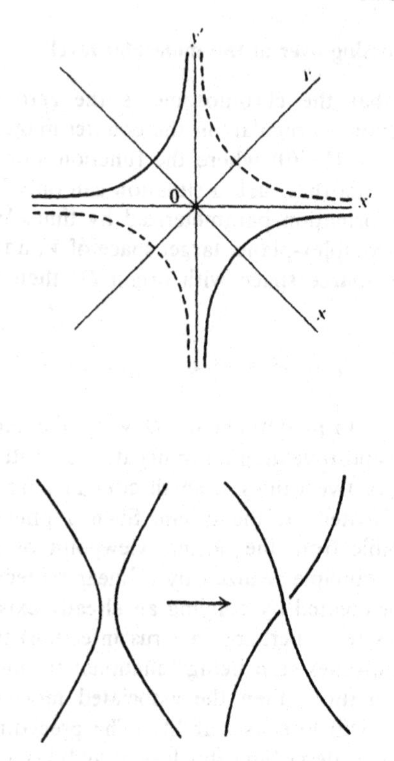

FIGURE 11.6 AND 11.7. A scheme for crossing over.

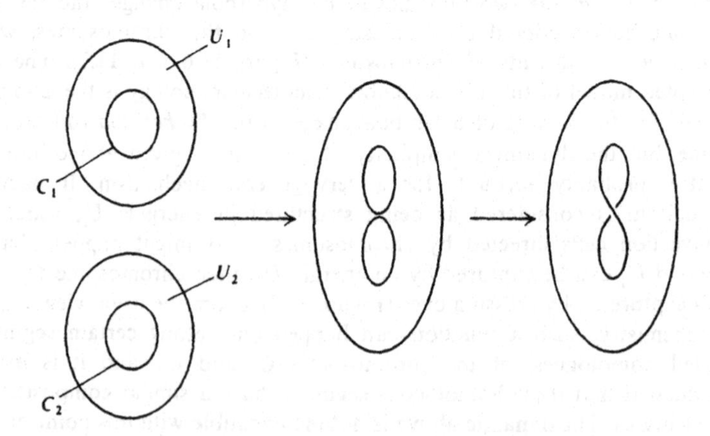

FIGURE 11.8. The union of two chromosomes during fusion of two energids.

the case of lambda phage, it is accepted that the phage can detach itself from the bacterial chromosome at a point other than the initial one, carrying away the gene *Gal* or other neighboring genes by transduction. Thus the points of attachment are not specified absolutely.

11.2. MITOSIS

A. Mitosis in internal coordinates

We can construct a reasonable satisfactory dynamical model of the phenomenology of mitosis by supposing that the local biochemical regime of the cell is defined by a map V of this cell into the complex plane $z=u+iv$. In interphase, there will be the following vector field X in the z-plane: X has two attractors, the origin O (corresponding to the nuclear regime) and the circle $|z|=2R$ (the cytoplasm regime), with the two basins of attraction of these attractors separated by the circle $z=R$, a closed, unstable trajectory of X. The counterimage under V of $|z|>R$ defines the cytoplasm, that of $|z|=R$ the nuclear membrane, and that of $|z|<R$ the nucleus (Figure 11.9). At the beginning of prophase, the nuclear attractor O undergoes a Hopf bifurcation and transforms into a small invariant circle c of radius r, and the origin becomes a repeller point (Figure 11.10). As a result, a curve (made up of very small, circular filaments), the counterimage of O under V, appears in the nucleus—this is the nuclear material that appears in the nuclear liquid by a filament catastrophe and is the set of points of zero metabolism. Then the radius r of c increases, the catastrophe simplifies topologically, the small nuclear circles capture each other by crossing over, thus initiating the condensation of chromosomes, and, when c coincides with the unstable trajectory $|z|=R$, the nuclear membrane ruptures, leaving only one attractor, the cytoplasmic attractor $|z|=2R$. At this moment, in full metaphase, the metabolism is quasi-zero (Figure 11.11); this means that the field X has a zero tangential component on $|z|=2R$, and therefore X has the configuration of the field grad $|z|$; the cell is in a state of *reversible death*.

The momentary degeneration of the internal metabolic field into a static gradient field is expressed by the formation of the mitotic apparatus between the two asters, which correspond to attractors of the field induced by V in X. After anaphase, which can be described only on a complete model with spatial coordinates (not just the z-plane of internal metabolism), there is the telophase: the origin in the internal space (u,v), a repeller point representing nuclear material, once again undergoes bifurcation and gives birth to a small circle m of radius r separating the cytoplasmic regime (again $|z|=2R$) and the origin (a new point attractor that will represent the new nuclear regime) (Figure 11.12); the counterimage $V^{-1}(m)$ is the new nuclear membrane, and the cycle starts again. Observe that this model requires that the new nuclear membranes of the successor cells form in some sense by delamination of successor chromosomes—a feature that apparently contradicts the observations of many cytologists who say that the new membrane forms from fragments of the old. Phantom structures in metaphase probably also play a part and return to service when the occasion presents itself. Pleuromitosis, in which the nuclear membrane divides by constriction but does not disappear, follows a slightly different scheme.

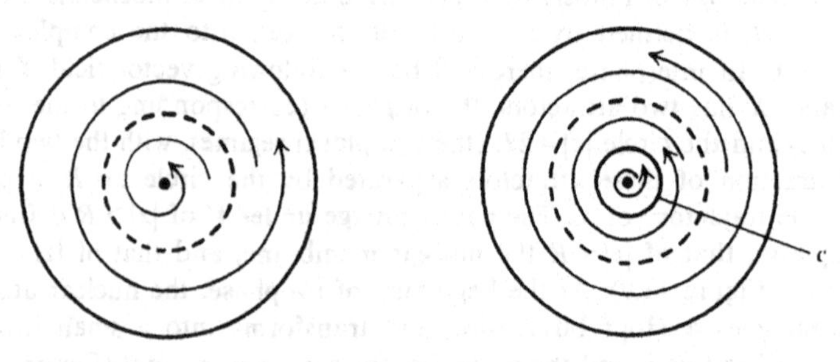

FIGURE 11.9 AND 11.10. A Hopf bifurcation during mitosis.

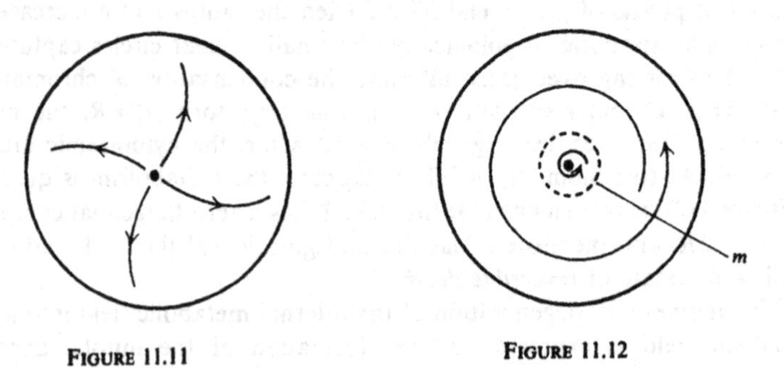

FIGURE 11.11 FIGURE 11.12

B. Mitosis with spatial coordinates

Only the fiber space of internal variables has been considered in the preceding, whereas a complete model needs spatial coordinates. Because of the very incomplete nature of the scheme, we will consider only a one-dimensional model with coordinate x, internal coordinate z, and two special points on Oz: a cytoplasmic attractor c, which corresponds to the centriole, and a higher point n representing nucleic material. Normally (in interphase) the state of the cell will be represented by the line $z = x$, with a small plateau with ordinate n corresponding to the nucleus (Figure 11.13). The first symptom of prophase is the duplication of the centriole, when an offshoot will go to the opposite pole of the cell. This cytoplasmic duplication has graphical representation of the type in Figure 11.14, where c has two counterimages at symmetric points of the cell. Then the nuclear plateau moves about, giving two neighboring counterimages that correspond to the two systems of chromatids paired in prophase. The arc thus formed develops first a cusp at ordinate n (see Figure 11.15), thus identifying the two chromatids, and then a loop leading to two new separate chromatids n_1 and n_2; but, as these points are on metastable branches of the state curve, there will be a discontinuity in the representative graph. In fact this discontinuity corresponds to the formation of the mitotic apparatus, represented by a dotted line joining c_1 to n_1, c_2 to n_2 (Figure 11.16). At anaphase the representative graph steps up from the metastable to the stable branch, situated higher at n_1' and n_2', respectively. The dotted line follows the point n_1 in its upward moment by pivoting about c, and so the counterimage of n associated with n_1 is carried to a neighborhood of c_1: this movement represents the anaphase separation of the chromosomes. Then the graph evolves toward a system of two normally joined graphs.

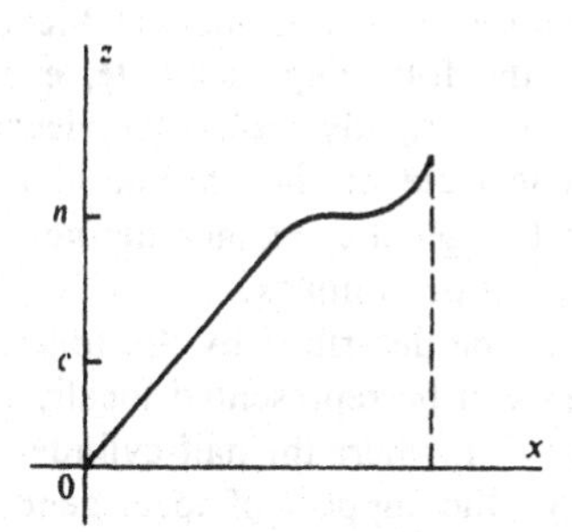

FIGURE 11.13

Of course this is only a somewhat idealized description of mitosis and does not pretend to be an explanation. Nevertheless, this analogy with the breaking of a wave has some merit worth noting.

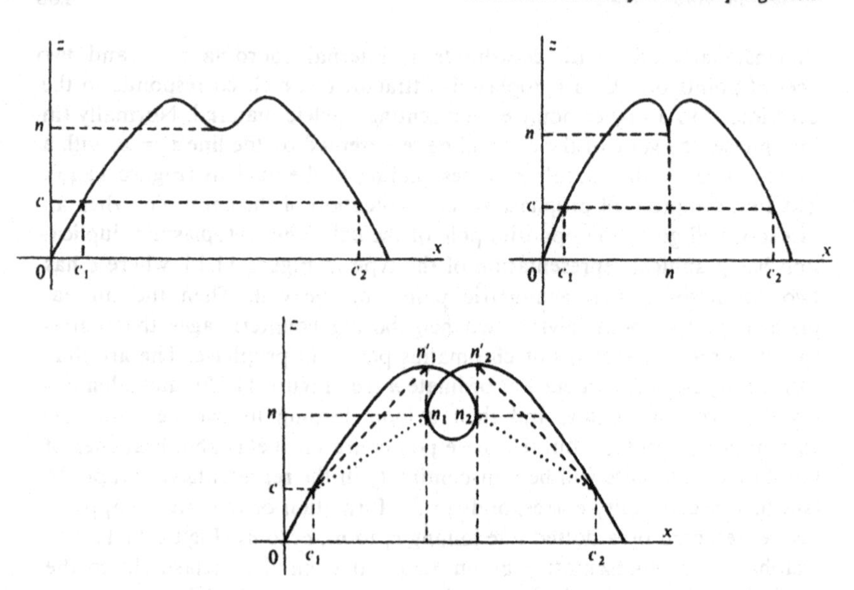

FIGURE 11.14, 11.15, AND 11.16. **The general scheme of mitosis in spatial coordinates.**

11.3. MEIOSIS

The model that we shall give of crossing over in meiosis is not founded on any explicit structure of chromosomes (in Metazoa, which we consider here). We need only the following postulate, even though it is perhaps difficult to defend from a rigidly molecular viewpoint: at the pachytene stage, the local genetic state of the chromatid tetrad is the result of a competition between the genetic regimes defined by the parent chromosomes joined in the synaptic complex.

This competition can be described by the following geometrical model. The synaptic complex will be represented locally in *Oxyz* by the cylinder with equation $x^2 + y^2 = 1$, where the half-cylinder $x < 0$ represents one of the chromosomes (e.g., the support of local gene a), and the half $x > 0$ represents the other, with local gene a^+, an allele of a. At the pachytene stage, the whole synaptic complex is cut in two by a shock wave which, in principle, divides each chromosome into two chromatids. Suppose that this dividing shock wave is represented in each plane z = constant by a curve passing through the origin of the *Oxy*-plane, and consider first the case in which it is a straight line (Figure 11.17). The angle α that it makes with *Oy* is a function of z, and it is natural to suppose that this function $\alpha(z)$ is continuous. When $\alpha(z) \neq 0 \pmod{\pi}$, the shock wave H is transversal to the plane $x = 0$ separating the two chromosomes; inevitably, the two chromatids into which a chromosome is cut will then inherit the genetical properties of their parents. Next suppose that in the plane $z = z_0$, separating the two genes (a, a^+) of $z > z_0$ from the two genes (b, b^+) of $z < z_0$, the angle $\alpha(z_0)$ is zero and is changing from a positive to a negative value as z increases. Then the chromatid associated with the acute angle at $z < z_0$ in $x < 0$ will extend, by continuity, to the chromatid for $z > z_0$ in $x > 0$, the positive acute angle. Thus this chromatid will have the combination (a, b^+); similarly the chromatid situated symmetrically with respect to Oz will give rise to the symmetric combination (a^+, b). Geometrically the global configuration of the four chromatids (a, b), (a, b^+), (a^+, b^+), (a^+, b) is that of a braid with four strands under a rotation of $\pi/2$. This gives rise to the classical configuration of chiasma, which occurs at the diplotene stage (see Figure 11.18).

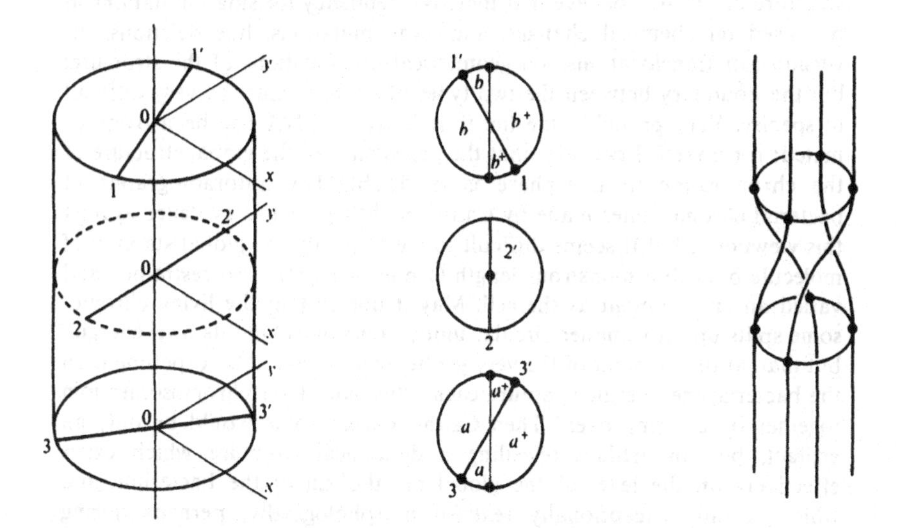

FIGURE 11.17 AND 11.18. The scheme of meiosis.

Only pure and simple crossings over, with symmetric recombinations, can occur in a situation such as we have just described. The hypothesis that the shock wave cuts each plane z = constant in a straight line is too restrictive; it is more natural to suppose that the shock wave cuts each plane in a convex curve through the origin $c(z)$, which, approximately speaking, rotates through an angle $\alpha(z)$. In the critical case ($z = z_0$), where the curve $c(z_0)$ touches Oy [$\alpha(z_0) = 0$], we can see that one of the chromosomes (e.g., $x < 0$) will divide into three chromatids whereas the other will remain unchanged (Figure 11.19), and there will be a gene conversion in which three of the chromatids of the tetrad will adopt the locus a of the chromosome $x < 0$. This situation can persist as long as the rotation $\alpha(z)$ is sufficiently small; but quickly, for example, in $z > z_0$, one branch of H will recut Oy, and this may give rise to competition for the nature of the corresponding chromatid (Figure 11.19). Hence our model of meiotic crossing over will give the following situation associated with a chiasma: a narrow zone of gene conversion bounded by two zones of multiple fractures of total odd order, each involving a pair of distinct chromatids. The methods of genetics are now so refined that such a fine structure must be observable experimentally if it exists.

One of the basic questions connected with nuclear material is whether a mutation arises from a chemical or a geometrical modification in the structure of DNA. I believe that there is a tendency for small mutations to be based on chemical changes, and large mutations, like deletions, inversions, or translocations, on geometrical modifications of the structure; but the boundary between the two types of changes must be very difficult to specify. Very probably the linear stability of DNA has been overestimated; for myself, I consider that the persistence of the global structure of the chromosome in interphase is in doubt. The autoradiographs of bacterial chromosomes made by Cairns might be used as evidence against this viewpoint, but it seems difficult to me to justify the global stability of molecule of such a monstrous length (1 mm or more) in so restrained and varied an environment as the cell. May it not be that the living chromosome splits up into smaller circular units, each controlling its own energid, but that, at the moment of the very gentle death required by experiment on the bacteria, the energids capture each other, and their chromosomes join together by crossing over? Then Cairns' chromosome would be only an artifact, but an artifact revealing a dynamical structure which exists effectively on the level of the global metabolism of the bacterium and which is only exceptionally realized morphologically, perhaps during mitosis or sexual conjugation.

11.4. MORPHOGENETIC FIELDS OF CYTOPLASM

There are many examples in animal and plant cells of very precise and very specialized morphologies. The most primitive animals, coelenterates in particular, compensate for their histological simplicity by a highly developed infracellular differentiation. To what extent is the building of these structures a result of morphogenetic fields of the same kind as those introduced in the epigenesis of larger animals? It seems to me that we may hazard the following reply: there may be, as in the epigenesis of tissues, a metabolic field in the cytoplasm that is capable of having different types of catastrophes; but this time a narrowly and specifically adapted macromo-

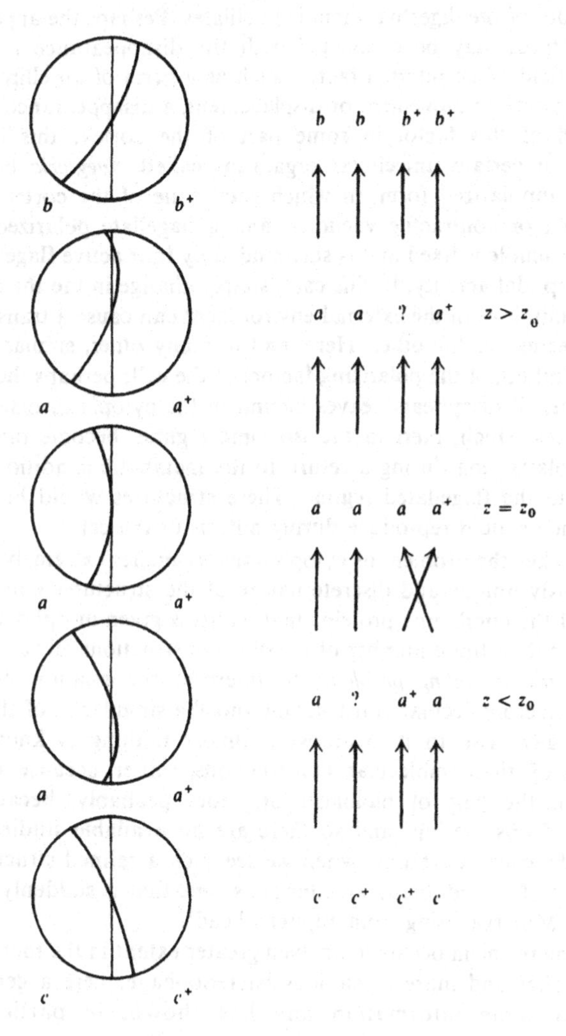

FIGURE 11.19. The scheme of meiosis.

lecular structure will correspond to a singularity of the metabolism, although the nature of this adaptation may remain mysterious to us. For example, we know that a hair or a flagellum is always connected with a macromolecular structure consisting of nine tubules, but at the present time nobody knows the reason for such a structure, and why it is so successful. On the other hand, the placing of these organelles appears to depend on relatively precise and explicit laws, and they are found, in general, at geometrical or functional singularities of the cortex, for example at the closure of the digestive vacuole in ciliates. Perhaps the appearance of these structures may be connected with the disappearance in the local metabolic field of an internal factor, such as a germ of an elliptic umbilic specifying a certain movement or displacement, a disappearance due to the localization of this factor in some part of the cortex; this is like the oscillation of certain unicellular organisms called *Naegleria* between an ameoboid unpolarized form, in which each zone of the cortex can form pseudopodia or contractile vacuoles, and a flagellate polarized form, in which the vacuole is fixed and is surrounded by four active flagella without any pseudopodal activity. In this case, simply change in the the concentration of certain ions in the external environment can cause a transformation in one direction or the other. Here, and in many other, similar examples, we know nothing of the polarizing factors of the cell; perhaps the polarized regime, when it disappears, leaves behind in the cytoplasm macromolecular structures which, inert in the isotropic regime, become oriented and provoke polarization during a return to the metabolic conditions that are favorable to the flagellated regime. These structures would be structure-germs, which should reproduce during mitosis of the cell.

What makes the problem of cytoplasmic organelles extremely difficult is the obviously unique and discrete nature of the structures constructed. It seems as if the combining proteins that realize a given morphogenetic field can make only a finite number of possible constructions, always the same. *This is really a coding problem*: to interpret the relation between an observed macromolecular construction and the singularity of the metabolism that gave rise to it. Moreover, almost nothing is known of the epigenesis of these molecular constructions, either because of lack of interest on the part of biologists or, more probably, because of the difficulty of observation; and so there are no available studies of *cytoplasmic embryology*. Certainly, when we see such a refined structure as the nematocyst of a cnidoblast, we cannot assume that is suddenly came into being like Minerva rising from Jupiter's head[1].

The same problem occurs to an even greater extent in the morphogenesis of the smaller and more mysterious bacteriophage; here a gene analysis has given some information and has shown, in particular, that morphogenetic genes exist whose mutations prohibit the attachment of certain pieces of the capsule to other pieces. But the problem raised by the obviously finalist nature of these structures remains untouched, and I can only repeat what I said in Chapter 8: that those who believe that they can discover the solution of morphogenesis more easily through the fine structure than the large are on a false trail; genetic and epigenetic constraints bear much more heavily on a discrete, quantified universe than on one in which continuous adaptions are possible, and for this reason reconstruction of the generating morphogenetic field from the form is a much harder process.

Apart from these highly specialized forms like the nematocyst and the phage capsule, ordinary cells contain many widespread structures: mitochondria, endoplasmic reticulum, Golgi apparatus, various membranes and vesicles, and so forth. The morphology of these organelles is fairly uniform and is governed by repeated variations on the theme of the double membrane and the sac. A morphological theory for these structures seems easier; they are already much more labile and respond rapidly to biochemical variations of the local cytoplasm. A double membrane may be considered as a shock wave between two phases or regimes, a sac as a preliminary form in the appearance of a new phase or the residue of a disappearing phase, and the stacked forms of endoplasmic reticulum as a local laminar catastrophe. We are now going to consider these further and to return to the central problem of the relationship between metabolism and macromolecular constructions.

11.5. THE THEORY OF CYTOPLASMIC STRUCTURES

A. The idea of an enzyme

An enzyme E is a substance with the property of rapidly transforming a system of substances A into a system B, either without itself being altered during, or by being reconstructed after, the reaction $A+E \longrightarrow B+E$. Let us, as usual, represent local biochemical states by the points of an internal space M, the state A being represented by an attractor localised at a, and B by another attractor at b. The presence of enzyme E has the effect, in M, of translating a neighborhood of a through $\overrightarrow{ab}$. More precisely, we may have to suppose that the speed of the reaction $A+E \longrightarrow B+E$ depends on a certain characteristic spatial relationship between the molecules of A and E, whose stereochemical properties allow the enzyme to exercise its catalytic properties. We shall apply this idea immediately.

B. The structure of a shock wave, and transitional regimes

Suppose that a cell contains two stable regimes A and B in conflict along the surface of a shock wave, and let a and b in M represent the attractors defining the two regimes. Let us postulate that there exists a flat enzymatic system H which, on one of its sides, will cause the transformation $A \rightarrow B$; and similarly that there is another system H' giving the inverse transformation $B \rightarrow A$. If the enzymatic configurations H and H' can combine, they will stabilize the shock wave S into a double membrane of the two configurations stuck together; a transitional regime will form between the two sheets, characterized by fluctuations and periodic changes between the two regimes A and B. Formally this situation is analogous to the formation of a vortex street in the theory of wakes in hydrodynamics, where the vortices form a transitional regime between the zone of dead water behind an obstacle and the exterior zone of laminar flow (Figure 11.20).

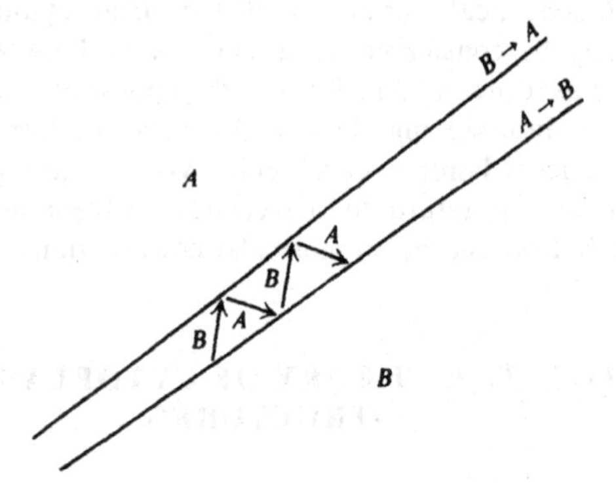

FIGURE 11.20. Duplication of a shock wave.

Whereas in hydrodynamics the transitional regimes are very unstable and depend almost entirely on variations in initial conditions, in biology these regimes are connected with supramolecular configurations which, once formed, can survive the metabolic conditions that gave birth to them. Expressing this in another way, we can say that the dynamic of transitional regimes is itself as quantified and rich in distinct attractors as the ambient dynamics. Once a transitional regime is established, it will tend to remain even though the ambient dynamics change; then the transitional regime associated with a shock wave S can, in its turn, act on the ambient dynamics. In this way the existence of a dynamical regime A can depend on the transitional regimes themselves associated with structures bounding the domain of regime A.

It is possible, however, that the transitions $\vec{ab}$ and $\vec{ba}$ that are necessary for the stabilization of a shock wave between A and B cannot be realized by flat enzymatic systems; other organelles, such as points or circles, may be necessary. Illustrations of this are perhaps the ribosomes on the ergastoplasm and pores in the nuclear membrane. These organites on the membrane of endoplasmic reticulum are arranged in periodic geometrical patterns; this suggests that the transitional regime can have a decomposition into convection cells, like a fluid with two conflicting gradients (Bénard's phenomenon). Perhaps this is a possible origin of the biochemical polarization necessary in all organogenesis (see Section 10.5).

C. The rule of three states

A macromolecular structure can occur in three different states.

At stage 1, *formation*, the structure is generated by a conflict of ambient dynamics or, more generally, a singularity of the ambient metabolism.

At stage 2, *functional activity*, the structure has acquired a supramolecular order with internal stability and properties enzymatic for the ambient dynamics; therefore it acts on these dynamics.

At stage 3, *rest or death*, the molecular structure remains but is completely passive in the environment; the structure has no influence on the local dynamic, and its enzymatic effect is zero.

In principle, we can investigate the state of a form by trying to destroy it artificially by micromanipulation.

At stage 1, a partial destruction will result in the regeneration of the structure without causing serious perturbations of the environment.

At stage 2, a partial destruction will not be followed by regeneration and may result in serious disturbances, often lethal, in the ambient metabolism.

At stage 3, destruction of the structure will cause neither regeneration nor disturbances in the environment. The form is now only a phantom, a relic of its past activity.

The normal evolution of a structure is from stage 1 via stage 2 to stage 3, but the transitions $1 \longrightarrow 2$ and $2 \longrightarrow 3$ are often reversible, for examples, in mitochondria and nuclear membranes.

This outline raises a major question in biology: if it is true that a conflict between two regimes can stabilize only by constructing a macromolecular structure with an enzymatic effect facilitating the transition between the two regimes, how can such a complex arise? It may be that, if the precursor molecules are in the environment in sufficient concentration, the stabilizing configuration can arise by a simple coupling effect; the presence of these enzymes decreases the formation of entropy along the shock wave, and, by a generalized Le Chatelier's principle, we may imagine that the complex can then form more easily. However, a sufficiently large region of the complex will probably be necessary to initiate the process; furthermore we come up against a sterochemical constraint: every enzyme is actually a protein, and every protein molecule has a complicated three-dimensional structure which, if it has an essential part in the enzymatic action, will prevent the molecule from playing any part in a duplication mechanism. The living dynamic has resolved this difficulty, no doubt after many attempts, by allotting the germ-part to nuclear material which is well adapted to duplication. Hence we must think of the free boundary of the finite shock wave as being captured by a chromosome (or a fragment of a chromosome); then this chromosome emits filaments of RNA messages in the boundary of S, which capture the ribosomes connected with S and synthesize the necessary molecules of protein. Thus the chromosome appears as a biochemical caustic, which beams information just as an optical caustic beams light.

The spatial relationships given here must be refined because they are far too naive. In fact, these double-sheeted shock waves occur in the topological product $M \times D$ of the internal space M and the cell D (Figure 11.21). One sheet Δ_1, the least eccentric in M, has a sensory function. When the local metabolism has a point of excessive variation which crosses Δ_1, this variation propagates by diffusion to the nucleus, where it releases a correcting mechanism by polarizing the nucleus in a regime g; this corre-

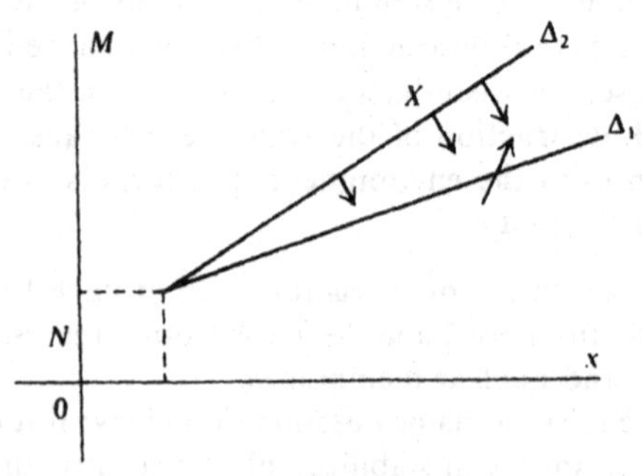

FIGURE 11.21

sponds to providing proteins to the external shock wave Δ_2, represented by correcting vector field X. This spatially bounded corrector mechanism probably occurs only inside the nucleus; there are probably no more spatially well-defined polarizing factors beyond the nuclear membrane, apart from perhaps very general natural gradients (and probably the local morphogenetic cytoplasmic fields, which we have already discussed).

D. The nucleus as a chemostat

With a little imagination we might describe this correcting mechanism as follows. The cellular nucleus plays the part of a chemostat working in the following way: let m denote a fundamental metabolite, whose concentration must remain close to an optimal value c_0, and suppose that, after a shock from outside, the concentration of m in the cytoplasm falls below a critical level $c_0 - k$. Then the metabolite will tend to diffuse from the nucleus into the cytoplasm. Suppose also that there is a relatively narrow specialized region W in the nuclear membrane whose outward permeability with respect to m is much greater than that of the rest of the membrane; this zone W will be the base of a cyclonic outward current of metabolite m, which will form both a kinetic and biochemical cylindrical shock wave T. Then there will be the ordinary nuclear regime n outside, and a perturbed regime a, characterized by an excessive concentration of m, inside. According to the earlier model, the only equilibrium position of the free boundary T will be the chromosome C, which codes the protein complexes carrying the biochemical transition $n \longrightarrow a$, and it is likely that this chromosome is relatively close to the area W, with which it is functionally associated. In these conditions the diffusion current of metabolite m across W will itself carry the enzyme molecules needed for making m, and the process will continue as long as the difference in concentration of m between the nucleus and the cytoplasm exceeds a certain threshold. Once the correction is completed and the cytoplasmic concentration of m is sufficiently close to the optimal value c_0, the diffusion across W will stop, and the shock wave T will be reabsorbed, involving perhaps the condensation of chromosome C (see Figure 11.22 and Note 2).

The least explicable part of such a model is the adaptation of the chromosome to the permeability properties of the membrane. It is worth noting in this respect that, according to some cytological observations, a pore in the nuclear membrane is the joining point of a nuclear fiber in interphase. In addition, one might reasonably object that homeostatic regulation also exists in bacteria, which have no nucleus. In this case the regulation is probably managed through the intermediary of labile and reversible structures like the hypothetical energids described at the beginning of this chapter; the nuclear membrane will then appear, as in the general model of all organogenesis, as a final realization and a definitive localization of a chreod that was at first labile and reversible.

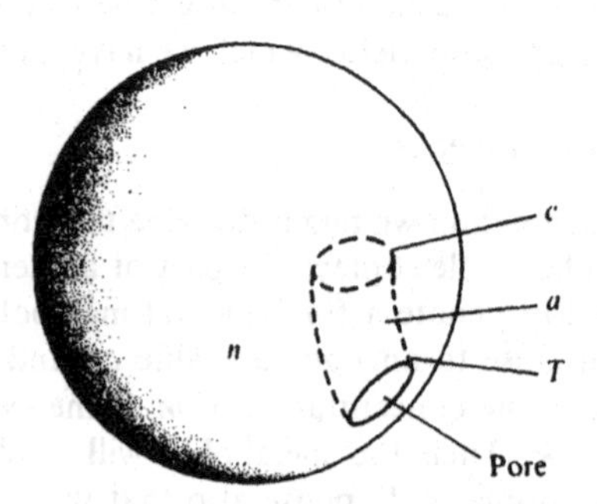

FIGURE 11.22. Local scheme of a regulation structure associated with the nuclear membrane.

11.6. FORMAL ASPECTS OF SPATIAL DUPLICATION

None of the local models in the list of ordinary catastrophes in $\mathbf{R}^4$ (cusp, swallow's tail, butterfly, and umbilics) shows the phenomenon of duplication; therefore let us impose the following condition: a singularity s, at $t = -1$, is replaced, after passing through the origin $x = t = 0$, at $t = +1$ by two singularities s_1 and s_2, each isomorphic to s. Suppose first that s is a point singularity; then it is not possible for the total graph of the point s to be in the form of a Y, because each algebraically defined singularity gives rise to a set forming a cycle mod 2, and this forces an even number of segments through the branch point. Therefore the stable couple s_1, s_2 will create or destroy a purely virtual, unstable regime s', which is in general unobservable. The simplest model in $\mathbf{R}$ (coordinate x) is of the function $f: x \to u$, defined by $u = x^4 + tx^2$. When t is positive, $x = 0$ is a simple minimum of u; when t is negative, u has two symmetrically placed simple maxima s_1 and s_2, and $x = 0$ is a simple minimum, the unstable regime s' (see Figures 11.23 and 11.24).

The interesting point about this model is that it is not structurally stable. A perturbation of u of the form $v = u + ax$ transforms the singularity into that of Figure 11.25, where the regime s extends directly, without branching, into s_1 and the other stable regime s_2 seems to loom up anew by the side of regime s, which acts as its parent. The model may be represented in terms of chreods as follows: the chreod W defining the field s has an umbilical zone in a neighborhood of the origin, where a chreod W', isomorphic to W, is grafted (Figure 11.26). More precisely, suppose that the evolution of a one-dimensional being is described by a growth wave with values in a torus having coordinates (x, t), represented by $0 < x < 1$, $0 < t < 1$, and suppose also that there is a conflict of regimes g_1 and g_2 on this torus, bounded by a line of conflict with two free ends, forming a Riemann-Hugoniot catastrophe (Figure 11.27). When g_1 is a regime of rapid growth, and g_2 a regime of slow growth, this will give a model for a budding system, with the part in g_1 forming the bud. The bud will not always be spatially separated from the parent in this oversimplified model.

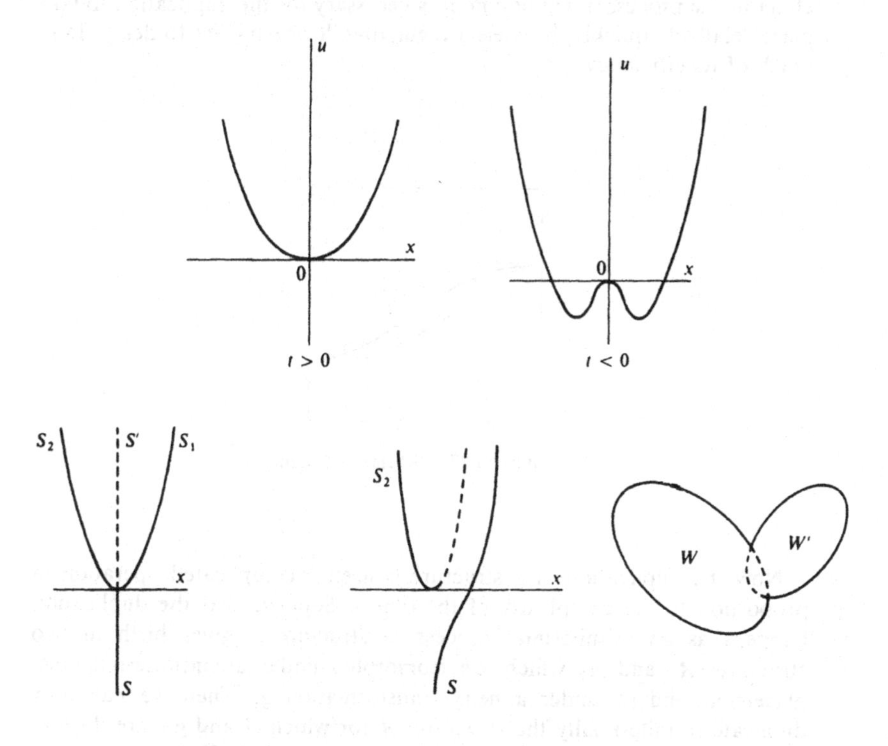

Figure 11.23, 11,24, 11.25, and 11.26. Scheme of duplication.

Almost every elementary phenomenon of duplication that has been observed in biology, particularly at the level of subcellular organelles, is a budding phenomenon (e.g., the case of the centriole). Pure symmetric duplication into two isomorphic models seems to be very exceptional.[3] Mitosis of a cell is too complicated to be considered an elementary phenomenon; it seems to me to be difficult to judge the value of the splitting of the double Watson-Crick helix, but the thermodynamical and topological difficulties of this model are, a priori, considerable. There still remains division of a chromosome into two successor chromatids; here again the fine structure of the process is not known. But it seems certain, a priori, that the linear structure of genetic material is the result of a topological constraint. In fact duplication of a structure in biology is a regulation, a homeostatic effect; whether the duplication of the structure has a favorable effect on the metabolism (by decreasing the local entropy), or, as we conjectured in Section 11.1, duplication retards a heterocatalytic effect in the process of extinction, it is necessary for the duplication to take place relatively quickly, because a regulating device subject to delays loses much of its efficiency.

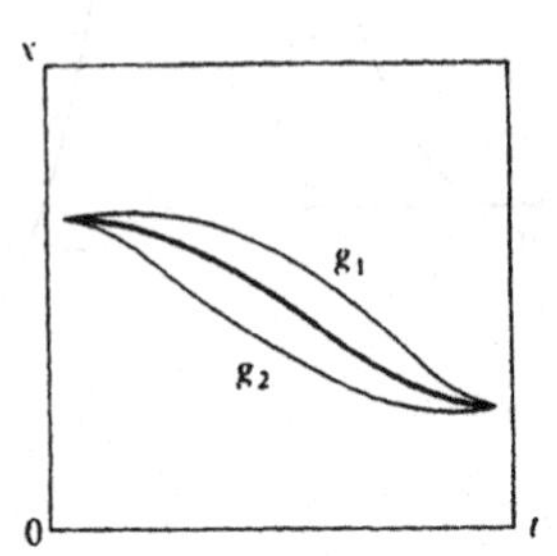

FIGURE 11.27. Scheme of budding.

Now the duplication of a structure is itself a complicated operation in proportion to the complexity of the object. Suppose that the duplication happens as an infinitesimal process; a structure A gives birth to two structures A_1 and A_2 which are isomorphic under an infinitesimal displacement and so under a helix transformation g. Then we can only duplicate infinitesimally the structures A for which A and gA are disjoint for all g belonging to a one-parameter subgroup $\{g\}$. If A has a simple structure it is, in practice, a curve or surface; it cannot be a three-dimensional object without A and gA meeting for sufficiently small g, and for the same reason it cannot be a compact surface without boundary. Hence the only simple geometrical objects that can duplicate quickly are curves or perhaps small surfaces with boundaries, like discs or ribbons. The necessity of filling these structures with important chemical information which is easily transmitted by a mechanism of local nature eliminates the latter possibility. Of course this is an a posteriori explanation, and the role of genetic material remains one of the enigmas of the dynamic of life.

NOTES

1. There are, in fact (T. L. Lentz, *The Cell Biology of Hydra*, North Holland, 1966, Figures 83–95), some interesting pictures of the formation of the nematocyst. The triangular nature (characteristic of elliptic breakers) of the stylets of the thread of nematocyst is particularly interesting (Photographs 26a and b).

2. In the same book, the section of a neuroganglion cell (Photograph 27) has many microtubes ending in certain pores of the nuclear membrane. Might these not be cytoplasmic extensions of an intranuclear polarization?

3. As we said in Section 10.4.C, there are perfectly symmetric duplication loops in the universal unfolding of the parabolic umbilic. These loops which simulate the haploid-diploid alternation of sexual reproduction, clearly require a threshold stabilization or, equivalently, a hereditarally fixed structure much more developed than any we are considering here.

CHAPTER 12

THE BASIC PROBLEMS OF BIOLOGY

La vie n'a pas le temps d'attendre la rigueur.

PAUL VALÉRY
L'idée fixe.

We shall discuss four topics: finality in biology, irreversibility of differentiation, the origins of life, and evolution.

12.1. FINALITY IN BIOLOGY

A. Finality and optimality

When a biologist finds an organ or behavior that is obviously well adapted, his first concern is to ignore this adaptive character and to emphasize the factors immediately responsible for the process. For example, in the well-known study of the orientation of leaves toward light, he isolates a substance, an auxin, produced by light rays, which inhibits the growth of tissues. The immediate mechanism of the process is then explained perfectly, and usually, to a biologist, that is sufficient. But if we, goaded by an understandable feeling of intellectual dissatisfaction, ask him how it comes about that the process is so obviously beneficial to the plant's metabolism, he will certainly invoke a principle of natural selection: plants in which an accidental mutation established this process enjoyed an advantage that eliminated those without it through selection. This lazy and entirely unverifiable answer is at present the only interpretation of biological finality, even though the process presents a challenge worthy of further explanation.

The mathematician von Neumann [1] commented that the evolution of a system can be described in classical mechanics in two ways: either by local differential equations, for example, Lagrange's or Hamilton's equations, or by a global variational principle, like Maupertuis' principle of least action;

and these two descriptions are equivalent, even though one seems mechanistic and locally deterministic, whereas the other appears to be finalistic.

The same is probably true in biology: every epigenetic or homeostatic process is susceptible of a double interpretation, deterministic and finalistic. We must not forget that the essential object of study in biology is not the isolated individual but the continuous form in space-time joining parents to descendants (the regulation figure); more precisely, when two or more species have some functional interaction between each other, such as predation or being an auxiliary in the fertilization process, etc., it is necessary to consider the total figure in space-time, the union of all the forms associated with each species. Then, for each adaptive process, we can probably find a function S of the local biological state expressing in some way the local complexity of the state with respect to the process considered, and the configuration will evolve between two times t_0 and t_1 (e.g., the parent at age A and the descendant at the same age) in such a way as to minimize the global complexity $\int_{t_0}^{t_1} |S|\, dt$. In this way the minimum complexity and hence the most economical adaptation of the process will be realized. Natural selection is one factor in this evolution, but I myself think that internal mechanisms of Lamarckian character also act in the same direction. However, in contrast with classical mechanics, we should not expect that this evolution will be differentiable, or even continuous, on the individual level because the global continuous configuration must conform to the boundary conditions of a system restricted by spatial reproduction in a given chemical and ecological context. Hence there will be not a continuous deformation but a finite chain of relatively well-determined, subtly interrelated, local processes (or chreods); even the variation of the global figure can introduce qualitative discontinuities into the structure of this chain—this is called mutation. The effect of the global variational principle will be too weak for the local mechanisms to show no random fluctuations, and only the resultant of these local variations will finally be oriented by the variational principle.

Although the teleological nature of organs and behavior in living beings will be immediately apparent to us (with reference to what we ourselves are and to our own behavior as human animals; see Appendix 1), the deterministic and mechanistic nature will escape our attention because it operates on a very long time scale and has a statistical character inherent in evolution, and its decisive factors (the influence of metabolism on the statistic of mutations) are probably very tenuous. Let me be more precise.

B. Chance and mutations

One of the dogmas of present-day biology is the strictly random (if this means anything) nature of mutations; however, it seems to me that this dogma contradicts the mechanical principle of action and reaction: of two possible mutations m and m', the one with the better effect on metabolism (i.e., the one that minimizes the production of entropy) must have a greater probability of happening. In the classical diagram of information theory,

$$\text{source} \rightarrow \text{channel} \rightarrow \text{receptor}\,,$$

it is clear that the source has an effect on the receptor; therefore the receptor has an inverse effect on the source, usually unobservable because the energy of the source is very large with respect to the interaction energy. This is certainly not the case, however, of nucleic acid, where the binding energy is much less than the energies of the metabolism. One might object that here the receptor is an open system, in the language of thermodynamics; it is possible that DNA has a directing action on the metabolism not requiring the introduction of a large interaction energy. In systems in catastrophe, a very slight variation in the initial conditions can cause large modification of the final state, and the interaction of the DNA chromosome could give rise to very small initial variations amplified later to large effects, a situation similar to that of a point determining the route of a train whereas the train has no effect on the point. But this comparison is specious, as are all examples taken from human technology; they can occur only in a state of zero metabolism. The effect of a signalman altering slightly the points under a moving train is disastrous, whereas it seems that most spontaneous mutations occur in interphase, during full metabolic activity. The breakages and displacements of chromosomes observed in metaphase are only the visible results of earlier metabolic accidents in the interphase which have upset the course of the anaphasal catastrophe.

Most mutations are attributed to chemical modifications in the DNA sequence in nucleotides, due to errors in the duplication process of DNA. I am reluctant to subscribe to the current belief that a point mutation, affecting just one nucleotide, is sufficient to inhibit the activity of a gene; this seems to me to repeat on another plane the error of the morphologists who believed that the destruction of one neuron in the brain would stop the process of thinking. To suppose the strict validity, without some random noise, of the genetic code amounts to making the basic regulation mechanism of the cell fully dependent on a process in a state of permanent catastrophe. Even if life is only a tissue of catastrophes, as is often said, we must take into account that these catastrophes are constrained by the global stability of the process and are not the more-or-less hazardous game of a mad molecular combination. Even adopting the anthropomorphic point of view that there is a mechanism for reading the DNA that is perturbed by errors, might we not push this anthropomorphism to its full extent and admit that the errors are oriented, as in Freudian psychology, by the "unconscious" needs and desires of the ambient metabolism? It seems difficult to avoid the conclusion that the metabolism has an effect, probably very weak, which in the long run can dominate the statistic of mutations, and the long-term results of the effect explain the variational principle of minimum complexity and the increasing adaptation of biological processes leading to the finality.

12.2. IRREVERSIBILITY OF DIFFERENTIATION

A. The main types of differentiation

I shall give here three geometrical models for the different transformations observed in cellular differentiation.

1. The silent catastrophe. A local dynamic defined by the attractor V of a differential system (M,X) has a silent catastrophe when the dimension of this attractor increases. This can happen, formally, in two different ways: either by bifurcation in the sense of Hopf, where, for example, a closed trajectory gives birth through bifurcation to a two-dimensional invariant torus; or by coupling with another system (M',X') whose regime diffuses, leading to the replacement of V by the product $V \times V'$ of attractors of the two dynamics, which can then couple.

Biochemically the silent catastrophe can be interpreted as the beginning of new reaction cycles, the synthesis of new products (RNA?) and new organelles (ribosomes), and the freeing of previously repressed degrees of freedom. In embryology the silent catastrophe usually denotes a gain of competence. But since it concerns a continuous transformation, it does not give rise, in principle, to morphogenesis.

2. The catabolic catastrophe. This is characterized by an abrupt decrease in the dimension of an attractor V of the local dynamic; it can be interpreted as the setting up of a resonance between two or more cycles of hitherto-independent reactions. It corresponds biochemically to a sudden loss of competence (by differentiation), to the massive synthesis of specific proteins, and to the end of syntheses not specific to the new regime. Morphogenetically there will be a catastrophic spatial differentiation of the new tissue (by a generalized catastrophe if the competent tissue has not undergone a sufficiently long preliminary polarization).

3. Aging, or the sliding catabolic catastrophe. The loss of dimension of an attractor V can happen almost continuously. The new attractor V' will be of lower dimension than V but embedded in V in a complicated topological way, as if V were covered ergodically by V'; then the embedding will simplify topologically through an infinite number of discontinuous changes whose individual thermodynamical effect is unobservable. There is a good picture of this phenomenon with the vector field $\dot{x}=1, \dot{y}=m$ on the torus T^2, x and y considered modulo 1 (when m is rational, we must replace this field by a structurally stable field with an attracting trajectory); then there is a deformation of this type as m tends to a simple rational value. In another model, the new attractor V' is not fixed but undergoes fluctuations, making it vary within an open set U of V (initially all of V), which decrease in amplitude until V' comes to rest at V_0'.

In both cases we can take as coordinate a function v for which V_0' is defined by $v=0$ in V, and express the entropy (the logarithm of the Liouville measure given by the probability function of the system) as a function of v; there will initially be a plateua-like curve at level $g = g_0 > 0$, for which $\int g\, dv = 1$, and finally g will evolve toward the Dirac measure at the origin $v=0$. Referring to the model of Chapter 4, we see that the system will have less and less interaction with an inductor system coupled with it for the variable v. Such a model explains the progressive loss of competence in the course of time: if the inductor is itself polarized, the induced zones that the new coupled regime influences become narrower in time as the function $g(v)$ tends to a steeper and steeper peak. Although this scheme may illustrate loss of competence, it is not certain that senescence, the global aging of the organism, can be represented by a model of this type.[1]

B. Sexuality

The periodic global nature of the living dynamic demands that each of the transformations described above be reversible—this is so for transformations 2 and 3, which can be reversed by a transformation of type 1. It is possible, for example, that the field degenerated into an attractor $V' \subset V$, returns, under the effect of a shock, to its primitive situation as an ergodic field on V. This may be the situation in the "dedifferentiation" of muscular or bony cells in the formation of blastema in the regeneration of a limb, but it must be remarked that this phenomenon can happen only locally, as if its realization placed on the remaining structures of the organism a relatively great effort. However, there is one exception: the generalized lysis of larval structures at pupation in the metamorphosis of insects; but here also the transformation is not a strict return to the past, and besides the rather exceptional structure of the polytene chromosomes observed in salivary glands shows the great tension of the genetic material at this instant.

The reversal of silent catastrophes (type 1) presents a very difficult problem because it cannot occur as a result of a transformation of type 2 or 3. This follows because, if the two systems of attractors V and V' are coupled, a later catabolic catastrophe will give rise to an attractor G embedded in $V \times V'$ which will, in general, project essentially onto the two factors; the regimes V and V' will be mixed in a way comparable thermodynamically to the diffusion of two gases when their containers are joined together. It is necessary to use some radical process, like fractional condensation, to separate the gases; similarly the living dynamic must undergo a quasi-complete halt of its metabolism, followed by a suitably arranged readjustment to reverse the effect of a catastrophe of type 1, and it is the characteristic of meiosis and gametogenesis in general to realize this inversion. In diploid species with normal sexual reproduction, the operation can perhaps be realized as follows. The local dynamic provides a weak coupling between two isomorphic dynamics (M_1, X_1) and (M_2, X_2), where M_i is of the form $K_i \times T_i$, $i = 1, 2$, and T_i denotes an n-dimensional torus representing, in some way, the purely kinetic part of the biochemical dynamic (the part that escapes biochemical analysis). Meiosis consists of a special coupling allowing components of X_1 and X_2 on T_1 and T_2 respectively to act on each other by addition; this makes them equal and allows chromatic reduction, and this period of interaction corresponds to the formation of a synaptic complex between homologous chromosomes. Sexual reproduction appears, from this viewpoint, as a dynamical necessity imposed by the periodical nature of biological morphogenesis. The old idea that sexuality is necessary in some way to recharge the morphogenetic potential (to express it in vague terms) has been abandoned in favor of finalistic arguments on the benefits of genetic exchange; perhaps this merits re-examination. Might not *sexuality be the slumber of morphogenesis*? It is a slowed-down period of life, made necessary by the accumulation of a mixture of dynamics following differentiation, on the one hand, and by the transport morphogenesis in space-time of the gametes, on the other. Very generally gametes or spores appear in the organism where the morphogenetic capacities of the species are all displayed and seem to be exhausted; this is particularly striking in fungi like the amanites, which generate spores that seem incapable of germination.[2]

I am not unaware of the fact that many species (parthenogenetic or not) have left behind sexual reproduction. In these cases, however, it would be surprising if the mitosis giving rise to future oocytes or spores has not special cytological or other properties that may have escaped observation.

C. Irreversibility and death

None of the transformations that we have encountered is fundamentally irreversible; they are only difficult to reverse. Then why is an individual not immortal, as he often is in lower organisms like Hydra? The aging and death of an individual are usually interpreted as a result of somatic differentiation, and I find it difficult to see why a fully differentiated being could not be immortal. Here is another, hypothetical explanation: if the individual became immortal, his dynamic would, after a finite time, have an attracting state g. There would be a temporal evolution from birth and embryonic development which must be considered as an aging. It would then be necessary for this aging to stop and the dynamic to tend toward, and reach, state g after a finite time; for this to happen, this state must be strongly attracting and must not allow any later evolution to other states: in the language of topology, the manifold of these final states of g must be compact. Thus it seems to me that the immortality of the individual and the possibility of later evolution of the species are incompatible; the death of the individual is the price that must be paid to preserve all the possibilities of future perfection of the species.

12.3. THE ORIGIN OF LIFE

A. Synthesis of life

Just as death is the generalized catastrophe determined by the transformation of a metabolic field into a static field, so the synthesis of living matter demands a truly anabolic catastrophe which, starting from a static field, will lead to a metabolic field. As we have seen in differential models, this requires the execution of an infinite number of ordinary catastrophes controlled by a well-established plan before the stable metabolic situation can be established, and so *an infinite number of local syntheses in a well-defined spatiotemporal arrangement*. This seems to exceed the possibilities of traditional chemistry.

Another method, which seems at first sight to be more accessible, consists in realizing a simulation of a living metabolism: creating a continuous flux of energy in a container and canalizing it into kinetic structures with properties of autoreproduction may cause these structures, which were originally labile, to acquire genetic structures that will stabilize them. We are going to try to imagine, in a highly speculative way, the primitive forms of life and their regulating dynamics.

B. The three-regime soup

We start with a primeval soup, an open set U in $\mathbf{R}^3$ containing a field of local dynamics in a state of generalized turbulence; U is irradiated by a flux of energy, light which is then degraded to heat. This soup is far from being homogeneous: there are stable local dynamical regimes $(r_1, r_2, \ldots, r_n)$ forming phases in U, and these phases are separated by narrow fluctuating and turbulent transitional regimes, which are essentially labile. These regimes can be ordered (at least partially) by increasing topological complexity corresponding, intuitively, to the amount of information, of negentropy, contained in a unit volume of their dynamic. Suppose, for simplicity, that there are three such regimes, written in the order ρ, σ, τ, with the property that τ is a self-duplicating regime in interaction with σ and admitting a stable transitional regime $V_{\rho\tau}$ with respect to ρ. The system can then evolve qualitatively as follows. There might be globules of regime τ, in an open set U_σ of σ, which either existed at the beginning or were formed spontaneously by a local excitation of the dynamic σ; these globules will then multiply and occupy the whole set U_σ. In doing so, they will use up the complexity of the ambient material; and, when the density of τ reaches a certain level, it will return to the less excited dynamic ρ. The globules of τ that are freed in this way in U_ρ will be preserved by a membrane of the stable transitional regime $V_{\rho\tau}$. This situation will continue until the globules τ are forced by the ambient agitation into a new open set U_σ, and again start to proliferate and repeat the process.

This is still only a preliving state, because of the labile, reversible nature of all the regimes, but the following modification will give rise to typically living phenomena. Suppose, in addition, that τ is a budding dynamic whose successor dynamic, controlling the bud, is a transitional stable regime $V_{\rho\sigma}$ between ρ and σ, where this regime $V_{\rho\sigma}$ is connected with a supramolecular organization having a certain geometric rigidity and directional enzymatic properties that stabilize σ. In this case, as soon as a globule τ meets an open set U_σ, it will implant itself first on the limiting surface, the boundary of U_σ in U_ρ, and then, budding, will excite the transitional regime $V_{\rho\sigma}$. This region will impose its own geometric structure, for example, a sphere S, on the bounding surface; the regime σ will be protected inside S by the enzymatic barrier on S of the ambient random dynamic ρ, and the globules τ can proliferate regularly inside S. Then after having used up the ambient dynamic σ, which cannot indefinitely support an exponential development of regime τ, the dynamic inside S will return to the less excited state ρ, and the sphere S, now only a phantom, in the third state, independent of the ambient environment and unstable, will disintegrate and release globules of τ, which can then colonize another free U_σ region.

Now suppose that the regime σ decreases, or even disappears, through an unfavorable change in the exterior conditions, such as a decrease in the incoming energy; it might be expected that this would imply the disappearance of τ, which depends on σ for its reproduction. In fact, however, this will not be so because of the phenomenon of induction associated with an organizing center. Suppose that we have a very complex regime, defined by a representative function g whose support is a small ball b, coupled with a much less complex regime g_0 controlling the ambient environment; then we get a structurally stable situation, after the local polarization of a neighborhood of b by g as follows. The shock wave bounding the ball b will exfoliate, and the separate regions isolated by the leaves will have regimes $g_0, g_1, \ldots, g_k$ of increasing complexity from the outside to the inside. This development will be governed by a chreod of space-time that depends only on g_0 and g, the bounding regimes, a process that can be called the germination of the globule g (Figure 12.1). It can also be supposed that the transitional regime $V_{\rho\tau}$ weakens under the effect of a local excitation of regime ρ and gives birth by exfoliation to an infinitesimal region of regime σ, intermediate between ρ and τ in topological complexity and from the point of view of the function-space topology. This is sufficient for implanting the transitional regime $V_{\rho\sigma}$ in the boundary of the region σ in U_ρ; then a sphere S will form, inside which the regime σ will survive because of the enzymatic properties of $V_{\rho\sigma}$, although it would disappear naturally without this barrier. In this sphere the globules of τ can proliferate, and the cycle can start again. One expression of the final state of this situation is that τ maintains a regime of ρ-duplication, and the essential difference from the previous case is the following: spontaneous generation of τ-globules is no longer possible, and the presence of these globules is connected with the presence of parent τ-globules and the information they contain. The same procedure allows the possibility of being viable in regimes $\upsilon, \mu, \ldots$ of lower complexity than τ at the cost of an increase in the topological complexity of τ. In this way one of the important processes of evolution—the enlargement of the conditions of existence and the conquest of the outside environment—appears in an elementary form.

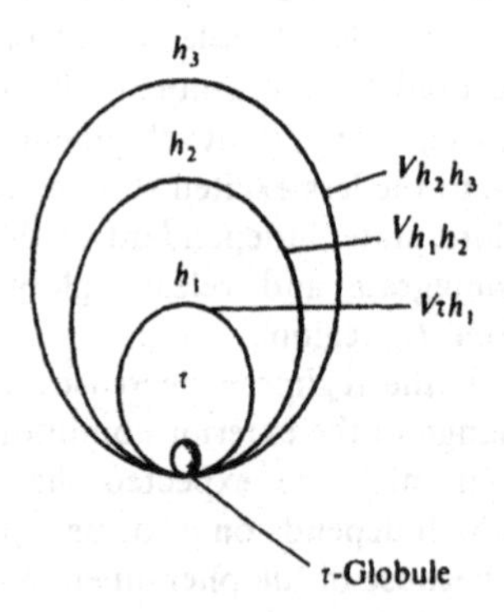

FIGURE 12.1

This is, of course, only a formal, conjectural explanation; the change from a σ-duplicating regime into a ρ-duplicating regime in a larger environment requires an increase in the topological complexity of the regime, and this increase will be attributed to a mutation, truly a process of the invention of life. It is tempting to believe that this process is favored by forces of necessity; as soon as the duplication of the τ-globules is slightly hindered, by unfavorable extremal conditions, the appearance of aborted or reversible mitosis, of the type of meiosis, might be expected. Now in such a situation of the successor elements may have greater complexity than their parents, precisely because their separation is more hesitant and unstable; the shock wave of mitosis could divide them in a topologically more delicate way. Of course only mutants with favorable mutations will survive, and the fact that such mutations exist appears as a postulate imposed, a posteriori, by the success of the living dynamic.

This three-regime soup, with regimes ρ, σ, and τ, gives a reasonably good idealization of cellular behavior, ρ representing the regime of cytoplasm, σ that of the nucleus, τ that of a gene (or nucleic material in general). Viruses are typically σ-duplicating τ-globules, but they also have a protean device (shell, tail, etc.) that protects them from the outside world or facilitates their attachment to the cells on which they are parasitic.

The scheme described above has been largely inspired by the theory of the virus-fossil, which is of great conceptual interest, even if it is difficult to justify because of the complexity of cytoplasm.[3]

C. Recapitulation

The preceding scheme affords a good explanation of the classical law of recapitulation according to which the embryo retraces its phylogeny during development. To simplify, let us consider the static global model of Chapter 10: the global dynamic of the species will always be represented by τ. The confrontation with the more and more hostile environments (represented by regimes $h_1\, h_2, \ldots, h_m$) during evolution will result in foldings, or successive "complexifications," of the dynamic τ, which permit it to create successive transitional regimes $V_{\tau h_1}$, $V_{h_1 h_2}, \ldots, V_{h_{m-1} h_m}$; the structures associated with these transitional regimes will be the organs which permit adaptation to the successive environments h_j. Then the egg of the species may simply be considered as a τ-globule, bounded by the transitional regime $W_{\tau h_m}$, where h_m denotes the external world (in mammals, this transitional regime is at first in the hands of the mother organism; in ordinary eggs, it consists of the vitelline membrane and extraembryonic structures); the germination of the egg is caused by a weakening of the transitional regime $W_{\tau h_m}$, which then becomes incapable of maintaining the permanence of the inside regime τ, and so this regime will pass in order through all the states h_1 to h_m. The τ-globule reacts to this variation of ambient regime by forming transitional regimes $V_{\tau h_1}$, $V_{h_1 h_2}, \ldots, V_{h_{m-1} h_m}$, which construct the associated organic structures, and so these structures appear temporally in the order in which they appeared in the course of evolution, and spatially in a centrifugal order (from inside toward outside), as Uexküll has already remarked (epigraph to Chapter 10). In fact, as we saw in Chapter 9, this is only an idealized scheme, because the epigenetic dynamic has achieved short cuts allowing it to leapfrog certain steps that have become unnecessary.

12.4 EVOLUTION

A. Eigenforms of duplication

Let us start with the very basic objection of the finalists to a mechanist theory of evolution: if evolution is governed by chance, and mutations are controlled only by natural selection, then how has this process produced more and more complex structures, leading up to man and the extraordinary exploits of human intelligence? I think that this question has only a single partial answer, and this answer will be criticized as idealistic. When the mathematician Hermite wrote to Stieltjes, "It seems to me that the integers have an existence outside ourselves which they impose with the same predetermined necessity as sodium and potassium,"[4] he did not, to my mind, go far enough. If sodium and potassium exist, this is so because there is a corresponding mathematical structure guaranteeing the stability of atoms Na and K; such a structure can be specified, in quantum mechanics, for a simple object like the hydrogen molecule, and although the case of the Na or K atom is less well understood, there is no reason to doubt its existence. I think that likewise there are formal structures, in fact, geometric objects, in biology which prescribe the only possible forms capable of having a self-reproducing dynamic in a given environment.

Let us express this in more abstract language. Suppose that the evolution of a biochemical soup in a given environment, depending on parameters W, can be described by the trajectories of a vector field in an appropriate function space, and denote by F_t the operation of moving along a trajectory through a time t. If at the beginning the data specified by a form S are given, this form will be self-reproducing if the image form $F_t(S)$ is the disjoint topological union $S_1 \cup S_2$ of two forms S_1 and S_2, each isomorphic to S; hence we must find the spectrum of the duplicating operator for each value of t, that is, the set of eigenforms S such that $F_t(S) = S_1 \cup S_2$. This spectrum is very probably discrete, and it may be possible to define operations such as sum, topological product etc. on the set of eigenforms. Each eigenform will have its domain of stability on the space W of external parameters, and the set of these stability domains, separated by conflict or catastrophe strata, will form a phylogenetic map decomposed into chreods, just like Waddington's epigenetic landscape. Evolution is then the propagation of an immense wave front across this map W. Phenomena of induction, regression, and catastrophes will occur exactly as in the development of the embryo and for the same formal reasons.

Global finalists (like Teilhard de Chardin) push the analogy between evolution and epigenesis of the egg to the extent that they believe that, just as the embryo develops according to an established and unchanging plan, so must the wave of evolution in the space W; but this neglects the important difference that the development of the embryo is reproducible and thus an object of science, whereas the wave of evolution is not. To assert that a unique and unrepeatable phenomenon occurs according to a plan is gratuitous and otiose.

Might not the same criticism be leveled against the idea of the phylogenetic map W? Dealing, as it does, with a unique and non-reproducible phenomenon, is it not a gratuitous attitude of mind? In fact, the only possible merit of the scheme is to point out analogies and local mechanisms. It is not impossible to provoke, experimentally, local evolutionary steps in desired directions (particularly in lower species which reproduce rapidly; see, e.g., Waddington's experiments on *Drosophila*). At bottom, it must be observed that each point of an embryo is close to a cellular nucleus containing almost all of the information needed for the local realization of the individual; the only factors common to living beings in the wave of evolution are the elementary biochemical constituents (such as DNA, RNA, proteins, the genetic code), which do not seem to form sufficiently complex structures to support the global plan of life. This is why, until more information is available, I prefer to think that evolutionary development happens according to a local determinism.

Attraction of forms is probably one of the essential factors of evolution. Each eigenform (one might even say each archetype if the word did not have a finalist connotation) aspires to exist and attracts the wave front of existence when it reaches topologically neighboring eigenforms; there will be competition between these attractors, and we can speak of the power of attraction of a form over neighboring forms, or its malignity. From this point of view it is tempting, with the present apparent halt in evolution, to think that the human attractor is too malignant. Of the theoretically possible living forms only very few are touched by the wave front and actually come into being

B. A hypothetical mechanism of the attraction of forms

The idea that we have proposed of the attraction of archetypes invites an obvious objection: how can a structure, an archetype, with only a virtual existence exert a *real* attractive force on existing beings? Here is a possible mechanism, depending on a further analysis of genetic material. An eigenform g is defined, in principle, entirely by its chromosome spectrum, that is, by the wild genotype of the species considered. Now this genotype, at least in Metazoa, is in practice determined by the structure of the chromosome material; however, it is possible that the data alone of the genotype are insufficient to characterize completely the metabolic state of the animal, which can vary under the effect of internal or external perturbations. This situation can be described, as in quantum mechanics, by saying that the average state of the metabolism is a mixed state u, a linear combination of eigenforms g_i,

$$u = \sum c_i g_i, \quad c_i > 0 \quad \text{and} \quad \sum c_i = 1,$$

where all weights c_i are zero or very small except for c_0, the weight of the archetype g_0, and perhaps c_1, of a form g_1, topologically close to g_0. Then, during relatively frequent fluctuations, the metabolism of the animal will be in regime g_1 even though the chromosomal material is in state g_0; the transition $g_0 \rightarrow g_1$ cannot happen without producing metabolic singularities (discontinuities, shock waves, etc.) at certain times and at certain places in the organism, and large amounts of entropy are generated at these singularities. We know that these singularities can provoke the creation of macromolecular structures m_0^1, with the effect of facilitating the transition $g_0 \rightarrow g_1$ and decreasing the production of entropy, and these particles m_0^1 could have an internal stability allowing them to survive the metabolic conditions of their birth, and moreover have a tendency to excite the presence of the singularity of the transition $g_0 \rightarrow g_1$ responsible for their birth. In this way they would seem to have the property of self-duplication as previrus or plasmagene-like particles.

The thermodynamical conflict between two regimes g_1 and g_1' close to g_0 might then be interpreted as a struggle between two plasmagene populations m_0^1 and m'^1_0, and from this point of view natural selection acts within the metabolism of the individual, just as it acts between individuals and between species. When form g_1 is sufficiently malignant, particles m_0^1 will multiply and facilitate further the reappearance of form g_1. The weight c_i of g_i will increase at the expense of c_0, and when c_1 exceeds a certain level, a tiny perturbation may provoke the appearance of genotype g_0 in form g_1; the transitional previrus m_0^1 will then move into the chromosome position and may have phenotypical effects. What will be the influence of the environment on a transformation of this kind?

C. Unusual stimuli

Let us consider the effect on a living organism of an unusual stimulus S; the representative point in the regulation figure will then move to an unknown point s of large amplitude. When the stimulus is excessive, there will be no further possible movement, and this, with the death of the individual, removes any problem. For smaller stimuli, the individual will react and bring correcting reflexes r_j into action. Suppose that there is a sequence of reflexes $r_1, r_2, \ldots, r_k$ correcting the perturbing effects of the stimulus; in fact, in most cases there are several such sequences of reflexes, and the choice between the sequences is initially structurally unstable. In any case, repeating the stimulus S on the individual under consideration will lead to the establishment of one of the correcting sequences. If the same stimulus is inflicted on the descendants of the individual, might it happen that the habitual use of this correcting sequence could implant itself in the genetic endowment of the species to the point of stimulating physiological or morphological effects? What should we think of the Lamarckian dream according to which acquired characteristics can become hereditary? I shall now show how the model could envisage this problem.

Each correcting reflex r_i is associated with a stratum X_i of the bifurcation set of the function space of the regulation figure. Hence it could happen that a correcting sequence $r_1, r_2, \ldots, r_k$ defines a starred stratum $X_1, X_2, \ldots, X_k$; and if this stratum is of a sufficiently low codimension, and not too far from the strata through which the growth wave passes during development, there could be a stabilization process of the organizing center of the sequence $r_1, r_2, \ldots, r_k$ around this stratum. Once this process is initiated, the animal could even investigate the stimulus S for itself. When this process is sufficiently advanced, a mutation could occur that would correspond to a permutation of the chromosome material, and the geometry of this permutation, involving dissociation and subsequent recombination of the chromosome fragments after possible replication, will echo the topological position of the new organizing center with respect to the old. In our hydraulic model, such a permutation might be represented as a permutation between saddles, geographically comparable to a capture of one of the water courses of the landscape by a lateral affluent.

In conclusion, I am tempted to think that the influence of the environment plays only a relatively secondary part in the important evolutionary changes—at most, that the frequency of an external stimulus could have a role in starting an evolutionary process.

D. Bacteria and metazoa

In respect to evolution, there is a great difference between bacteria and Metazoa. The bacterial mechanism is relatively rigid, and a modification of any importance in the environment requires, almost *ipso facto*, a modification of the genome, a mutation. The situation is different with Metazoa; they can undergo large physiological adaptations without having to modify their chromosome stock. Another way of saying this is that there is no distinction between soma and germ-plasm in bacteria. If my ideas are correct, the Metazoa have, located in the kinetic configuration of the metabolism of their gametocytes, a model of their actual conditions of existence, and even, in a broader sense, a representation of the external environment and other living species with which they are functionally associated (e.g., as prey or predators). Now these models have many common geometric features, at least among animals, which are expressed as through a global isomorphism of the epigenetic regimes, and therefore it is natural to think of these models as derived by specialization (as is said in algebraic geometry) from a universal model (see Appendix 2) containing all regulation figures of living species. The topology of this universal model will reflect less the phylogenetic relationships than the functional interaction between species, so that the distance between bee and snapdragon will be less than that between bee and butterfly. The big evolutionary advances of history will be described by global deformations of this universal model; the metabolism of gametocytes will appear as a kind of research laboratory, a device simulating conditions of existence close to the actual ones, and an evolutionary advance in this device will provoke unstable virtual catastrophes that will correspond to a new functional morphology, expressing the organic and physiological adaptations made necessary by this evolutionary change. Only when this new morphology is sufficiently stabilized, can evolutionary advance begin, and it will manifest itself as a mutation, rearrangement of chromosome stock, and appearance of new organic forms. According to these ideas, the fish already "knew", before they became amphibious, that a life on land would be possible for them, and what new organs they would need.

It may also be the case that a widespread epidemic affecting a species is a sign of disequilibrium between its genotype and its metabolic conditions, and that the effect of this shock can accelerate any necessary transition. I cannot hide the extent to which these ideas are speculation. My only ambition here its to propose an conceptually acceptable mechanism to explain one of the most formidable problems of science.

APPENDIX 1

Finality and archetypal chreods. A first question on the finality of a behavior is: Is there an intrinsic criterion (connected with the geometry of the process) identifying its finalist character? It might be expected that a finalist process is characterized by the existence of an aim, a final state toward which the organism tends along some convergent funneled route (cusped canalization), and most finalist processes in biology show this behavior. But, as we saw in Section 6.2.C, with mean coupled fields, this cusped canalization behavior can appear structurally stably in a natural way and therefore does not automatically characterize a finalist process. Homeostasis (global regulation) of living beings—in particular, of man—depends on the use of a small number of regulating functions (food, respiration, excretion, reproduction, etc.) which define several breaking hypersurfaces in the global model considered in Section 10.2.C, and each of these hypersurfaces begins with an epigenetic chreod which creates the corresponding organs, and finishes by constructing function fields in the nervous system regulating the behavior of these organs. Thus they create organizing centers of global functional fields (archetypal chreods) in our mental organization.

A very contracting process with attracting final state in an animal is called finalist if its observation sets up, by resonance, one of these archetypal chreods in the mind of the observer. As the main regulating functions are the same for most living beings, it is not surprising that most processes observed in living beings are called finalist. However, it often happens in biology that certain organs or certain behaviors, though very canalized, do not have any visible finality. In some cases there may even be suicidal chreods, leading to the destruction of the individual. This happens, for instance, if there is conflict between the finality of the individual and that of the social group or species or, even more mysteriously, if the chreod is geared to the finality of a predator, as in the case of the bird fascinated by the serpent.

APPENDIX 2

The universal model. The following metaphor gives the idea of this universal model.

In the last analysis, whence can life on our planet come but from the continuous flux of energy from the sun? The solar photons arriving in contact with the soil and seas are immediately stopped, and their energy abruptly degraded into heat; in this way the discontinuity of the earth and water surface is also a shock wave, a cliff down which the negentropy of the sun's rays falls. Now, life can be considered as some kind of underground erosion of this cliff, smoothing out the discontinuity; a plant, for example, is nothing but an upheaval of the earth toward the light, and the ramified structure of its stem and root is the same as that found when a stream of water erodes a cliff and produces a mound of debris. Plastids, veritable photon traps, are the minuscule orifices where this subterranean circulation begins. The energy stored in the noble form of chemical energy begins its slow decline. It flows underneath the cliff like a fluid, and its circulation echoes the inverted pyramid of the ecology of living beings. Each living species is a structurally stable singularity, a chreod of this circulation. As in hydrodynamics the energy of a turbulent regime flows from low-frequency oscillations toward higher frequencies, finishing in thermal chaos, so in life those with slow metabolism (plants) are the prey of the faster-metabolizing (animals).

NOTES

1. It is very natural to consider that aging is essentially the local evolution toward a more stable local state. As the global stability of the organism depends on the possibility of adapting to aggression through local qualitative variations, this easily explains how an increase in local stability leads to a decrease in global stability.

2. We might perhaps consider sexuality as arising from a smoothing of the self-reproducing catastrophe. For a complex being E, the topological separation of a bud E', isomorphic to the parent organism E, gives rise to a catastrophe whose complexity is proportional to that of E. This catastrophe can be smoothed into two intermediary steps by the intervention of sexuality. In the male, the spatial separation of the gametes is made easier both by their cellular and therefore small nature and also by the presence of the female, at the same time attractor and recipient; in the female, the topological separation of the infant from the maternal organism is facilitated by the fact that the infant has a different genetical constitution from the mother, and hence the mechanisms of immunological rejection can act.

3. Perhaps it is better to consider a virus as a self-destructive process of life taking on a living formalism, suicidal chreods of cellular metabolism.

4. "Les nombres me semblent exister en dehors de nous et en s'imposant avec la même nécessité, la même fatalité que le sodium et le potassium."

REFERENCES

1. J. von Neumann, *Collected Works*, Vol. VI, Pergamon, 1963 (The role of mathematics, p. 484).

CHAPTER 13

FROM ANIMAL TO MAN: THOUGHT AND LANGUAGE

13.1. A FUNDAMENTAL CONTRADICTION IN BIOLOGICAL REGULATION: THE PERSISTENCE OF THE SUBJECT AND PERIODICITY OF ACTIONS

A. The predation loop

The essential regulation constraint on an animal is feeding, which alone allows it to replace its loss of chemical energy and to restock its reserves. Now feeding implies predation, the presence of another living being (animal or plant), the prey, which is to be captured and ingested. The corresponding morphology is thus the capture morphology (Figure 13.1), whose simplest algebraic realization is given by the Riemann-Hugoniot catastrophe. As we want specifically to express the asymmetry between subject and object, between predator and prey, we associate to each living being (or, in linguistic terminology, each *actant*) a minimum of a potential on the space of internal variables. Then the spatial capture of one actant by another is to be interpreted as the capture of the basin representing the prey by the basin of the predator. This leads to representing the capture morphology as a typical section of the cusp by an oriented line L of the universal unfolding of the catastrophe (Figure 13.2). For example, on a line parallel to the Ou-axis, the point K where this parallel meets the discriminant curve $4u^3 + 27v^2 = 0$ denotes the catastrophe point where the prey is ingested by the predator. Now such an interaction is fundamentally irreversible. If we want to arrange things so that the animal can again be in a state of capture of a new prey, we must replace the oriented line L by a closed cycle C which restores the organism to its original state, a requirement of the periodicity of actions. Such a closed cycle must necessarily be centered on the origin, the organizing center of the catastrophe (Figure 13.3). However, if we lift this closed cycle to the space of internal variables, we find that, often having traversed arc *123* of C, the predator ***becomes its prey***. Thus the periodicity of the action implies an identification between predator and prey, a *confusion of actants*.

Faced with so paradoxical a conclusion, one might be tempted to reject the model; however, I think, to the contrary, that this difficulty of principle

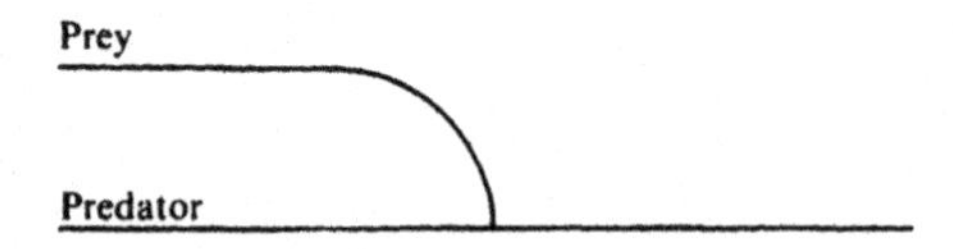

FIGURE 13.1. The capture morphology.

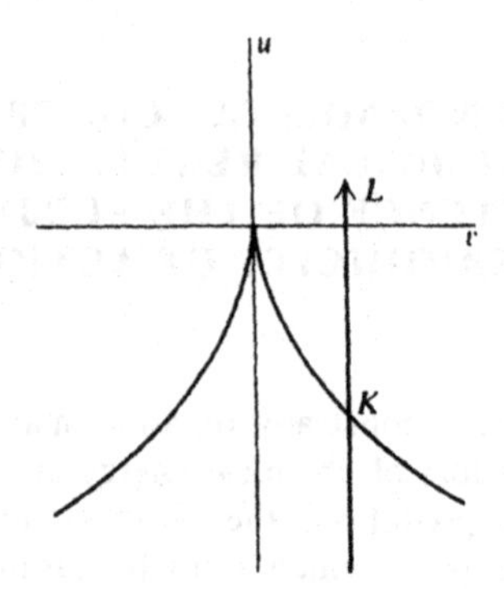

FIGURE 13.2

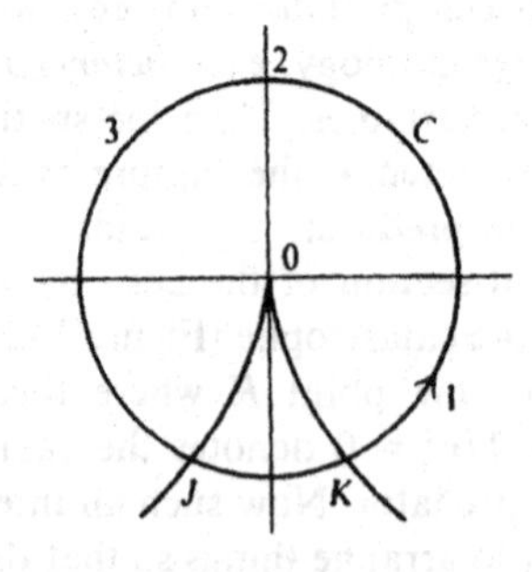

FIGURE 13.3. The predation loop.

must occur in all models. Even more, I think that the progressive elimination of this difficulty sheds unexpected light on many aspects of biological morphogenesis and on the origins of magical ritual in man (on *participation*[1] in the sense of Lévy-Bruhl). We can suppose that the spatiotemporal continuity of the organism, of the domain of space-time occupied by it, is the very basis of the unity of the organism. However, if we extend this requirement to the "semantic spaces" describing the internal properties of an organism, we might easily encounter situations that contradict this requirement of spatiotemporal connectivity. It could be that the essential function of the mind and cerebral organization is to overcome this contradiction. What does this mean?

The hungry predator having traversed arc *123*, is its prey until it reaches point *J*, where the cycle *C* meets the upper branch of the bifurcation cusp. There is a catastrophe at *J*, the *perception* catastrophe. If the predator *is* its prey before *J*, this means that the mind of the predator is dominated, alienated by the image of its prey. In some sense, the nervous system is an organ that allows an animal to be something other than itself, an *organ of alienation*. As soon as the external prey *p* is perceived and recognized by the predator, it becomes itself again, and it jumps from the surface corresponding to the prey to its own surface in an instantaneous *cogito*. At this moment, the chreod of capturing the prey is triggered, this chreod having several modalities according to the behavior of the prey (fight, flight, etc.); thus this is a motor chart focalized, at least at the beginning, not on the organism but on the prey. The ingestion catastrophe occurs at *K*; and a spatial smoothing of this catastrophe gives rise to the digestive tube, a tubular neighborhood of the trajectory of the prey in the interior of the organism of the predator (Figure 13.4).

How should we interpret this catastrophe which, at *J*, ensures the organic continuity of the predator despite the mental discontinuity? Consider the hysteresis loop associated with the folding defined by the predator-prey conflict (Figure 13.5); it has a simple interpretation in terms of energies. The *Ov*-axis denotes the amount of reserve chemical energy, and in the capture chreod the predator must reach into his energy reserves (on the lower sheet, decreasing *v*) in order to convey the prey *p* to the point

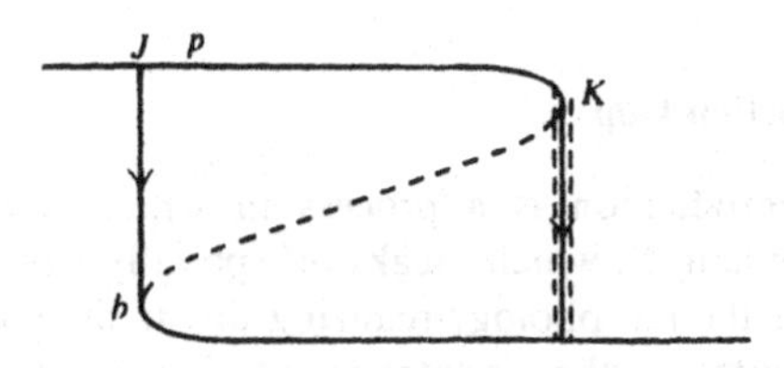

FIGURE 13.4. The digestive tube.

K of ingestion. We might regard the vertical arc *Jb* as the central nervous system in vertebrates; or, considering the stable sheets as rivers, the arc *Jb* might be interpreted as a capture process of the upper river by a particularly active affluent of the lower river. This capture catastrophe is none other than the neurulation producing the invagination of the nervous tube in the epidermis. After capture, the arc *bd* (Figure 13.6) will be a dead valley where the hydraulic circulation will become ramified, leading eventually to a generalized catastrophe; this corresponds to the dissociation of neural crests which will mainly contribute to the organogenesis of sensory organs. On the other side of this threshold *d*, there will be another generalized catastrophe *p* describing the spatial indeterminism of the position of the prey. Recognition of the analogy between the perceived form *p* and the typical form, the "genetic" form, initiates the perception catastrophe by a phenomenon of resonance, for how can we ever recognize any other thing than ourselves?

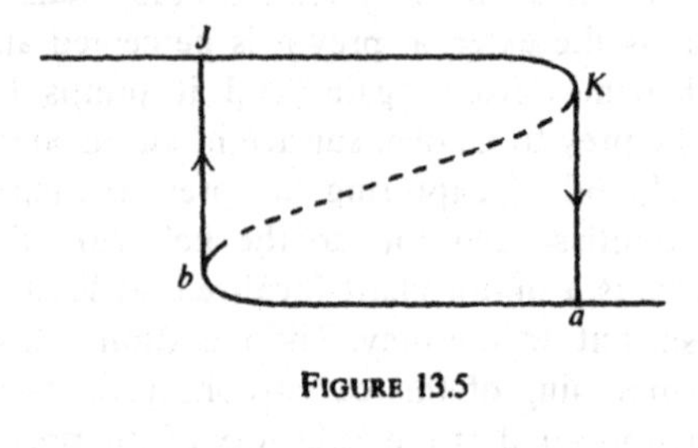

FIGURE 13.5

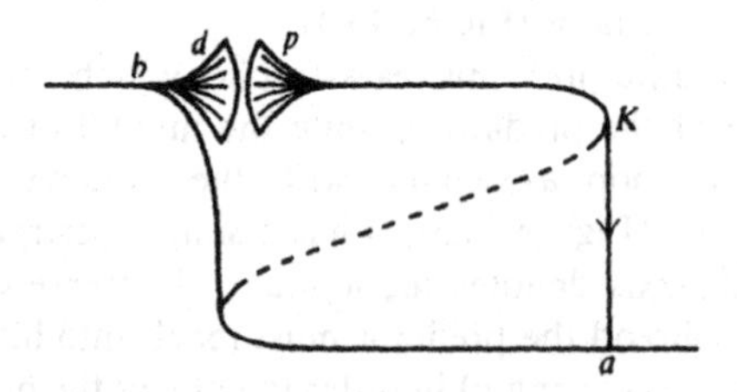

FIGURE 13.6. The dissociation of neural crests.

B. The reproduction loop

Biological reproduction is a process in which a parent organism *A* creates a descendant *A′*, which breaks off spatially from the parent (Figure 13.7). Therefore the morphology realizing this is the emission morphology obtained by reversing the orientation of line *L* of Figure 13.2 in the universal unfolding of the Riemann-Hugoniot catastrophe.

If again we form a cycle C, as in the previous case, we find this time that we arrive at a less paradoxical conclusion; the parent organism A perishes at point K (Figure 13.8), captured by its descendant A', and after traversing arc *123*, the descendant becomes a parent, as is normal after a lapse of a generation. Here the confusion of actants is natural, with daughter becoming mother after puberty. But if we want the parent to have several descendants in order to ensure the permanence of the subject, it is necessary to proceed as with the predation loop: we must guarantee continuity with the upper sheet by a process of capture (Figure 13.9).

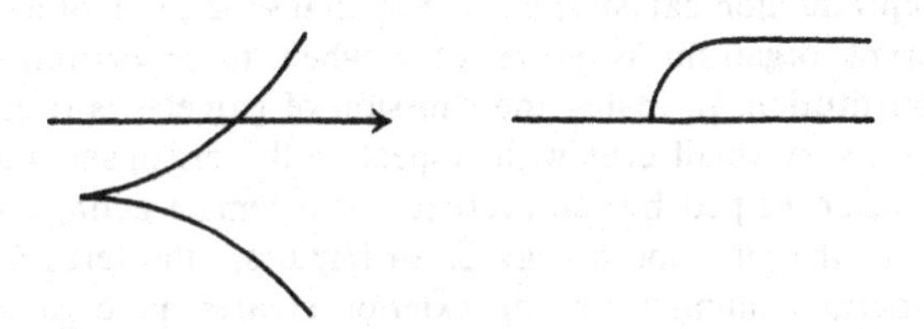

FIGURE 13.7. The emission morphology.

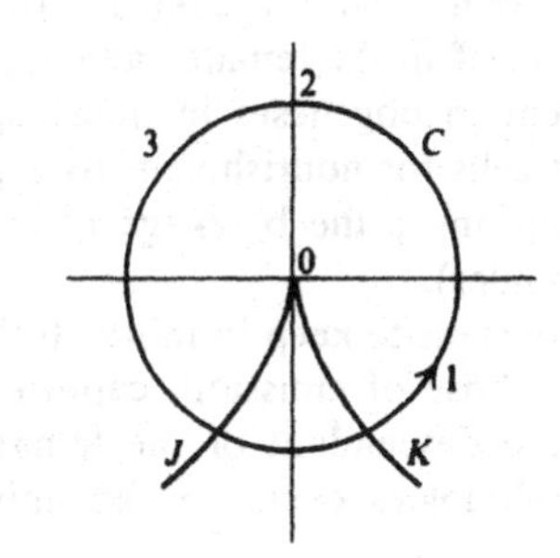

FIGURE 13.8. The reproduction loop.

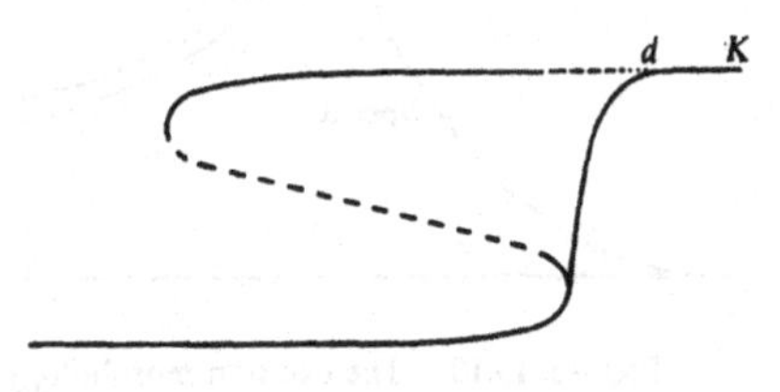

FIGURE 13.9

Hence the descendant sheet A' must perish at the threshold d in a generalized catastrophe, expressing the reduction of the corresponding tissue into a set of dissociated cells in the process of multiplication, and then the elimination of these cells from the organism by rejection toward the outside.

C. Sexuality

It is possible to give an interpretation of sexuality in this plan, if we suppose that one step in reproduction must necessarily comprise a capture process. The profound motivation of sexuality is probably not the exchange of genetic material, as is usually supposed, but rather the smoothing of the reproduction catastrophe. The spatial separation of a descendant from a parent organism is never easy when these organisms have a complex constitution. In males, the emission of gametes is relatively easy, since they are very small cells with respect to the organism, and also this emission is often helped by the presence of a female acting as attractor, conforming to the gift morphology: ⌶ In females, the fertilization of an ovum by sperm coming from the exterior creates an organism that is genetically and therefore metabolically different from the mother, thus allowing her to reject this organism by the formation of a shock wave (or an antigene reaction, biologically speaking). In this way the capture morphology manifests itself in the female gamete in relation to the sperm; it is also often apparent in oogeneses in many species where the future ovum engulfs adjacent cells for nourishment by a process of phagocytosis. Another probable symptom is the blockage of meiosis in the ovum (the formation of polar globules).

At any rate, it is important to keep in mind all the morphologies present in sexual interactions: those of emission, capture, gift, and excision (the male gamete excises the descendant of the female organism; see Figure 13.10). All these morphologies occur in the universal unfolding of the double cusp.

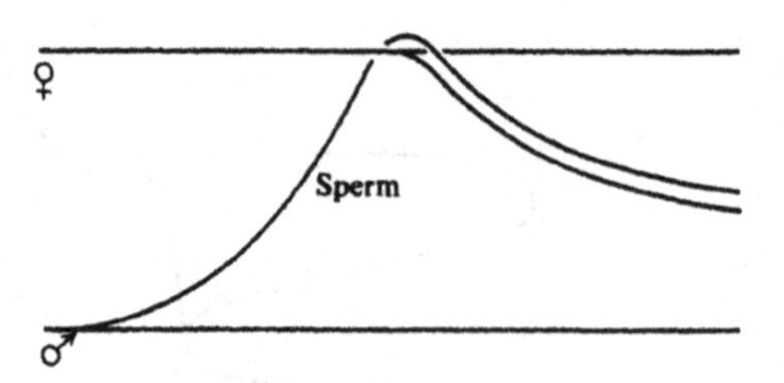

FIGURE 13.10. The excision morphology.

13.2. THE ANIMAL MIND

A. Genetic forms

It is striking to observe how efficiently the capture of prey is realized, even at the lowest levels of the phylogenic tree. This forces us to postulate that each animal has an organic chart of its mobility that enables it to control its own actual movements with remarkable precision so as to capture its prey or flee from its predators.[2] However, there is no reason to believe that this chart is permanent; it is more often associated with an object of fundamental biological importance (e.g., a prey or predator) and is focused on that object. We shall designate the innate forms, handed down in the endowment of the species and determining a well-defined motor action, by the name of *genetic forms*. Then, as soon as an external form is recognized as a genetic form, a perception catastrophe takes place, and the "ego" is recreated in an action, in the motor chreod (of capture or flight) that the genetic form projects onto the external form.

This poses an important theoretical problem: how can these genetic forms be represented, and on what space are they defined? Although it is not easy to reply to these questions, one thing is certain: it would be wrong to represent these forms as permanently fixed engrams, like prints on a photographic plate. In reality, they are defined dynamically, by a kind of never-ending embryology which extends into the motor chreod focused on the form. From this comes the fact that the metrical control of genetic forms is in general not very rigorous and often *supranormal releasers* trigger off the reflex more effectively than the normal biological form. Besides, the space on which these forms are defined can only exceptionally be identified with the outside space, and when the identification happens without external cause we say that there is a *hallucination*. Very probably this biologically disastrous phenomenon of hallucination is as rare in animals as in man, and it can be realized only in pathological states like those of sensory deprivation. In states of normal vigilance the permanent input of the senses into the mind represses the intrusion of genetic forms in the cycle of sense and motor activities.

Thus the ego of the animal is not, in principle, a permanent entity, any more than his vision of space is global. The *territory* of an animal is, in reality, an aggregate of local charts, each associated with a well-defined motor or psychological activity (areas for hunting, congregating, sleeping, etc.), and passage from one chart to another takes place through well-defined visual or olfactory markers. For certain animals some of these charts can extend over enormous distances, as with migrating birds, but here again the charts are centered on a territory and have a well-defined physiological vocation.

These, then, are the characteristic signs of the animal mind: impermanence of the ego, alienation by genetic forms, and decomposition of space into local charts, each associated with a partial ego. However, in the higher animals at least, there are mechanisms acting to remedy this fragmentation.

B. Animal in quest of its ego

One of the most evident manifestations of this impermanence of ego, still clearly apparent in man, is the circadian alternation of sleep and wakefulness. It is not unreasonable to see this cycle as a realization of the predation loop, synchronized to the alternation of day and night. In animals the wakeful period is appropriate for hunting—night time for nocturnal predators, and day time for diurnal; and sleep is then the ill-defined period covering the transition between the satisfied predator and the hungry predator identified with the prey, as we have seen. Then, beginning with reptiles, we see the appearance of a mechanism smoothing to some extent the transition from sleep to wakefulness: dreaming.

C. Dreaming

It is well known that the relative length of dream or paradoxical sleep increases as one climbs the phylogenic tree. It is natural to see this paradoxical sleep as a kind of virtual spatialization of the genetic forms; dreaming gives rise to a partial ego, without come-back to the dreamer, without substance or liberty—a veritable prey of its preys or its predators. Thus we could define dreaming as a constrained activity, dealing fictitiously with fictitious objects. As such, sleep permits a considerable temporal extension of the ego during this period of unconsciousness.

D. Play

Another factor having an important role in stabilization of the animal ego is play. Observe, for example, a young cat in the process of playing; it will behave as if attacking genuine prey in front of an object like a ball of wool or string, having only a distant morphological analogy with its prey. The animal is certainly not deceived; it has created a playful ego that is not at all disturbed by the lack of final reward when the pseudoprey is found to be inedible. Thus play is a dual activity to dreaming; whereas the latter is a constrained activity dealing fictitiously with fictitious objects, the former is a spontaneous activity practiced in reality on real objects. Of course play may follow rules: in fact, it is often a highly structured combination of rules. But the absence of effective reward makes it into a free activity that the animal may abandon or take up at any moment.

Also this predatory playful activity gives valuable experience in distinguishing between edible and inedible objects; this exploration goes on almost continuously in human babies between 8 and 11 months of age. In this sense wakefulness in animal and man is a state of continuous virtual predation. Every perceived object is treated as a virtual prey, but only objects of sufficiently promising form can initiate the capture process, while the others are weighed by perception, which in man behaves in this respect like a virtual hand. The etymology of *percipere* is "to seize the object continually in its entirety."

13.3. *HOMO FABER*.

A. Organs and tools.

The capture chreod typically includes two phases: a somatofugal phase of throwing out an organ to seize the prey, and a somatotrophic phase of carrying the prey, once seized, toward the mouth.

Roughly speaking, the use of tools corresponds to a smoothing, a *threshold stabilization*, between these phases. For example, when we want to gather fruit from a tree, it is often useful to grasp first the branch carrying the fruit, so that the characteristic of being prey extends from the fruit to the branch on which it is growing. Playful activity adds further to this extension. This stabilization at the extreme point of the somatofugal extension has brought about an organogenesis based on *opposition*: the opposition of thumb and index finger, the opposition between the two hands, and so forth. The *grip* archetype is probably only an organic realization of the hysteresis cycle that we met in the study of a bone joint (Section 9.6.B); the vertical segments correspond to the fingers, the upper sheet is an arbitrary object, and the lower sheet is the wrist (Figure 13.11). The hand appears as a universal rotula with an exogenous tibia of arbitrary form.

Man, the omnivore, must kill and tear up his animal prey; therefore outside objects are to be considered, in play, not only as prey but also as tools. This has implied a global transfer of organogetic fields into fields of tool manufacture. In addition to the fields associated with the elementary catastrophes (such as splitting, pricking, and piercing, etc.; see the table in Figure 13.12), there are fields, such as polishing and planing, associated with the conflict between two solid objects that we met in the epigenesis of a bone. Each tool has it own space T, in which it is created by a kind of embryology; this space has its own epigenetic gradients unfolding the functional catastrophe that the tool provokes. Thus almost all everyday objects (tools, furniture, etc.) have a bilateral symmetry, just like animals, and a proximal-distal gradient. In use, the characteristic space T of the tool is canonically identified with the local motor chart of the organism and often realizes an important extension of it.

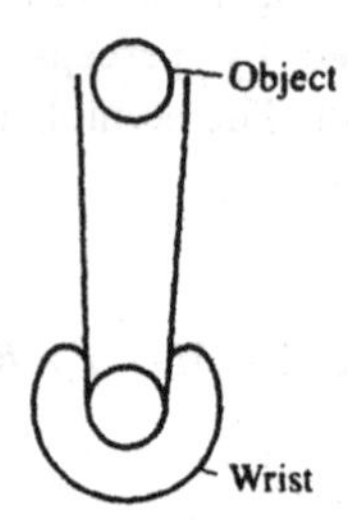

FIGURE 13.11. The grip.

B. An example: The construction of a club as a chreod

The *projectile chreod* is known throughout the animal kingdom; it consists of throwing an object in order to hurt an enemy. To begin with, the projectile is often produced by an appropriate organ of the animal that throws it (in a hydra, its nematocyst; in a cuttlefish, its ink; some monkeys bombard their attackers with their conveniently produced excrement). This chreod has been highly specialized in man by the use of a club; he tries to split open his enemy's head and provoke a swallow-tail catastrophe, and this requires the projectile to have a typical beveled form. The mental picture of this catastrophe to be wrought on the adversary creates a secondary field, that of the manufacture of the club. This process can be described as follows. Unlike those of many fields found in animals, the initial data, namely, the form of the block to be fashioned, are very varied; from the variety of different forms of the basic material, the worker must produce a unique form. Thus the construction of the club is a highly attracting chreod and demands the continual presence in the worker's mind of the required form. In this sense it is the earliest example of a rigorously finalist process in which instantaneous direction at each moment is the result of a conflict between the end required and the present condition of the material being worked. Such a process can be described as follows.

When the worker takes up the block of rock B that he has chosen to work, he compares it in his mind with the required result O. This comparison produces a mental shock wave which stimulates a work process, for example, to file from B along a face f. When this face is sufficiently flat, a

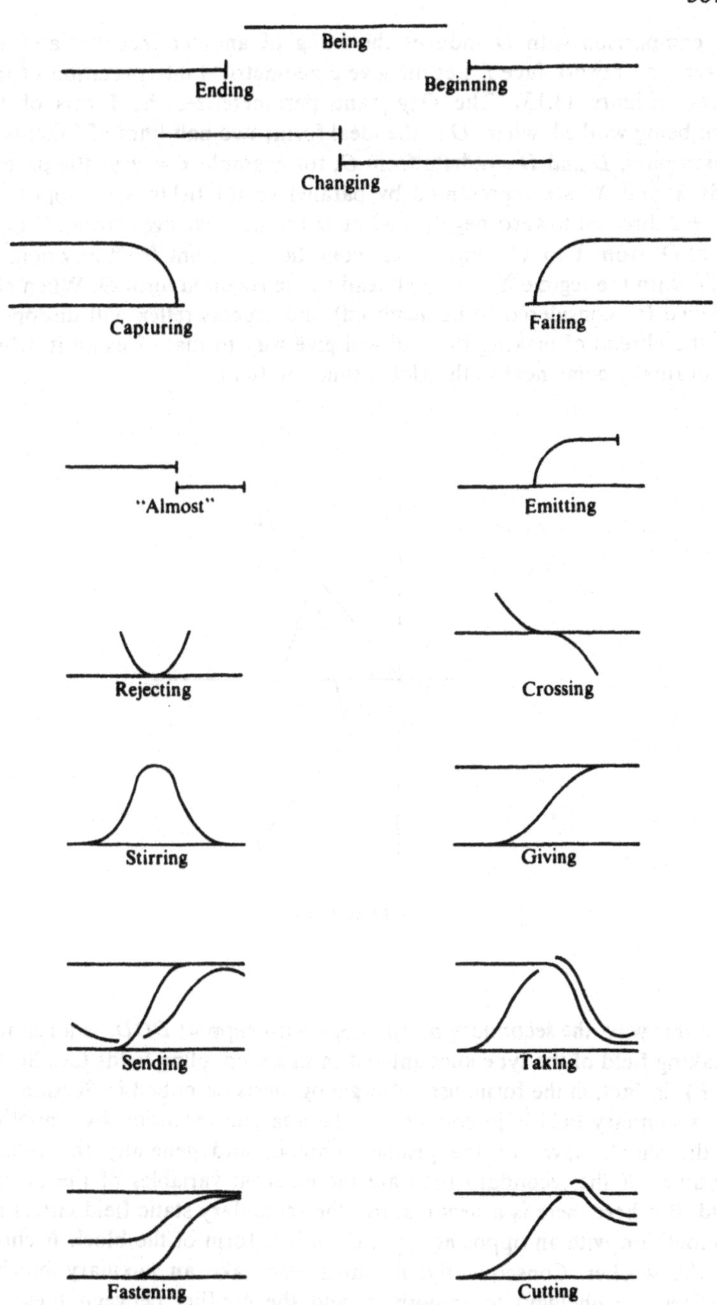

FIGURE 13.12. The table of archetypal morphologies.

new comparison with O induces the filing of another face f'; later the worker may rework face f. Let me give a geometrical interpretation of this process (Figure 13.13). The Oxy-plane parameterizes the forms of the block being worked, where O is the ideal form; two half-lines of functional catastrophes, D and D', radiate from O, for example $x = \pm y$; the process fields X and X' are represented by parallel vector fields with slopes -2 and $+2$ directed toward negative x; at D the field changes from X' to X, and at D' from X to X'. Under these conditions, starting from any point in DOD' with the regime X or X' will lead to the required form O. When O is achieved (or considered to be achieved), the process reflex will disappear, and the chreod of making the tool will give way to that of using it, which will normally come next in the global function field.

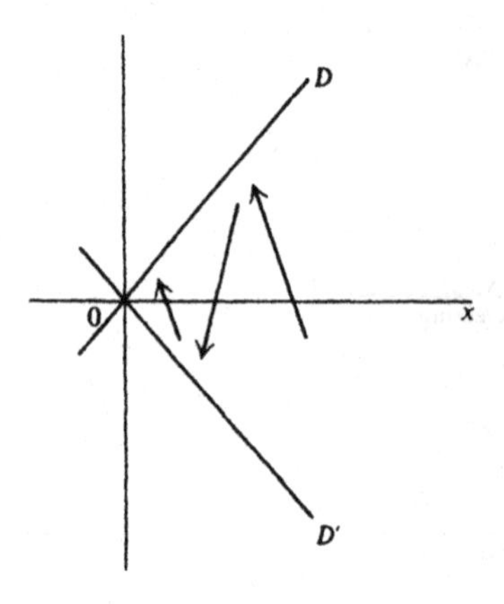

FIGURE 13.13

In this way, the secondary morphology with support DOD' is a funneled breaking field of the type encountered in mean coupling fields (see Section 6.2.B); in fact, in the formalism of organogenesis described in Section 10.4, this secondary field is in some sense the analytic extension by smoothing of the shock waves of the primary chreod, and generally the internal variables of the secondary field are the external variables of the primary field. But here there is a new feature: the secondary static field enters into competition with an opposing field, the initial form of the block B chosen by the worker. Consequently, he must also take an auxiliary block A (tertiary morphology) to smooth B, and the conflict between these two fields produces the funneled configuration of the secondary static field (and observe that the bounding shock waves D and D' give, by analytic continuation, the edges of the wound to be inflicted on the adversary). This gives a secondary catastrophe whose external variables characterize the relative position of B with respect to A and whose nonpunctual organizing center has for support the shock waves D and D'. The tertiary field formed by the smoothing of these shock waves is a static field, indicating the attraction by B of an auxiliary block A. To sum up, the set of the motor fields of the hands forms a semantic model of at least twelve dimensions, the dimension of a product of two copies of the group of Euclidean displacements.

It is the same hand that fashions and brands the club, and this functional subordination between chreods is evidence of a syntax already almost as refined as that of language. In particular, the formal process of organogenesis (Section 10.4), transforming a mean coupling field on an external space U into a static field with internal space U, is already realized in mental activity.

These considerations raise the question of how it is that the human mind can achieve this complicated architecture, this hierarchy of fields, of which animals appear incapable, even though the brains of the higher vertebrates are so little different, anatomically and physiologically. Personally, I believe that this capability depends on a topological discontinuity in the kinetics of nervous activity, that in the human brain there is a device simulating self-reproducing singularities of epigenesis allowing, in the presence a catastrophe with internal space Y and unfolding U, the unfolding space to be mapped back into the internal space, causing the confusion of internal and external variables. Such a device would not require any great modification of the anatomical or physiological structures.

13.4. *HOMO LOQUAX*

A. The double origin of language

The appearance of language in man is a response to a double need:

1. For a personal evolutive constraint, aiming to realize the permanence of the ego in a state of wakefulness.
2. For a social constraint, expressing the main regulating mechanisms of the social group.

The first constraint fulfills the need to virtualize predation. Man in a state of wakefulness cannot pass his time like an infant of 9 months, seizing every object and putting it in his mouth. He has greater things to do; he must "think," that is, seize the things lying between exterior objects and genetic forms, namely, *concepts*.

The second constraint expresses the need for the social body to disseminate the information necessary for its survival, like the presence of nearby enemies or prey. Language then works as a *sense relay*, allowing one individual X to describe to another Y what he, X, is in a position to see but Y, less well placed, cannot see.

The social constraint will create structures in the most unstable zone of the individual by an effect of interaction between hierarchical levels of organization; these structures will be on the shock waves separating sleep and wakefulness, or genetic and spatial forms, and these shock waves will exfoliate. A baby is equipped at birth with a stock of sensory-motor schemas, genetic forms that manifest themselves in the so-called *archaic* reflexes. Later, toward 6 months, these schemas undergo a kind of melting, a generalized catastrophe, coinciding with the onset of infantile babbling. This babbling seems like the need to expel by the process of articulation some of these alienating genetic forms, clearly a playful emission of forms, not a capture of forms. Recall that a genetic form is not fixed, but rather is equipped with mechanisms of self-regulation analogous to those of a living being; a concept occurs by superposition, by projecting the regulation schemes of the subject onto a spatial form, an exterior image. By a geometrical analogy, we might say that the concept forms by exfoliation from the spatial image, where the normal coordinate along which the exfoliation takes place is associated with the direction of an articulated emission whose phonetic structure has little to do with the genetic form that gave its regulation to the concept. (This is the Saussurian idea of the arbitrariness of sign.) This association forms by habituation, with the emission of the sound occurring with the (playful or biological) use of the corresponding object. On the other hand, the laws of combination, the syntax, of these words are not arbitrary since they are imposed by the semantic interactions between the concepts, themselves defined by the regulation schemes of the subject and thus of the concepts.

If a child spends the time between 1 and 3 years of age without other human verbal contact, the articulatory emission catastrophe (the babble) rapidly degenerates into the production of a few crude sounds (the verbalization of "wolf children"). The exfoliation of the semantic support space of the concept is inhibited by the incoherence or absence of the sounds associated with the object; this results in the mental retardation or idiocy to which these children are condemned. Those who have studied these wolf children have observed their extreme reactions to some noises like the cracking of a nut. There is little doubt that their minds are still dominated by a small number of alienating forms of genetic origin. Man gets rid of these alienating forms by giving them a name and so neutralizing their hallucinatory powers by fixing them on a semantic space distinct from space-time.

If man has escaped from the fascination of things through the use of language, he remains under the fascination of action incorporated as the grammar of the language. (a verb conjugates, etc.). Only the Oriental philosophers have tried to withdraw the subject of this fascination by reducing virtual action to pure contemplation.

B. Syntax and archetypal morphologies

It is well known that all speech can be decomposed into elementary phrases, each phrase being characterized by the fact that it contains precisely one verb, ignoring here the difficulties (about which specialists are still debating) of the definition of the traditional grammatical categories: noun (substantive), adjective, verb, preposition, and so forth. The fact that any text can be translated from one language to another confirms the belief that these categories are almost universal. Now, given a spatiotemporal process which we are to describe linguistically, are there any formal criteria relating to the intrinsic morphology of the process that enable us to predict the decomposition into phrases?

To this end, we must start by "objectively" describing a spatiotemporal morphology. In fact every linguistically described process contains privileged domains of space-time bounded by catastrophe hypersurfaces; these domains are the *actants* of the process, the beings or objects whose interactions are described by the text. As a general rule, each actant is a topological ball and hence contractible; this is the case, for example, for animate beings. At each moment t we contract each actant to a point, and when two actants interact this implies that their domains come into contact in a region of catastrophe points which we also contract to a point of intersection of the lines of the two contiguous actants. In this way we associate a graph with every spatiotemporal process.

I then propose that the total graph of interactions describing the process can be covered by sets U_i such that the following conditions are met:

1. The partial process with support U_i is described by one phrase of the text.
2. The interaction subgraph contained in U_i belongs to one of the sixteen archetypal morphologies of the table of Figure 13.12.

In principle each of the morphologies is generated by a construction of the following type. In the universal unfolding of each elementary singularity, take a ray emanating from the organizing center and having contact of maximum order with the discriminant variety, and then displace this ray parallel to itself to avoid the confusion of actants at O. Then lift this ray to the space of internal variables, with each actant being represented by the basin of a minimum, to give the corresponding interaction. It may be necessary to cheat a bit by bending the rays in order to ensure the permanence of the actants to times $t = \pm\infty$. Furthermore, certain verbs called iteratives require repetitions of an action, and in this case the basic cell is described by the archetypal morphology. The occurrence of excision morphologies, characteristic of sexual reproduction, is noteworthy.

This theory of the spatial origin of syntactical structures accounts for many facts, for example, the restriction to four actants in an elementary phrase and the origin of most of the cases in a language with declension: the nominative, for the subject; the accusative, for the object; the dative, with verbs having the morphology $\mathcal{I}$, the instrumental, with verbs having the excision morphology of cutting, or ablative. The only classical case that cannot be interpreted by this tableau is the genitive, which is an operation of semantic destruction, dislocating a concept into its regulating subconcepts in a kind of inverse embryology (see [1]).

C. The automatisms of language

It remains to account for the palpably automatic character of the formation of syntactical structures. To this end we suppose that each main verb type (each archetypal morphology) is represented mentally by an oscillator, which vanishes at the organizing center at a certain critical energy level E_c. When $E < E_c$, it typically describes a cycle in the universal unfolding with almost a stationary point on a sheet or a branch of the discriminant variety; the corresponding arc describes one of the typical sections generating the associated archetypal morphology. For example, in a verb of the capture morphology there will be a cycle C in the universal unfolding of the Riemann-Hugoniot catastrophe with the stationary point K on the capture branch. When $E = E_c$, the radius of C tends to zero and the cycle vanishes at the organizing center O. This gives the unstable potential $V = x^4$ in the internal space, and then the situation evolves toward a generic situation with the representative point in the uv-plane following a curve close to the capture branch K inside the cusp. This results in the formation of two dummy actants corresponding to the minima of potential, and then these dummy actants will play an instrumental role in the capture of the concepts of the meaning. Each dummy actant excites the concept and reduces it to its splitting into image + word; finally the dummy actant unites with the word and is emitted as a word.

Such a process then describes the emission of a phrase of type SVO, subject-verb-object. First the vanishing of the cycle C leads to the emission of the verb V, and then the packet of two dummy actants liberates the subject-object; each of these actants interacts with the corresponding concept, which it maps on the subject image axis while the dummy actant is re-emitted as a word (substantive). The total morphology is that of the verb "to take" (see Figures 13.12 and 13.14). Thought is then a veritable conception, putting form on the dummy actant arising from the death of the verb, just as the egg puts flesh on the spermatizoid; thus thought is a kind of permanent orgasm. There is a duality between thought and language reminiscent of that which I have described between dreaming and play: thought is a virtual capture of concepts with a virtual, inhibited, emission of words, a process analogous to dreaming, while in language this emission actually takes place, as in play.

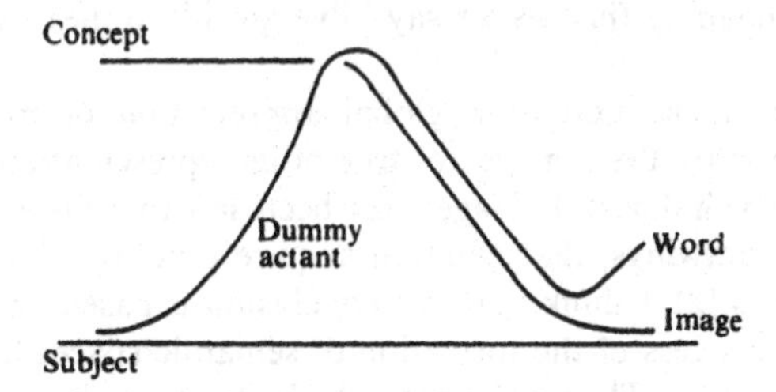

FIGURE 13.14. The take morphology.

More details of the problem of grammatical categories and the question of the order of words in a sentence are given in Appendix 2.

13.5. THE ORIGIN OF GEOMETRY

In the state of wakefulness, of continuous virtual predation, man can reach out to any point sufficiently close to his body; and these voluntary movements give rise to proprioceptive sensations which permit a rigorous metrical control of the displacement of limbs far more precise than the control of induced movements. Thus we can assert that, very early in human development, there are local charts associated with the organism that describe all the metrical structure of Euclidean space.

The use of tools allows us to extend this chart by adjoining to it the chart associated with the tool. Even better, the disappearance of alienating forms, which center the chart on an external prey, enables these charts to be adjoined indefinitely to each other. For example, the operation of measuring, by adjoining a yardstick to the end point of a previously measured length, is typically an extension of a chart by the chart associated with the yardstick. This is how geometric space is made up: a space of pure unmotivated movement, the space of all the play-movements of which we feel we are capable.

It is striking to observe how most spaces, even in pure mathematics, are not homogeneous but are well and truly endowed with a base-point, an origin, the mathematical equivalent to a subject in a state of continual predation in the space. Even the typical picture of Cartesian axes *Oxy* (Figure 13.15) conjures up irresistibly a mouth about to close on the typical point *p*, the prey. It is surely not by chance that the most of the catastrophes of organic physiology are idempotents ($T^2 = T$) like projection onto an axis, measurement in quantum mechanics, or the capture by the mind of a meaning (just as we say "I've got it" to mean that we have understood).

Although the acquisition of a global appreciation of metrical space appears early in man, the same is not true of its representation by pictures. Although the psychologist J. Piaget has been led to believe, through an examination of drawings, that children acquire topological structures before metrical ones [2], I think that this conclusion is based on a misunderstanding of the process of the formation of semantic spaces by exfoliation of the spatial image. Thus a 3-year-old child knows how to recognize whether his piece of cake is smaller than his neighbour's; but when he tries to draw, he clearly wants above all to draw objects and not the ambient space in which they lie, which is very difficult from him. Now each object is generally seen from a privileged direction, giving it a simple apparent contour and revealing best the three-dimensional form of the object. Generally this direction is perpendicular to the plane of bilateral symmetry for an object, like an animal or a tool, having such a symmetry; only man is represented first face-on, rather than in profile. For a complicated object, like a car, there are subobjects that may be necessary for recognition, like wheels. In a global chart of the object, the subobjects may be correctly located but drawn with their own privileged perspective, thus explaining the phenomenon of flattening that Luquet has described in many designs by children (e.g., see Figure 13.16). In such cases there is a conflict between the global perspective and the requirement of semantic dominance of an object over its subobjects. The relationships of surrounding, touching, and so on, which are considered by Piaget as "topological," are in fact semantic relationships of dominance between concepts, relationships expressed precisely by the genitive case (the wheels of the car, the eyes of the head, etc.).

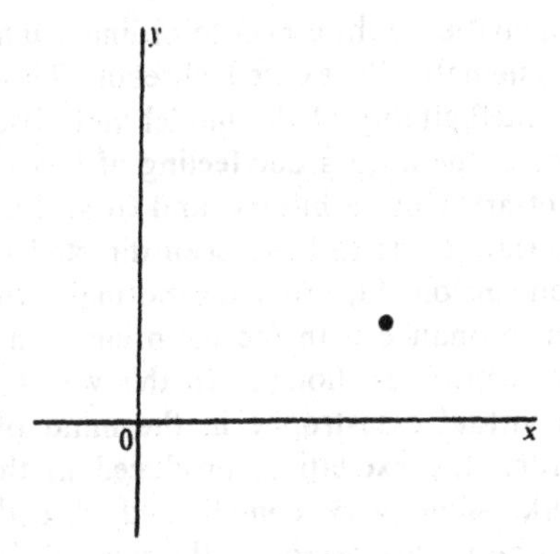

FIGURE 13.15. Cartesian axes.

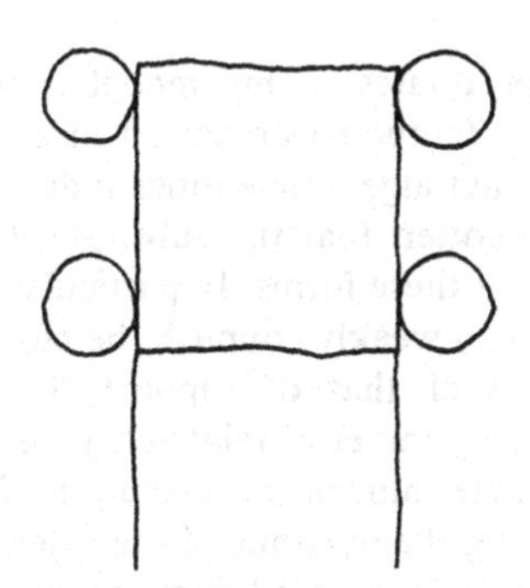

FIGURE 13.16. Flattening (rabattement) in children's drawing, according to Luquet.

13.6. THREE IMPORTANT KINDS OF HUMAN ACTIVITY

A. Art

Can artistic activity appear in our semantic models? Of course this is so with poetry, which uses ordinary language, and music, which requires the use of a discrete notation. For visual arts, however, the case is less clear, even though the recurrence of typical motifs and forms suggests that the idea of the chreod and, with it, semantic models may be valid; but what characterizes these activities with respect to ordinary language is the use of concepts, that is, systematically excited chreods. This has the effect of harming the global intelligibility of the model and obscuring or removing the sense. From where, then, does our feeling of beauty come? From the idea that the work of art is not arbitrary, and from the fact that, although unpredictable, it appears to us to have been directed by some organizing center of large codimension, far from the normal structures of ordinary thought, but still in resonance with the main emotional or genetic structures underlying our conscious thought. In this way the work of art acts like the germ of a virtual catastrophe in the mind of the beholder. By means of the disorder, the excitation, produced in the sensory field by looking at the work, some very complicated chreods (of too great a complexity to resist the perturbations of the normal thought metabolism) can be realized and persist for a moment. But we are generally unable to formalize, or even formulate, what these chreods are whose structure cannot be bent into words without being destroyed.

B. Delirium

One of the basic postulates of my model is that there are coherent systems of catastrophes (chreods) organized in archetypes and that these structures exist as abstract algebraic entities independent of any substrate, but it must not be forgotten that the substrate does have a part fundamental in the dynamic of these forms. In particular, if the substrates A and B are independent or only weakly coupled, the morphology of support A is practically independent of that of support B. During normal mental activity, there are a large number of relatively independent substrates, each undergoing its own determinism or evolution. In this respect mental activity is only simulating the dynamic of the external world, which admits many independent or weakly coupled domains, made possible in particular by spatial separation. But such a separation of substrates inside a unique functional system as highly interconnected as the nervous system is very difficult to maintain; it is unstable and can break down under crude physiological influences (fever, drugs, etc.). Then the coupling interactions between two substrates A and B can increase to the extent that an archetype defined on A can evolve on B and upset the normal evolution of B's own dynamics; furthermore, this mixture of substrates destroys the more refined chreods, whose organizing centers have high codimension, in favor of more primitive fields that are more stable and more contagious. This will give rise to a syncretistic mental dynamic with oversimple structures—what is usually called *delirious thinking*.

In this connection, a good doctrine in the use of analogies in science remains to be established. Local isomorphisms of dynamical situations over independent substrata are, I think, very frequent, and many of them have yet to be recognized; because they are so frequent, however, these analogies have neither the weight nor the significance that the incautious might be tempted to attribute to them. It is an enormous step from noticing the presence of isomorphic morphological accidents on different substrates, to establishing some fundamental coupling between these substrates to explain these analogies, and it is precisely this step that delirious thinking takes. If some of my arguments, particularly in biology, have seemed to the reader to lie on the boundaries of delirium he might, by rereading, convince himself that I have, I hope, at no point made this step.

C. Human Play

In play the mind is given over to an eminently combinatorial activity; it constructs and uses a semantic model according to rules that it knows and respects. In the simplest case the rules are such that the strategy is determined; there are only a finite number of possible evolutions, which can be considered in turn. Frequently, however, the rules are not powerful enough to determine the system completely, and play appears as a kind of artistic activity in which the player, motivated by some esthetic sense, tries to form the most attractive and effective moves. But playing games is fundamentally different from artistic activity in that the set of rules must be consciously kept in mind. It is completely incompatible with delirium because it requires a mental domain kept completely free and independent of any external coupling, in which the mind may realize high-codimensional chreods; thus formalized, or axiomatic, thought can be considered as a game whose rules, forming the organizing center, are codified as a system of axioms. In this way, formal thought can be regarded as magic that has been psychoanalyzed and made conscious of its organizing structures.

Playful activity, left to itself, is not slow in creating gratuitous, worthless examples of semantic models with no other semantic realization than their own combination, nor is it slow to decline into trivia. Although some of modern mathematics is perhaps guilty of this charge, it is no less true that mathematical activity among mankind has essentially been inspired by reality and finds there its constantly renewed fecundity. The axioms of arithmetic form an incomplete system, as is well known. This is a happy fact, for it allows hope that many structurally indeterminate, unformalizable, phenomena may nevertheless admit a mathematical model. By allowing the construction of mental structures simulating more and more closely the structures and forces of the outside world, as well as the structure of the mind itself, mathematical activity has its place in the warp of evolution. This is significant play *par excellence* by which man can deliver himself from the biological bondage that weighs down his thought and language, and can assure the best chance for the survival of mankind.

13.7. THE STRUCTURE OF SOCIETIES

We are now going to outline roughly the structures of animal and human societies as metabolic forms. A society is a metabolic form first and foremost because it survives the individuals that constitute it; there is a permanent flux of constantly renewed individuals assuring the continuation of the social form. Moreover, there is an interaction between the individuals maintaining the stability of this form. This interaction often materializes as a circulation of complexity, of information, through the social body.

A. Basic Types of Society

There are two basic types of society.

1. The military society. Here each individual occupies a specified position and regulates his own movement so that the global form of the society is preserved, as well as his position within the society. It is clear that global invariance of the spatial body requires a permanent interaction of each individual with the individuals surrounding him. As the circulation of information, considered as a fluid, must be structurally stable, the simplest process to give this effect is a gradient circulation: here a positive function u is defined on the social body, called authority, zero at the boundary, and each individual is constrained to control his movement toward the closest individual on the trajectory of grad u in the direction of increasing u. This function u must have at least one maximum, and the individual at this point is the chief, because he takes orders from no one. Because delays in the transmission of orders may have a disastrous effect on global stability, particularly in troubled times when very rapid changes of behavior are necessary, u can have no other critical point apart from the single maximum personified by the chief. Thus the social body is a ball under monarchical rule.

As an application of this proposition, we see that most animal groups with invariant structures are topologically balls, generally governed by a chief, for example, flocks of birds, shoals of fish, or herds. However, the existence of a chief is not always obvious: the queen is not the director[3] in a hive of bees, but more complicated morphogenetical fields are in operation. Another example is slime molds.

Although this is the simplest structurally stable structure, it is not the only one. Military societies with many, or even with no chiefs are possible, but then the social body must be at least a three-dimensional manifold in order to admit a structurally stable ergodic field without singularity.

2. The fluid society. Here the typical example is a cloud of mosquitoes: each individual moves randomly unless he sees the rest of the swarm in the same half-space; then he hurries to re-enter the group. Thus stability is assured in catastrophe by a barrier causing a discontinuity in behavior. In our societies this barrier is fixed and doubly realized by the conscience of the individual and by the laws and repressive organisms of the society, and so our societies are of an intermediary type: they are not rigorously fluid, for they are stratified into social classes separated by shock waves that are difficult for an individual to cross. The classical analogy between the society and the individual organism is well known: the distinction between primary (production), secondary (distribution), and tertiary (direction) is echoed by the three fundamental layers of an embryo, but societies, unlike individuals, rarely reproduce themselves by budding (swarming). Marxism, which would explain the structure and evolution of societies solely in economic terms, is equivalent to Child's metabolic theory of embryology and suffers from the same simplifications. A typical character of social morphogenetical fields is that they modify the behavior of individuals, often in a lasting and irreversible way, as in the case of those who are persuaded to give all, including their own lives, for the preservation of the global social form, considered as supreme. This is an effect, without parallel in inanimate nature, that has disastrous consequences as far as the continuance of social injustice is concerned.

B. Other Aspects of Societies

1. Money. As a token for bargaining, money circulates in the opposite direction to goods and services and therefore tends to move away from consumers and accumulate at the producers. But as it must follow a closed path in the social body, it is necessary for an artificial mechanism to carry it in the opposite direction. Such a mechanism can depend only on the structure of society, on authority; in this way money is imbued with the gradient of authority (grad u) by the process of taxes. When it arrives at the summit, it is divided up by the chief, who can, in a permanent (or at least annual) catastrophe, share out the portions where necessary; in fact, this power is one of the essential means of government. Thus, in all societies, the gradient of production and the gradient of authority have a tendency to organize themselves in an antagonistic manner and so bring about a sufficiently stable circulation of money.[4]

2. The mind of a society. In this respect we might ask whether a social group acquires a "mind" that could have an autonomous existence. It seems that the social mind has a fragmentary character very similar to that of the animal mind: society finds its identity only in the face of an urgent threat, like war, where its existence and stability are threatened; similarly, the spatial conscience of the society has a local character, focalized on certain threatened zones. On the other hand, one may see in large collective displays such as fêtes and celebrations an activity homologous to dreaming in the individual, a virtual manifestation of genetic "social" forms. Individual semantics is perhaps another example.

The view outlined above is basically pessimistic, for it shows that social injustice is inextricably bound up with the stability of the social body. Personally I believe that the only way of reducing oppression is to stop giving an ethical value to social forms, in particular, to nations. The celebrated dictum of Goethe, "Better injustice than disorder," can be justified only inasmuch as disorder may generate even worse injustices; but this danger is not likely if the members of a society have achieved sufficient moral values for them not to exploit a temporary lapse of authority to their own profit. In this case, a very labile situation, with a fluctuating authority, has the best chance of providing a regime optimal for each member.

It is tempting to see the history of nations as a sequence of catastrophes between metabolic forms; what better example is there of a generalized catastrophe than the disintegration of a great empire, like Alexander's! But in a subject like mankind itself, one can see only the surface of things. Heraclitus said, "Your could not discover the limits of soul, even if you traveled every road to do so; such is the depth of its form."[5]

13.8. CONCLUSION

Here, before the final discussion, is a summary of the main points that have been made.

A. Summary

1. Every object, or physical form, can be represented as an *attractor* C of a dynamical system on a space M of *internal variables*.

2. Such an object is stable, and so can be recognized, only when the corresponding attractor is *structurally stable*.

3. All creation or destruction of forms, or *morphogenesis*, can be described by the disappearance of the attractors representing the initial forms, and their replacement by capture by the attractors representing the final forms. This process, called *catastrophe*, can be described on a space P of *external variables*.

4. Every structurally stable morphological process is described by a structurally stable catastrophe, or a system of structurally stable catastrophes, on P.

5. Every natural process decomposes into structurally stable islands, the *chreods*. The set of chreods and the multidimensional syntax controlling their positions constitute the *semantic model*.

6. When the chreod C is considered as a word of this multidimensional language, the meaning (*signification*) of this word is precisely that of the global topology of the associated attractor (or attractors) and of the catastrophes that it (or they) undergo. In particular, the signification of a given attractor is defined by the geometry of its domain of existence on P and the topology of the regulation catastrophes bounding that domain.

One result of this is that the signification of a form C manifests itself only by the catastrophes that create or destroy it. This gives the axiom dear to the formal linguists: that the meaning of a word is nothing more than the use of the word; this is also the axiom of the "bootstrap" physicists, according to whom a particle is completely defined by the set of interactions in which it participates.

Let us now consider the prejudicial question of experimental control.

B. Experimental control

Are these models subject to experimental control? Can they give experimentally verifiable predictions? At the risk of disappointing the reader, I must answer in the negative; this is an inherent defect of all qualitative models, as compared with classical quantitative models. When the process being studied is entirely within a chreod C, all that the experiment can do is to confirm the stability of the chreod; when the process has several chreods $C_1, C_2, \ldots$ separated by indeterminate zones, the process is by definition structurally unstable, and no individual prediction is possible. The only benefit to be drawn from the model is a statistical study of the morphologies presented by a *set* of processes of the type under consideration; this is the method used in quantum mechanics and often in quantitative biology. In the case of a single chreod, it might be possible to construct, by internal analysis of the chreod, a quantitative model; but for this the dynamical properties of the substrate must be well understood—something that is rarely possible, except in the study of shock waves in hydrodynamics, where some partial progress has been made. In general, the way is blocked by the nonexistence of a quantitative theory of catastrophes of a dynamical system; to deal with this it would be necessary to have a good theory of integration on function spaces, and we saw in Section 7.3 the difficulties of such a theory.

The strict empiricist, faced with this deficiency, will tend to reject these models as speculative constructions, devoid of interest; and, as far as present scientific progress is concerned, he is probably right. On a larger scale, however, there are two reasons that might commend them to the scientist.

The first reason is that *every quantitative model first requires a qualitative isolation from reality* in setting up an experimentally reproducible stable situation. We take the main divisions of science, the taxonomy of experience, into physics, biology, chemistry, and so forth as given a priori, a decomposition bequeathed on us almost unconsciously by our our perception and used by every scientist, no matter who he is, rather like Monsieur Jourdain speaking prose. Should it not be of interest in this situation to reconsider this decomposition and integrate it into the framework of an abstract general theory, rather than to accept it blindly as an irreducible fact of reality?

The second reason is our ignorance of the limits of quantitative models. The enormous successes of nineteenth century physics, based on the use and exploitation of physical laws, created the belief that all phenomena could be justified in a similar way, that life and thought themselves might be expressed in equations. But, on reflection, very few phenomena depend on mathematically simply expressed ("fundamental") laws: scarcely three, namely, gravitation (Newton's law), light, and electricity (Maxwell's law). Their simplicity is only apparent, and only expresses how gravitation and electromagnetism are intimately connected with the geometry of space, the result of a statistical effect of a large number of isolated, independent, small phenomena. The situation is different at the level of quanta; the fundamental reasons for the stability of matter are still unknown, and the stability of the proton remains unexplained. Quantum mechanics, with its leap into statistics, has been a mere palliative for our ignorance. Furthermore, even when a system is controlled by explicit laws of evolution, it often happens that its qualitative behavior is still not computable and predictable; as soon as the number of parameters of the system increases, the possibility of a close calculation decreases—what Bellman [4] has called the curse of dimensionality. Those who sell electronic gadgetry would have us believe that the computer age will be a new era for scientific thought and humanity; they might also point out the basic problem, which lies in the construction of models. Since Newton's proud cry, "Hypotheses non fingo," it has been hoped that a happy intuition, a lucky guess, would be sufficient to reveal the fundamental laws underlying everything; but this method of blind groping without any intuitive support seems now to have produced as much as it is able. After all, it is not impossible that science is now approaching the ultimate possibility of finite description; then the indescribable and unformalizable will be on hand, and we shall have to face this challenge. We shall have to find the best means to approach the action of chance, to describe the symmetry breaking generalized catastrophes, to formalize the unformalizable. For this task, the human brain, with its ancient biological heritage, its clever approximations, its subtle esthetic sensibility, remains and will remain irreplaceable for ages to come.

So what I am offering here is not a scientific theory, but rather a method; the first step in the construction of a model is to describe the dynamical models compatible with an empirically given morphology, and this is also the first step in understanding the phenomena under consideration. It is from this point of view that these methods, too indeterminate in themselves, lead not to a once-and-for-all explicit standard technique, but rather to an art of models. We may hope that theoreticians will develop a quantitative model in the framework of a given substrate, just as quantum mechanics has been developed for elementary interactions. But this is only a hope.

C. Philosophical considerations

There is no doubt it is on the philosophical plane that these models have the most immediate interest. They give the first rigorously monistic model of the living being, and they reduce the paradox of the soul and the body to a single geometrical object. Likewise on the plane of the biological dynamic, they combine causality and finality into one pure topological continuum, viewed from different angles. Of course this requires the abandonment of a universal mechanism and Laplacian absolute determinism, but have these ever been anything but wishful thinking?

Our models attribute all morphogenesis to conflict, a struggle between two or more attractors. This is the 2,500 year old idea of the first pre-Socratic philosophers, Anaximander and Heraclitus. They have been accused of primitive confusionism, because they used a vocabulary with human and social origins (conflict, injustice, etc.) to explain the appearance of the physical world, but I think that they were far from wrong because they had the following fundamentally valid intuition: *the dynamical situations governing the evolution of natural phenomena are basically the same as those governing the evolution of man and societies*, profoundly justifying the use of anthropomorphic words in physics.[6] Inasmuch as we use the word "conflict" to express a well-defined geometrical situation in a dynamical system, there is no objection to using the word to describe quickly and qualitatively a given dynamical situation. When we geometrize also the words "information," "message," and "plan," as our models are trying to do, any objection to the use of these terms is removed.

And this will represent not a little progress for present-day molecular biology.

D. Epilogue

I should like to have convinced my readers that geometrical models are of some value in almost every domain of human thought. Mathematicians will probably deplore abandoning familiar precise quantitative models in favor of the necessarily more vage qualitative models of functional topology; but they should be reassured that quantitative models still have a good future, even though they are satisfactory only for systems depending on a few parameters. The qualitative methods considered here, which appeal to the ideas of the morphogenetic field and chreod associated with singularities of the bifurcation set of an infinite-dimensional function space, avoid this difficulty. In theory, they still provide only a local classification, and so only a local investigation of the singularities of morphogenesis; the problem of integrating these local models into a stable global structure (dynamical topology), although sketched out in the case of living beings, remains wide open.

An essential tool is still missing: a more precise description of the catastrophic process of the disappearance of an attractor of a differential system and its replacement by new attractors. This problem is not only of theoretical importance. Physicists, if they want one day to obtain information about very small processes at subquantic level, will need the intermediary of the interaction of a *highly controlled process with an enormous degree of amplification*. Such processes, in which an infinitesimal perturbation may cause very large variations in the outcome, are typically catastrophes.

Biologists will perhaps reproach me for not having spoken of biochemistry in precise terms. This is true, and I do not deny the importance of chemical constraints on the dynamic of life. But I believe that any such constraint, and any chemical bond, can be considered as a geometrical factor in an appropriate space. Writing the equation, in atoms, that connects two constituents of a chemical reaction is one, the coarsest, of these constraints; the topology of biochemical kinetics and its relation with the spatial configuration of macromolecules are others that are certainly more decisive. On the other hand, *what is it that assures us that the formal structures governing life as a stable process of self-reproduction are necessarily connected with the biochemical substrate that we know today*?

As far as life and social sciences are concerned, it is difficult for me to judge whether my present ideas may be of interest, but in writing these pages I have reached the conviction that there are simulating structures of all natural external forces at the very heart of the genetic endowment of our species, at the unassailable depth of the Heraclitean *logos* of our soul, and that these structures are ready to go into action whenever necessary. The old idea of Man, the microcosm, mirroring World, the macrocosm, retains all its force: who knows Man, knows the Universe. In this *Outline of a General Theory of Models* I have done nothing but separate out and present the premises of a method that life seems to have practiced since its origin.

A mathematician cannot enter on subjects that seem so far removed from his usual preoccupations without some bad conscience. Many of my assertions depend on pure speculation and may be treated as day-dreams, and I accept this qualification—is not day-dream the virtual catastrophe in which knowledge is initiated? At a time when so many scholars in the world are calculating, is it not desirable that some, who can, dream?

APPENDIX 1

A model for memory

Structure of memory. We propose the following structure for memory:

1. A relatively slow dynamic (P, ψ) representing consciousness and mental activity.
2. A rapid auxiliary dynamic (M, X).
3. A weak coupling between these two dynamics by a third system (Q, X), of type a product of N linear oscillators, each with the same period.

The configuration space of Q is then an N-dimensional torus T^N. The parallel constant field in Q is structurally unstable and evolves through resonance toward a structurally stable situation characterized by the presence of attracting cycles. There may be competition between several possible resonances corresponding to slightly different biochemical states of neurons. To fix the idea, suppose that the possible resonances are parameterized by the points of a set J of phase differences in the torus T^N, varying, in principle, according to the past history of the individual. Suppose, for example, that J is a tree with successive bifurcations; then the terminal points of the tree will correspond to specifically catalogued memories. When Q is in a stable regime corresponding to such an extremal point s, the system Q will impose a certain weak coupling between (P, ψ) and (M, X) with the effect of producing a mean coupling field in P; the unique attractor of this mean field will then capture the dynamic ψ, recalling the memory belonging to s. The internal evolution of Q is itself governed by a secondary coupling with (P, ψ); when mental activity requires the recollection of a memory, Q will become unstable and the representative point in J will return to the origin of the tree J and then move toward an extremal point.

Two factors come into play at each bifurcation of J.

1. The coupling with (P, ψ), which can direct the choice in one direction or another.

2. The local chemical memory affecting the probabilities of each of the branches of the tree at the corresponding vertex.

Very probably the structure of J is not absolutely determined but depends on the genetic information and the experience of the lifetime of the individual. On the other hand, the extremal points probably cover only a very small part of all the possible couplings leading to mean fields with stable attractors. In this sense it can be said that our brains contain not only our actual memories, but also all virtual memories that we could have but never shall have.

The mechanism of acquiring a memory. Each branch point of the tree J can be represented by a partial oscillator D with three states: two stable regimes a and b corresponding to the branches of J issuing from the point, and an indifferently excited state. Each stable state synthesizes substances (RNA?) m_a and m_b, respectively, in the affected neurons, and these substances catalyze the return to the corresponding state. However, active synthesis of these substances, and their deposit in the neurons, cannot start until the excitation of D has died down. The recollection of the memory s will usually lead to the excitation of a reflex $r(s)$ by the organism. This reflex $r(s)$ may have agreeable or painful effects on the organism. When the effect is painful, there will be a generalized excitation of the local oscillators D in Q and the substances giving rise to this choice will be suppressed; this will leave almost no trace of the chemical of the choice just made. On the other hand, if the effect is agreeable, the excitation of the local oscillators D will be reduced and material m_a or m_b, according to the choice just made, will be synthesized. It could also be that the pain wave in Q generates the synthesis of substances unfavorable to the return to the corresponding regime.

Temporary loss of memory is a well-known and curious phenomenon. There are two possible explanations: either the dynamic Q cannot reach the required terminal point s because the local chemical memory at one of the bifurcations of J prevents it (this is the Freudian interpretation, in which we forget memories connected with disagreeable sensations), or Q does reach s, but this was actually not a point, so that Q ends up a little to one side and gives a coupling that recalls the general structure of the bifurcations of the mean field, but, because of an auxiliary perturbation, the horizontal component Y does not give rise to any attracting cycle. The memory is there, virtual, in the consciousness, and needs only the field that excited it. In this case the best way of recalling the memory is to start again from the beginning, if possible some time later.

Whatever else are the virtues of this model, it shows that there is little hope of localizing memories either spatially, in specific neurons, or chemically, in well-determined substances.

APPENDIX 2

Grammar, languages, and writing

Grammatical categories and the typology of languages. We have seen that concepts have a regulation figure, a *logos*, analogous to that of living beings. We might regard a grammatical category (in the traditional sense) as a kind of abstract *logos*, purified to the point that only the rules of combination and interaction between such categories can be formalized. From this point of view, we say that a grammatical category C is semantically denser than a category C' if the regulation of a concept of C involves mechanisms intervening in the regulation of C'. For example, take a name of an animate being, say a cat: this cat must make use of a spectrum of physiological activities for survival—eating, sleeping, breathing, and so forth; once these are satisfied, he can then indulge in less necessary but quite normal activities—playing, purring, and the like. Similarly each substantive has a spectrum of verbs describing the activities necessary for the stability and the manifestation of the meaning of the concept. Since the verb is indispensable for the stability of the substantive, it is less dense than the noun. The adjective shares in the stable character of the noun, but it is defined on a space of qualities, deeper than space-time, the support of the verb.

When a category C is denser than C', there is, in general, a canonical transformation from C to C'. The inverse transformation, however, is generally not possible.

These rules lead to the following order, in decreasing semantic density, for the traditional grammatical categories: noun-adjective-verb-adverb-affixes and various grammatical auxiliaries. In the emission of a sentence, the meaning is analyzed and the elements are emitted in the order of increasing density. In the model of Section 13.4.C the density of the concept is, in practice, the time required by the dummy actant to reduce the concept to the representative sign. It is much longer for a complex being like man than for an inanimate object, whose regulation is much simpler. As an example, the normal order of emission of a transitive phrase, subject-verb-object, would be verb-object-subject; the object is less stable than the subject, since in such a transitive process the subject survives the whole interaction whereas the object may perish. (The cat eats the mouse, the morphology⊐.) The reception order, the one most favorable to the best reconstitution of the global meaning, is generally the opposite order: subject-object-verb. Now researches on the universals of language have shown that the pure emissive typology V-O-S is practically nonexistent (see [5]), whereas the receptive typology S-O-V is well represented (e.g., Japanese, Turkish, and Basque). This reflects a fundamental fact in the dynamic of communication: the act of speaking is initiated by the speaker, and in general he has a greater interest in being understood than the listener has in understanding. Consequently the transposition of the emissive order into the receptive order is generally carried out by the speaker, and this gives predominance to the reception typology. However, the mixed emissive typology S-V-O is the most common.

An elementary sentence generally contains other ancillary elements that go to make up a nuclear phrase; these are the *adjuncts*. The principal kinds of adjuncts are the epithet adjectives (A-N or N-A) and genitives (G-N or N-G). The adjective is semantically less dense than the noun; therefore the receptive typology of the epithet is N-A, the emissive A-N; and similarly for the genitive: receptive N-G, emissive G-N. Since a preposition is less dense, an adjunct of type Pre-N is an emissive type, in harmony with the order V-O, while a postposition N-Post is in harmony with O-V.

The second principle governing the typology of languages is this: the free adjuncts (those not tied to the central verb, e.g., A and G) have an inverse typology to the verb-object nucleus. This leads to the two main types of languages:

Emissive	*Receptive*
V-O	O-V
Pre	Post
N-A	A-N
N-G	G-N

English is not typical, since it has preserved from an older receptive stage the typology A-N for the epithet adjective and the partial type G-N in the Saxon genitive (John's house). For further details see my article [1].

The origin of writing The mental reconstruction of the organizing centers of elementary fields spreads, by a very natural contagion, to the functional fields of the hand. The external variables of the elementary catastrophes will be realized as spatial variables. The stylization of an action is nothing more than a return to the organizing center of this action. To the extent that the Riemann-Hugoniot cusp is conceptually stabilized by the concept of division or separation, the catastrophe can be realized by writing in clay with a stick the symbol $\prec$. Similarly the sign 入 is, like the previous one, an old Chinese ideogram meaning to enter or penetrate, and this could well be a stylization of the elliptic umbilic. In this way we cannot but admire the suitability of the Chinese system of writing; the dominant influence of the the spoken word in the West has resulted in an alphabetical or syllabic system of writing, and the expression (*signifiant*) has violently subjugated the meaning (*signifié*).

In conclusion, we have seen that an analysis of the grammatical structures of language requires a subtle mixture of algebra, dynamics, and biology. Without pretending to have a definitive answer to a problem whose difficulty can scarely be measured, I venture to suggest that these ideas may contain something of interest for many specialists.[7, 8]

NOTES

1. The belief that man can transform himself into an animal is very widespread even today, and the animals into which he believes that he can change himself are universally those that are in the relationship of predator on man (wolf, tiger, shark, etc.) or his prey. In particular, in many primitive tribes the hunt begins with a ceremony of a virtual dance in which the hunter, dressed in a skin of the prey, imitates its movements and behavior. In biological morphology, on the other hand, the presence in many predators of bait organs, like the end of the tongue of a lamprey with its wormlike form, can scarcely be explained other than by the existence in the genetic endowment of a simulating structure of the mind of the prey (or sometimes of the predator; for example, the ocellate designs on the wings of some butterflies, which have the effect of keeping birds off).

2. The idea that mental activity must realize some model of the environmental space of the animal—so ineluctable to my profane mind—seems very uncommon among physiologists. It does occur, however, in J. Z. Young, *A Model of the Brain*, Oxford University Press, 1964.

3. Note that in a beehive the place (i.e., the function) of an individual in the society varies with his age. As the worker bee grows older, her activities move away from the germinal center (care of recently enclosed larvae) toward somatic roles (wax-worker, then honey-gatherer).

4. In this model of society, the intellectuals (artists and thinkers) figure as the axis of the money circulation; in their central position, freed from the need to produce, deprived of the advantages and responsibilities of power, they are the eye of this continual hurricane. In terms of the model of Section 11.1.C we might regard them as the chromosome of the social body.

5. I have twice allowed myself to translate the Heraclitean λογος as "form." Allowing that to Heraclitus the *logos* is the formal structure that assures for any object its unity and its stability, I am convinced that this particular use of the word "form" (meaning the equivalence class of structurally stable forms) in this book is a reasonably good approximation.

6. It is striking how all past and present techniques of foretelling the future depend on the following principle: a generalized catastrophe (tea leaves in a cup, lines on the palm of a hand, drawing of cards, the shape of a chicken's liver, etc.) is studied and its morphology is then associated, by a suitable isomorphism, with the preoccupations and difficulties of the client. This method is not absurd insofar as the dynamic of morphogenesis may contain local accidental isomorphisms with the dynamic of human situations, and often a gifted soothsayer may well elicit some valuable conclusions from this examination. To classify these isomorphisms in some definitive manner would be to embark on the characteristic form of delirious thought.

7. Suppose now that the number of these categories organizing the meaning was reduced and that a typology of the events so instituted was possible; then such a typology, based on an exhaustive description of the structures of the message, would form the objective framework within which the representation of the contents, identified with the semantic micro-universe, would be the only variable. The linguistic conditions of the understanding of world would then be formulated. Translated from A. J. Greimas, *Sémantique structurale*, Larousse, 1966, p. 133.

8. It thus appears that the future of syntactical research lies in intranuclear investigation, which alone will allow an investigation of the interior of the nucleus and the phenomena based there, and which will produce, in the intellectural order, structures at least as complicated as those of the cell, the molecule, and the atom of the material order. Translated from L. Tesnière, *Eléments de syntaxe sıructurale*, Klincksieck, 1966, p. 157.

REFERENCES

1. R. Thom, Sur la typologie des langues naturelles: essai d'interprétation psycholinguistique, in *The Formal Analysis of Natural Languages*, Proceedings of the First International Conference, eds. M. Gross, M. Halle, and M.-P. Schutzenberger, Mouton, 1973.
2. J. Piaget and B. Inhelder, *The Child's Conception of Space*, Routledge and Kegan Paul, 1956.
3. G. H. Luquet, *Les dessins d'un enfant*, Alcan, 1913; and *Le dessin enfantin*, Alcan, 1927.
4. R. Bellman, *Adaptive Control Processes, a Guided Tour*, Princeton University Press, 1961, p. 94.
5. J. H. Greenberg, *Universals of Language*, M.I.T. Press, 1966. p. 104.

MATHEMATICAL SUMMARY

CONCEPTS AND NOTATIONS OF DIFFERENTIAL TOPOLOGY AND QUALITATIVE DYNAMICS

The following are assumed to be familiar to the reader:

1. The symbols of set theory: $\in, \subset, \cup, \cap$. The Cartesian product of two sets: $E \times F = \{(e,f) : e \in E, f \in F\}$. The idea of a map, its source and target; injective, surjective, and bijective maps. Equivalence relations.

2. Elementary theorems of analysis: the notions of derivative, partial derivative, integral. Differentiation of the composition of functions. Taylor series. The existence and uniqueness of solutions of a system of differential equations.

3. The rudiments of linear and multilinear algebra: real vector space, linear map, rank of a linear map, determinant, linear functional, dual vector space, transpose of a map. Exterior derivative and exterior product.

1. REAL EUCLIDEAN Q-DIMENSIONAL SPACE

Suppose that the *state* of a physical system A can be completely characterized by q parameters $x_1, x_2, \ldots, x_q$; then every state of A is represented by a system $(x_1, x_2, \ldots, x_q)$ of q real numbers, that is, by a point of *real q-dimensional Euclidean space* $\mathbf{R}^q$. The *Euclidean distance* between two given points

$$x = (x_1, x_2, \ldots, x_q) \quad \text{and} \quad y = (y_1, y_2, \ldots, y_q)$$

of $\mathbf{R}^q$ is defined by the formula

$$d(x,y) = \left[\sum_{j=1}^{q} (x_j - y_j)^2 \right]^{1/2} ;$$

the set $\{y;\ d(x,y) < r\}$ is the *open ball* with center x and radius r. The *triangle inequality*

$$d(x,y) \leqslant d(x,y) + d(y,z)$$

holds for any three points x, y, and z of $\mathbf{R}^q$.

Given a set $a = (a_1, a_2, \ldots, a_q)$ of q real numbers, it will not, in general, be possible to prepare a state of A corresponding to a (in particular, if the values of the a_j are too large). These points of $\mathbf{R}^q$ representing physically realizable states of A form the *defining set* M_A of A in $\mathbf{R}^q$. Since the values of the parameters x can never be known with precision, then, if $a = (a_1, a_2, \ldots, q_q) \in M_A$ (i.e., a is a physically realizable state of A), all sufficiently close points $b = (b_1, b_2, \ldots, b_q)$ will also be realizable. Hence M_A has the following property: if $a \in M_A$, then M_A contains a ball with center a and sufficiently small radius $r(a)$. In topology subsets of $\mathbf{R}^q$ having this property are called *open*.

If any two points a, b of M_A can be joined by a continuous path lying in M_A, then M_A is called *connected* (connected open sets are also called *domains*); every open set in $\mathbf{R}^q$ can be decomposed into a union of at most a countable number of disjoint domains.

If any two paths c and c' joining any two points a and b of M_A can be continuously deformed into each other, M_A is called *simply connected*; otherwise the set of paths joining a and b that can be deformed into c forms a *homotopy class of paths*. A path whose end points coincide is a *loop*.

The set of points in $\mathbf{R}^q$ not belonging to a set M is the *complement* of M, written $\mathbf{R}^q - M$. The complement of an open set is called *closed*.

Any union of open sets is open, and any intersection of closed sets is closed. For any open set M, the intersection of all closed sets containing M is a closed set $\overline{M}$, the *closure* of M. The set $\overline{M} - M$ is the *boundary* of M, and any neighborhood of a point $c \in \overline{M} - M$ contains points both of M and of $\mathbf{R}^q - M$, but c does not belong to M. If M_A is the set of definition of a physical system A, new, discontinuous phenomena appear at each point of the boundary of M_A to prevent the realization of A at this point: this is the idea of the *catastrophe set* of Chapter 4.

A set $M \subset \mathbf{R}^q$ is *bounded* if it is contained in some ball of finite radius. A closed and bounded set is called *compact*.

2. MAPS

Suppose that the data of a physical system A determine completely the data of a second system B (e.g., when B is a subsystem of A); and suppose further that A is parameterized by points $(x_1, x_2, \ldots, x_n)$ of $\mathbf{R}^n$, and B by points $(y_1, y_2, \ldots, y_p)$ of $\mathbf{R}^p$. Then to each point x of M_A, the defining set of A, there will correspond a point $y = (y_1, y_2, \ldots, y_p)$ of M_B, the defining set of B. This defines a *map* F from the source M_A to the target M_B given by formulae of the form

$$y_k = f_k(x_1, x_2, \ldots, x_n), \quad 1 \leqslant k \leqslant p.$$

It is important not to confuse the *target* of a map with its *image*; in this case the image of F is the set

$$F(M_A) = \{b : \exists a \in M_A \text{ such that } b = F(a)\}.$$

The image is a subset of the target; when the image and the target coincide, the map is *surjective*. For a subset $C \subset M_B$, the *inverse image* of C by F is the set

$$F^{-1}(C) = \{a \,;\, a \in M_A \text{ and } F(a) \in C\}.$$

When the inverse image of each point is again a point, or empty, F is called *injective*. An injective and surjective map is called *bijective*.

A map $F : M_A \longrightarrow M_B$ is *continuous* if the inverse image of any open set in M_B is an open set in M_A; it is sufficient that the inverse image of any open ball of M_B be an open subset of M_A, or, equivalently, that each f_k be continuous in the usual sense that the distance $d(f_k(a), f_k(a'))$ can be made arbitrarily small by making $d(a, a')$ sufficiently small.

When U and V are two open subsets of Euclidean space, a bijective and bicontinuous (i.e., both F and its inverse continuous) map is a *homeomorphism* and U and V are called *homeomorphic*. The central problem of topology is to know whether two topological spaces X and Y are homeomorphic; when they are open subsets of Euclidean spaces, a necessary condition for them to be homeomorphic is that the dimensions of their ambient spaces are equal (the theorem of *invariance of domain*).

A continuous map $F : U \longrightarrow V$ is *proper* if the inverse image $F^{-1}(K)$ of each compact subset $K \subset V$ is compact in U. Intuitively this means that $F(u)$ cannot be in the frontier of V unless u is in the frontier of U; when U and V are the domains of definition of physical systems A and B, respectively, this implies that when B is not in a state of catastrophe (and so is well defined) A is also well defined.

3. DIFFERENTIABLE MAPS

When all the formulae

$$y_k = f_k(x_1, x_2, \ldots, x_n)$$

defining the map $F : M_A \longrightarrow M_B$ have continuous partial derivatives up to and including the rth order, F is said to be *r-times continuously differentiable*, or to be of *class* C^r. Two maps of class C^r compose to give a map of class C^r, that is, if $F : U \longrightarrow V$ and $G : V \longrightarrow W$ are C^r, then $H = G \circ F : U \longrightarrow W$ is of class C^r.

Let F and G be two differentiable maps of $\mathbf{R}^n$ into $\mathbf{R}^p$, defined by formulae $y_j = f_j(x_i)$ and $y = g_j(x_i)$, and such that $f_j(0) = g_j(0) = 0$. Also, let $\mu = (\mu_1, \mu_2, \ldots, \mu_n)$ be a multi-index of order

$$|\mu| = \mu_1 + \mu_2 + \cdots + \mu_n, \quad \mu_i \geqslant 0.$$

When

$$\frac{\partial^{|\mu|} f_j}{(\partial x_1)^{\mu_1} \cdots (\partial x_n)^{\mu_n}} = \frac{\partial^{|\mu|} g_j}{(\partial x_1)^{\mu_1} 1 \cdots (\partial x_n)^{\mu_n}}$$

for all μ with $0 \leqslant |\mu| \leqslant r$, where all the partial derivatives are evaluated at 0, the local maps F and G (i.e, regarded as maps of some neighborhood of the origin) are said to have the same local *jet of order r* at the origin. The set of local jets of order r of $\mathbf{R}^n$ into $\mathbf{R}^p$ forms a vector space parameterized by the values a_j^μ of the partial derivatives

$$\frac{\partial^{|\mu|} f_j}{(\partial x_1)^{\mu_1} \cdots (\partial x_n)^{\mu_n}} .$$

This idea of local jet systematizes the classical idea of the expansion of order r of a differentiable function. When $H = G \circ F$ is the composition of two local maps of class C^r, the jet of order r of H is determined (by polynomial formulae) by the jets of F and G of order r.

For a point $a \in \mathbf{R}^n$, a *differentiable path*, starting from a, is a differentiable map $f : I \rightarrow \mathbf{R}^n$, where $I = [0, 1]$, and $f(0) = a$. This map is defined by formulae

$$x_j = f_j(t), \quad 1 \leqslant j \leqslant n.$$

For any differentiable map $F : \mathbf{R}^n \longrightarrow \mathbf{R}^p$, the composition $F \circ f : I \longrightarrow \mathbf{R}^p$ defines a differentiable path in $\mathbf{R}^p$ starting from $b = F(a)$. Furthermore, if two paths f and g starting from a have the same jet of order 1 at 0 [i.e., $f_j'(0) = g_j'(0)$], the image paths $F \circ f$ and $F \circ g$ will also have the same jet of order 1 at 0 (this follows immediately from the rule for differentiating composed functions); thus the set of jets of order 1 of differentiable paths starting from a forms a vector space [parameterized by the coordinates $(f_r'(0))$] of dimension r. This space is the *tangent space* to $\mathbf{R}^n$ at a, written $T_a(\mathbf{R}^n)$, and every differentiable map $F : \mathbf{R}^n \longrightarrow \mathbf{R}^p$ defines (by composition) a linear map $TF : T_a(\mathbf{R}^n) \longrightarrow T_b(\mathbf{R}^p)$, where $b = F(a)$, the linear *tangent map* to F at a. As this linear map depends differentiably on the point a, there is, in fact, a field of linear maps of $\mathbf{R}^n$ into $\mathbf{R}^p$ above the source-space $\mathbf{R}^n$.

Let F be a local C^r-map of $\mathbf{R}^n$, $(x_1, x_2, \ldots, x_n)$, into $\mathbf{R}^n$, $(y_1, y_2, \ldots, y_n)$, defined by $y_j = f_j(x_i)$, with $f_j(0) = 0$. If the linear tangent map $TF : T_x \to T_y$ is of rank n, that is, surjective [so that the Jacobian $\partial(f_1, f_2, \ldots, f_n)/\partial(x_1, x_2, \ldots, x_n) \neq 0$ at 0], there is a local C^r-map G inverse to F, so that $G \circ F$ and $F \circ G$ are each the identity map on some neighborhood of 0 (the *inverse function theorem*).

For two open sets U and V in $\mathbf{R}^n$, a C^r-map $F : U \to V$ with a C^r-inverse $G : V \to U$ is called a *C^r-diffeomorphism*. The inverse function theorem states that a map $F : \mathbf{R}^n \to \mathbf{R}^n$ is locally a diffeomorphism in a neighborhood of each point where the rank of F is maximum. More generally, if $F : \mathbf{R}^n \to \mathbf{R}^p$ is of maximum rank (the larger of n and p), then

(1) If $p > n$, F is locally an *embedding* in a subspace of the target space, that is, a change of coordinates (a diffeomorphism) mapping $\mathbf{R}^n$ locally onto a linear submanifold of dimension n; or

(2) if $n > p$, F can be transformed locally, by a diffeomorphism of the source space, to a linear projection $L : \mathbf{R}^n \to \mathbf{R}^p$.

4. DIFFERENTIAL MANIFOLDS

Let us return to the case of a physical system A parameterized by q parameters $x_1, x_2, \ldots, x_q$. The evolution of A as it varies in time will be described by functions $x_j(t)$, which define a path in the defining set M_A of A. It frequently happens that this variation is not arbitrary but must be such as to keep certain functions $G_1(x_i), G_2(x_i), \ldots, G_s(x_i)$ invariant; then $G_m(x_i(t)) = G_m(x_i(t_0))$ for all t. These functions are called *first integrals* of the evolution of the system. Geometrically, this is to say that the point $x(t)$ cannot move freely about M_A, but is constrained to lie in the subset $\{x; G_m(x) = G_m(x(t_0)), 1 < m < s\}$. If at all points of this subset, the rank of the map $G : \mathbf{R}^q \to \mathbf{R}^s$ $(s < q)$ is maximum, then at all points there is a minor of maximum rank s in the matrix of partial differential coefficients $(\partial G_m/\partial x_i)$ (although this minor is not necessarily the same for all points). Then at all points of this subset there is a local diffeomorphism of $\mathbf{R}^q$ mapping this set into a linear submanifold of codimensions s. Such a set is called an *embedded manifold* of dimension $q-s$.

The idea of an embedded manifold is thus a generalization of the idea of a curve embedded in the plane, of a surface in $\mathbf{R}^3$, and so forth.

A differential manifold can be defined as a Hausdorff topological space covered by charts U_i, where each U_i is an open subset of $\mathbf{R}^n$, and on the (nonempty) intersection of two sets U_i and U_j each point has two representatives, p_i in U_i and p_j in U_j, such that the relation between p_i and p_j is given by a diffeomorphism h_{ij}. The h_{ij} must also satisfy patching conditions $h_{ik} = h_{ij} \circ h_{jk}$ where this composition is defined. It can be shown (Whitney) that all paracompact (i.e., countable union of compact) n-dimensional manifolds can be embedded in $\mathbf{R}^{2n}$; however, there does not in general exist a system of n equations

$$G_1 = G_2 = \cdots = G_n = 0,$$

with $dG_1 \wedge dG_2 \wedge \ldots \wedge dG_n \neq 0$ everywhere, defining the manifold globally.

Let U^{n+k} and Y^n be two differential manifolds, and $p : U \to Y$ a differentiable map (i.e., differentiable in systems of charts of the source and target) with maximum rank n at each point. Then, by the implicit function theorem, the counterimage $p^{-1}(y)$ is a submanifold of codimension n for each $y \in Y$. Next observe that the topological product $Y \times F$ of two differential manifolds Y and F is a differential manifold (take as charts the product of a chart in Y and a chart in F), and the canonical projection of $Y \times F$ onto Y is a map of maximum rank, equal to the dimension of Y. A differentiable map $p : E \to Y$ is called a *differentiable fiber bundle* when each point $y \in Y$ has an open neighborhood V_y satisfying the condition of *local triviality*: there exists a diffeomorphism

$$h : p^{-1}(V_y) \overset{\text{onto}}{\longrightarrow} V_y \times F,$$

where F is some manifold (the *fiber*) such that $p \circ h^{-1}$ is the canonical projection of $V_y \times F$ onto V_y. When Y is a domain, all the fibers F of the fibration $p : E \to Y$ are diffeomorphic.

Every map $p : U \to Y$ that is proper and has maximum rank (equal to dim Y) is a differentiable fibration whose fiber is a compact manifold (Ehresmann's theorem).

Another very important example of a fiber bundle is the *tangent vector bundle* to a manifold V. When V is an n-dimensional manifold defined by charts U_i and coordinate transformations h_{ij}, the set of tangent vectors to V can be made into a $2n$-dimensional manifold [each vector of $T(V)$ is defined by its starting point y and its n scalar components of $(u_i'(0))$ in a local chart U_i], where a local chart on $T(V)$ is of the form $U_i \times \mathbf{R}^n$; where $\mathbf{R}^n$ has the components of $(u_i'(0))$ as coordinates, and the diffeomorphisms h_{ij} extend to diffeomorphisms of $U_i \times \mathbf{R}^n$ defined by linear invertible maps $c(u) : \mathbf{R}^n \to \mathbf{R}^n$. This situation is described by saying that the space $T(V) \to V$ is a *vector bundle* over the manifold V.

For technical reasons the vector bundle of cotangent vectors $T^*(V)$, and not tangent vectors $T(V)$, is used in mechanics. The fiber of the former bundle is the vector space T^* dual to the tangent vector space at the point considered.

A *section* of a fiber space $p : E \to Y$ is a map $s : Y \to E$ such that

$p \circ s : Y \to Y$ is the identity map; that is, to each point y of the base space Y corresponds a point $s(y)$ of the fiber $p^{-1}(y)$ over y. When E and Y are differential manifolds, the section s is called r-times differentiable when it is a map of Y into E of class C^r.

5. VECTOR FIELDS

A section of the vector bundle $T(V) \to V$ of tangent vectors to a manifold V is a *vector field* on V, and the points $y \in V$ for which $s(y)$ is the zero vector are the *singularities* of the field. A section of the cotangent bundle $T^*(V) \to V$ is a *differential* 1-*form* V.

When $F : U \to V$ is a differentiable map between the two manifolds U and V, and a is a 1-form on V, the 1-form a^* on U defined by

$$(X, a^*) = (F(X), a)$$

is the form *induced* by the map F. In particular, there is a canonical differential 1-form j on $\mathbf{R}$ defined on the vector X to be the algebraic measure of X. When f is a map of a manifold M into $\mathbf{R}$ (i.e., a numerical function), the induced form $f^*(j)$ is precisely the differential df of f.

6. DYNAMICAL SYSTEMS

Let X be a vector field on a manifold M. When this field is of class C^1, the differential system $dm/dt = X(m)$ has solutions with the following property of local uniqueness: there exists one and only one solution curve in the product $M \times \mathbf{R}$ of M with the time axis through each point $y \in M$. When M is *compact*, there is a one-parameter group of diffeomorphisms $h(t)$ of M such that $dh/dt = X(h(t))$; write $p_t = h_t(p)$ for the trajectory through p ($p \in M$), and write $\alpha(p)$ and $\omega(p)$ for the limit sets of p_t in M when t tends to $-\infty$ and $+\infty$, respectively. Then an *attractor* of a dynamical system is a closed set K, invariant under h_t, which contains the ω-limit set of all points of a neighborhood of K in M and is such that, if $\alpha(p) \subset K$, then $p \in K$ (but note that a more general definition is often used; c.f. Section 4.1.B). Birkhoff called a *wandering point* a point p that has some neighborhood B such that $h_t(B) \cap B = \phi$ for all sufficiently large t; then the *nonwandering* points are, in practice, the recurrent points, points from every neighborhood of which are trajectories returning infinitely often to that neighborhood. This qualitative distinction is very important, for only the *nontransient*, *stationary properties* of a dynamic have their support in the set of nonwandering points.

On any differential manifold, we can define a Riemannian metric ds^2 that is, a positive definite quadratic form Q defined on the tangent space at each point. Then, for any differentiable path $f: I \to M$, the integral $\int \sqrt{Q(df/dt)}\,dt$ defines the length of the path in the metric ds^2; and, if V is a real-valued function defined on M, the field X defined by $\langle X, Y\rangle = dV(Y)$ (where $\langle , \rangle$ is the scalar product associated with Q) is the *gradient* of V. A gradient dynamical system can only have wandering points because the function V is monotonically increasing on each trajectory, with the exception of singular points of the field, where dV is zero. These gradient systems are, in some sense, the simplest of dynamical systems: almost all trajectories tend to an attractor which is, in general, an isolated maximum of V.

We can define the following 1-form A on the cotangent bundle $T^*(M)$ to M. Let $p: T^*(M) \to M$ be the canonical projection, and (y,u) a covector at y. Let X be a vector tangent to $T^*(M)$ at (y,u), and write $A(X) = u(p(X))$. If q_j is a system of local coordinates f or M, then $T^*(M)$ has coordinates (q_j, p_j) on the associated chart, where $p_j = \partial Q/\partial q_j$ is the momentum associated with the variable q_j, and $A = \Sigma_i p_i\, dq_i$.

A *conservative Hamiltonian system* X is then defined by a real-valued function $H: T^*(M) \to \mathbf{R}$; if $dA = \Sigma_i dp_i \wedge dq_j$ is the exterior derivative of A, X is defined by $dA(X,Y) = dH(Y)$. In local coordinates the field X is defined by Hamilton's equations:

$$\dot{q}_i = -\frac{\partial H}{\partial p_i}, \qquad \dot{p}_i = \frac{\partial H}{\partial q_i}.$$

Hamiltonian fields have the Hamiltonian H (i.e., energy) as first integral, and leave invariant the 2-form dA, and hence also its exterior product

$$\wedge^n(dA) = dA \wedge dA \wedge \ldots \wedge dA = dp_1 \wedge \ldots \wedge dp_n \wedge dq_1 \wedge \ldots \wedge dq_n;$$

thus there is an invariant volume in $T^*(M)$ and also in every hypersurface $H = \text{constant}$. This implies that a Hamiltonian system cannot have strict attractors, and almost all points are nonwandering (Poincaré's theorem of return). Also, Birkhoff's ergodic theorem holds in every hypersurface of constant energy, $H = \text{constant}$: For a measurable function $f: T^*(M) \to \mathbf{R}$, the average $(1/T)\int_0^T f(m_t)\,dt$ converges as $T \to \infty$ for almost all points m. When this limit is the same at all points (and equal to the average value of f on the manifold $H = \text{constant}$, with its invariant measure), the field is called *ergodic*. Contrary to the frequent assertion in books on statistical mechanics, it is quite exceptional for a Hamiltonian field to be ergodic, for such a field can have, in structurally stable way, invariant sets of nonzero measure which do not fill out the manifold.

7. FUNCTION SPACES AND INFINITE-DIMENSIONAL MANIFOLDS

Let $f: U \rightarrow V$ be a map between two differential manifolds, and to each point u of U associate a variation δv of the image $v = f(u)$. Taking all possible variations $\delta v(u)$, depending differentiably on u, gives all maps g close to f, and these variations (regarded as vectors tangent to f) form an infinite-dimensional vector space. The function space $L(U, V)$ of maps from U to V can thus be considered as an infinite-dimensional manifold. I refer to books on functional analysis for definitions and properties of the topologies (Hilbert space, Banach space, Fréchet space, etc.) that are possible on infinite-dimensional vector spaces and, in turn, on infinite-dimensional manifolds. Here we are interested only in a closed subspace L of *finite* codimension, namely, the *bifurcation* set H. In the study of subspaces of this type, the actual choice of topology on the space of tangent vectors $\delta v(u)$ is, in practice, irrelevant.

INDEX

Printed in the United States
by Baker & Taylor Publisher Services